Game Theory
A Modeling Approach

T0256341

Textbooks in Mathematics

Series editors:
Al Boggess and Ken Rosen

ELEMENTARY DIFFERENTIAL EQUATIONS, SECOND EDITION

Charles Roberts

LINEAR METHODS FOR THE LIBERAL ARTS

David Hecker and Stephen Andrilli

CRYPTOGRAPHY: THEORY AND PRACTICE, FOURTH EDITION

Douglas R. Stinson and Maura B. Paterson

DISCRETE MATHEMATICS WITH DUCKS, SECOND EDITION

sarah-marie belcastro

BUSINESS PROCESS MODELING, SIMULATION AND DESIGN, THIRD EDITION

Manual Laguna and Johan Marklund

GRAPH THEORY AND ITS APPLICATIONS, THIRD EDITION

Jonathan L. Gross, Jay Yellen and Mark Anderson

A FIRST COURSE IN FUZZY LOGIC, FOURTH EDITION

Hung T. Nguyen, Carol L. Walker, and Elbert A. Walker

EXPLORING LINEAR ALGEBRA

Crista Arangala

A TRANSITION TO PROOF: AN INTRODUCTION TO ADVANCED MATHEMATICS

Neil R. Nicholson

COMPLEX VARIABLES: A PHYSICAL APPROACH WITH APPLICATIONS, SECOND EDITION

Steven G. Krantz

GAME THEORY: A MODELING APPROACH

Richard Alan Gillman and David Housman

https://www.crcpress.com/Textbooks-in-Mathematics/book-series/CANDHTEXBOOMTH

Game Theory
A Modeling Approach

Richard Alan Gillman
Valparaiso University, Valparaiso, Indiana
David Housman
Goshen College, Goshen, Indiana

CRC Press
Taylor & Francis Group
Boca Raton London New York

CRC Press is an imprint of the
Taylor & Francis Group, an **informa** business
A CHAPMAN & HALL BOOK

CRC Press
Taylor & Francis Group
6000 Broken Sound Parkway NW, Suite 300
Boca Raton, FL 33487-2742

First issued in paperback 2022

© 2019 by Taylor & Francis Group, LLC
CRC Press is an imprint of Taylor & Francis Group, an Informa business

No claim to original U.S. Government works

ISBN 13: 978-1-03-247559-2 (pbk)
ISBN 13: 978-1-4822-4809-8 (hbk)

DOI: 10.1201/9781315156880

Library of Congress Cataloging-in-Publication Data

Names: Gillman, Richard Alan, author. | Housman, David, 1957- author.
Title: Game theory : a modeling approach / Richard Alan Gillman and David Housman.
Description: Boca Raton : CRC Press, Taylor & Francis Group, 2019. | Series: Textbooks in mathematics | Includes bibliographical references and index.
Identifiers: LCCN 2019000626 | ISBN 9781482248098 (hardback : alk. paper) |
ISBN 9781315156880 (ebook).
Subjects: LCSH: Game theory. | Decision making--Mathematical models.
Classification: LCC QA269 .G4868 2019 | DDC 519.3--dc23
LC record available at https://lccn.loc.gov/2019000626

Visit the Taylor & Francis Web site at
http://www.taylorandfrancis.com

and the CRC Press Web site at
http://www.crcpress.com

Contents

Preface   vii

Acknowledgments   xi

CHAPTER   1 ▪ Models and Games   1

1.1	INTRODUCTION TO MODELING	1
1.2	INTRODUCTION TO GAME THEORY	2
1.3	EXAMPLES OF GAMES	4
1.4	RATIONALITY ASSUMPTION	7

CHAPTER   2 ▪ Player Preferences   13

2.1	ORDINAL UTILITIES	13
2.2	VON NEUMANN-MORGENSTERN UTILITIES	17
2.3	CONSTRUCTING UTILITIES	23
2.4	DETERMINING RISK	29

CHAPTER   3 ▪ Simultaneous Play   39

3.1	STRATEGIC GAMES	40
3.2	FINAL JEOPARDY	48
3.3	MIXED STRATEGIES	52
3.4	NON-MATRIX MODELS	61
3.5	LIMITATIONS	69

CHAPTER   4 ▪ Bilateral Agreements   81

4.1	NEGOTIATIONS	81
4.2	BARGAINING IN STRATEGIC GAMES	88
4.3	FAIRNESS PROPERTIES	92

CHAPTER   5 ▪ Sequential Play   103

5.1	SEQUENTIAL GAMES	104
5.2	SUBGAME PERFECT EQUILIBRIA	112
5.3	COMBINATORIAL GAMES	120
5.4	MORAL HAZARD	133

CHAPTER   6 ▪ Missing Information   145

   6.1   IMPERFECT INFORMATION   146
   6.2   ROMANS AND GERMANS   159
   6.3   INCOMPLETE INFORMATION   163
   6.4   BARGAINING GAMES WITH PRIVATE INFORMATION   170
   6.5   INTERNATIONAL COLLABORATION   175
   6.6   AUCTIONS   187

CHAPTER   7 ▪ Repetitious Play   205

   7.1   REPEATED SOCIAL DILEMMAS   206
   7.2   MODELING NOISE   215
   7.3   EVOLUTIONARY GAME THEORY   218

CHAPTER   8 ▪ Multilateral Agreements   239

   8.1   SHAPLEY ARGUMENT   240
   8.2   NUCLEOLUS ARGUMENT   247
   8.3   BARGAINING ARGUMENT   252
   8.4   VOTING POWER   257

CHAPTER   9 ▪ Resource Allocation   275

   9.1   RESOURCE ALLOCATION PROBLEM   275
   9.2   BARGAINING AND COALITION MODELS   280
   9.3   COMPLETE AND INCOMPLETE INFORMATION MODELS   284
   9.4   CONCLUDING REMARKS   291

APPENDIX   A ▪ Selected Answers   295

Bibliography   317

Index   325

Preface

Game theory is a fascinating field. It is deeply mathematical, yet frequently studied by people trained in other disciplines. These include economists, biologists, political scientists, psychologists, and many others. It is accessible and popular among the general public, where zero-sum games and prisoner's dilemma are often used as common, everyday phrases. Nonetheless it relies on deep mathematical arguments for its justification. The widespread interest in this field no doubt stems from its ability to help us understand human interactions in a manner grounded in mathematical logic and truth.

This volume is intended to fill the gap between the popular treatments of game theory and highly technical treatments of the subject. It is broad and introductory, but set at a level that will engage and challenge advanced undergraduate mathematics majors and graduate students in professional master's programs. Such students are capable of understanding the mathematical underpinning of game theory but do not necessarily need a full development of the underlying mathematical theory. Often, their primary interest will be finding applications of game theory to real-world problems that they will encounter in their work.

Hence this volume is inherently about modeling. It is, however, not quite a traditional modeling text. Because students typically only take one course in game theory, we must introduce the appropriate game theoretic concepts and tools as we engage in the modeling process. This is very different from a canonical mathematical modeling text in which the tools—calculus and differential equations—have been introduced and mastered in multiple prior courses.

Unlike many game theory textbooks that treat the real world only in the abstract, we take care to match conclusions to the real-world scenarios we set out to model. Throughout the book, scenarios are described inside boxes. These scenarios—real or fictitious but plausible— represent situations in which someone would want to make a strategic decision. They form the basis of the modeling process developed in the text. Chapter 1 contains a brief introduction to modeling that can be used as a review for students who have already had a modeling course or as a beginning point for new modelers. Figure 1 summarizes the prerequisite relationships among the chapters.

Chapter 1 also introduces the fundamental definition of a mathematical game and provides several basic examples. It closes with a brief discussion of the limitations and assumptions of our approach, specifically the assumption of rationality that game theorists make about game participants. This assumption is maintained throughout the book; however, we interweave challenges to the rationality assumption which guide readers in understanding both the limits and possibilities of game theoretic modeling.

Chapter 2 lays the foundation for mathematical game theory by introducing and developing the idea of a utility function. These functions transform vague personal preferences into quantifiable and manipulable objects. Understanding this process is essential to understanding the applicability of game theory. The key theorems are developed in detail and examples of many types of utility functions are provided. In particular, both discrete and continuous utility functions are considered. This is a key step in moving readers beyond

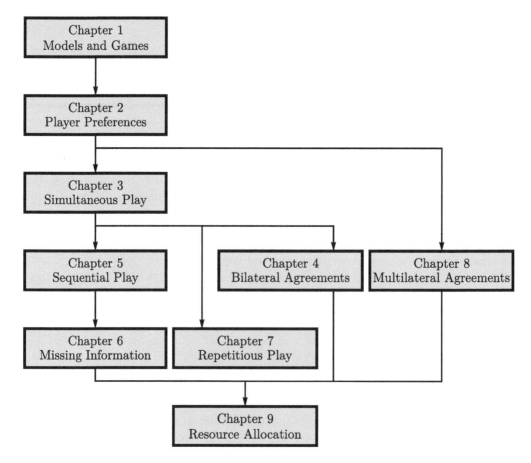

FIGURE 1 Chapter Prerequisite Relationships.

popular treatments of game theory which restrict themselves to finite cases. It is a dual perspective that is maintained throughout the volume.

Chapter 3 introduces strategic games, the simplest and most widely recognized game theoretic model. Here you will find Prisoner's Dilemma, Chicken, and Battle of the Sexes. We also devote an entire section to analyzing a model for Final Jeopardy! which demonstrates the difference between the colloquial and mathematical understanding of Prisoner's Dilemma.

Chapters 4 and 5 treat independent topics: bargaining and sequential play. Both of these require the foundation laid in Chapters 1–3. The later chapters build on sophistication of both topic and presentation. Chapter 6 relies on the study of sequential games in Chapter 5 to model missing information. Chapter 7, which covers repeated play, relies on the topics of strategic games covered in Chapter 3. Chapter 8 models multilateral agreements, which references material covered in Chapter 4.

Chapter 9 is truly a culminating chapter, drawing on the multiple approaches presented in the book to study the problem of resource allocation. Readers should finish all Chapters 1–8 in order to fully engage with the material in Chapter 9.

Each of the chapters begins with sections which provide definitions, explain important concepts, and provide illustrative examples. Sections 3.2, 5.3, 5.4, 6.2, 6.5, 6.6, and 8.4 are each devoted to a single scenario, providing an in-depth modeling example to illustrate key concepts. These could be included in core course material, used as independent studies or projects, or omitted. We also provide many citations throughout the text to suggest further investigation. Each chapter ends with exercises that follow the order of the sections in the text. Asterisks on exercises indicate problems that either move beyond the content of this book or could serve as a student project. Select answers and solutions are included at the back of the book, and an instructor's solution manual is available from the authors.

In our experience, a one-semester course for advanced undergraduate mathematics majors can expect to cover about the first six chapters of this book, which has been designed to be used in a classroom engaged in active learning, incorporating in-class discussions, interactive simulations, and group work. Specifically, we anticipate that courses will end with projects in which students research a new scenario of interest to them, and then build, analyze, and assess game theoretic models for their scenario. Examples from our previous courses have included investigations of pricing decisions in the video game console business, interactions between players and teams in crucial moments of basketball games, and negotiations between governments and protestors. The material in Chapters 7, 8, and 9 can be used as starting points for student projects that extend the material covered in the previous chapters.

About the Cover Images

A multiple stage bicycle race can be represented by a variety of game theory models. Identifying the individual cyclists as the players seems natural, but aggregating the cyclists on a team as a single player may make the resulting model easier to analyze without a significant loss of fidelity. Although there may be several teams, a model with only two teams could capture the essence needed given the goals of an analysis. The final placements provide a reasonable first estimate of player preferences, but a mixture of personal, team, and sports cultural motivations can refine and add intensities to the utilities assigned. Consideration of overall approaches to inter-team competition suggests a one-shot strategic game or a Bayesian game if differences in information are relevant, while intra-team planning and collaboration suggest a bargaining or coalition game, and day-by-day decisions with the mutual benefits of teams suggest a sequential or repeated game.

Acknowledgments

We wish to thank Audrey Malagon for her efforts on this project. After we had worked on this book project for a few years, we asked her to help with the final stages of the writing. Her input and advice improved the exposition significantly, and without her, we may never have finished.

Another professional colleague, Jennifer Galovich, at College of Saint Benedict and Saint John's University, and her student, Lydia Glen, should be recognized for the their careful review of the manuscript while Lydia was completing an independent study of game theory under Jennifer's guidance.

We also recognize the many other students at Goshen, Valparaiso, and Virginia Wesleyan whose studies evolved into content in this book, whose course work shaped how we developed the book, and who tested early versions of this book. Specifically, we would like to thank Savannah Carr at Valparaiso University for her assistance in editing and formatting the manuscript.

We would like to thank Goshen College, Valparaiso University, and Virginia Wesleyan University for their support of this project. In particular, this project was partially supported by Mininger funds from Goshen College.

Finally, this book would not have been possible without the constant encouragement and patience of our families. Rick thanks his wife Michele, and David thanks his wife Jeanne.

CHAPTER 1

Models and Games

Mathematical modeling bridges the mental distance between the real world in which we operate and the abstract world that provides guiding structures. Game theory has a collection of mathematical concepts and tools useful in modeling scenarios having multiple interacting decision makers: people, businesses, governments, animals, and genes. We review the basic components of mathematical modeling, learn the definition of a game, and explore two scenarios amenable to game theoretic modeling, a particular type of mathematical modeling. In addition, we reflect on the general rationality assumptions underlying game theory.

1.1 INTRODUCTION TO MODELING

While the development of a mathematical theory is interesting in its own right, the primary value of game theory, and the principal focus of this volume, is its ability to model real-world scenarios. Mathematical modeling helps provide a scientific understanding of real-world phenomena, using the tools of mathematics to clarify issues and provide precise answers. It is worth mentioning that mathematical modeling is one of many ways to perceive the world: one that is surprisingly accurate for describing the physical world. However, in certain situations, the output of a mathematical model should be considered in light of social, cultural, and ethical values. For example, Schiemann [93] builds a mathematical model to assess whether interrogative torture is efficient. While the mathematical model has value, it isn't designed to depict the ethical issues surrounding torture. Other types of models which represent torture pictorially or through fictional accounts, for example, may be more suited for assessing moral value.

In the natural sciences, modeling has descriptive, explanatory, and predictive goals: what is, why it is, and what will happen. In the social sciences and humanities, modeling can also have a normative goal: what ought to be. With these varied purposes of mathematical modeling in mind, we begin with an overview of the mathematical modeling process.

The first step in mathematical model building is to articulate a precise problem statement. Many, if not all, real-world scenarios are highly complex with many different components and peculiarities. By reviewing the appropriate literature, determining the data that is available, and reviewing the tools we have to work with, we can articulate a well-defined problem statement that we can reasonably hope to answer. This problem statement will enable us to determine what we want the output of our model to be by eliminating output that does not answer the question posed in the problem statement.

With the problem statement determined, the second step in the modeling process is to construct a mathematical structure (model) that represents the scenario and addresses the problem statement. To do this, we must identify the important factors to incorporate into our model and quantify those factors in the form of variables. We'll necessarily make assumptions to simplify the model we are building; it is important to identify and document these assumptions. Because of the complexity of the scenario, or the limits on our ability to use particular tools, it may take several iterations of the modeling process to complete this step, as we continue to refine the model and uncover more data relevant to the problem that we have posed.

The third step in the modeling process is to analyze the model. That is, we use the appropriate mathematical tools to determine solutions to the mathematical problems posed by the model. It is important to realize that the relative sophistication of the tools we are using does not reflect directly on the quality of our modeling process. Very simple tools can sometimes be used to describe complex relationships and, unfortunately, sometimes straightforward solutions are hidden in complex machinery. Much of the work of this volume will be devoted to helping us develop the tools necessary to find solutions to problems posed by game theoretical models.

The fourth step of the modeling process is to assess both the solutions and the model. How well do the solutions that we found correspond with the original scenario? Are they reasonable? Can they be validated by comparison to solutions that others have found in practice? How sensitive is our model to changes in the parameters that we established? How limiting were the assumptions that we made? What changes in the model might improve its accuracy?

Models may be assessed in terms of three traits: fidelity, cost, and flexibility. *Fidelity* assesses how well the model corresponds to the scenario being studied. Models that use real-world data typically have higher levels of fidelity than models which make significant assumptions about data. Secondly, we consider the *cost* to develop and implement the model. Considering real-world data again, we recognize that gathering that data has a much higher cost in time and money than models that use standardized formats. Finally, *flexibility* measures how sensitive the model is to variations in conditions. Again, models relying on real-world data are likely to be less flexible because of the effort required in gathering data while standardized models are more flexible with their various parameters being well-defined. The game-theoretic models described in this volume will typically have highly variable fidelity (because this depends significantly on the modeler's willingness to gather basic data), low cost (because their structure and implementation are straightforward), and reasonable flexibility (because of their internal structures).

These four steps may need to be iterated several times to refine and extend the model. Once we have reached a satisfactory model that accurately represents the real-world phenomena we set out to model, the fifth and final step in the mathematical modeling process is to implement the model and report the results. While it is not strictly part of the modeling process, we should expect and plan to share our results with some audience. For whom were we undertaking this modeling exercise? What do they expect us to report back to them? Answers to these questions may also have a impact on the type, complexity, and the level of detail that we might want to include in our model.

Figure 1.1 displays this process graphically. Adapted from [40], it emphasizes the iterative nature of the modeling process, with attention to the fact that we may loop through the process in different patterns.

1.2 INTRODUCTION TO GAME THEORY

People and organizations make decisions every day to take actions in complicated environments as they interact with others. Their goals are to influence the outcome of an interaction with others in ways that they perceive to be beneficial. Examples can include the following:

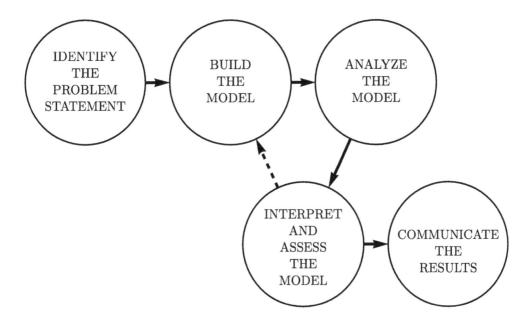

FIGURE 1.1 Mathematical Modeling Process. Adapted from *GAIMME: Guidelines for Assessment and Instruction in Mathematical Modeling Education*, Sol Garfunkel and Michelle Montgomery, editors, Consortium for Mathematics and Its Applications and Society for Industrial and Applied Mathematics, Philadelphia, 2016. Used with permission.

- **Office coffee sharing** In many offices, there is a shared coffee pot, and the employees voluntarily contribute to a pool of money to replenish the supplies. Each employee that wants to enjoy the coffee must decide how much, if anything, to contribute to the pool.

- **Negotiations between countries that share a border** Each of the countries is attempting to pick a course of action that will both ensure security as well as facilitate economic growth and the well-being of its citizens.

- **Representatives voting** Elected representatives may have to vote on policies that directly benefit themselves, like their own raises. Each may want the resolution to pass, but doesn't want to appear self-interested.

- **Sharing of water resources** In many regions, governments need to negotiate how to share common water resources. In these negotiations, each government needs to balance the personal needs of its citizens, needs of the agricultural community, and environmental sustainability.

Game theory is the mathematical field that has been intentionally developed to advise people and organizations on what their optimal strategy is and what outcome they might expect from following this strategy in situations such as these.

To begin building a game theoretic model, we have to answer some critical questions, namely:

1. Who takes action or makes decisions in the scenario?

2. What outcomes can arise?

3. What do the parties know about each other, how may they communicate, and what rules govern their actions?

4. What outcomes do each of the parties prefer?

These questions shape the central mathematical model of game theory, which we call a game.

Definition 1.2.1. A *game* consists of the following:

1. A set $N = \{1, 2, \ldots n\}$ of at least two players.

2. A set O of outcomes that can occur when the game is played.

3. Rules that specify how the joint actions of the players result in an outcome $o \in O$.

4. Utility functions $u_i : O \rightarrow \mathbb{R}$ that specify the preferences among the outcomes for each player $i \in N$.

One of the primary goals of this book is to show how we can model interactive scenarios as (mathematical) games, apply mathematical techniques to analyze the game, and interpret the mathematical solutions within the context of the original scenario. Scenarios/predictions/explanations are part of the real world and games/techniques/solutions are part of the mathematics world; these worlds are connected by assumptions and interpretations.

The above definition gives the precise, mathematical definition of a game. Colloquially we often say "play a game," which usually means "engage in an interactive scenario." Additionally, words chosen to describe scenarios can play to human emotions. For example, a common scenario asks people to choose either "cooperate" or "defect." If instead it had asked them to choose "conspire" or "resist," there are different moral connotations. The mathematical model should be analyzed based on the outcomes and preferences associated to these choices and shouldn't depend on the labels.

Even the precise definition of a game leaves lots of space for variation based on the rules, but there are two basic types of games. In one category are *non-cooperative games*. In these games, each player is focused on their own interests. Their interests may place them in competition with each other, or their interests may align, but there are no binding agreements made between players regarding their actions or choices. They may make their choices simultaneously (Chapter 3) or make choices in sequence (Chapter 5). The second category of games is *cooperative games*, where players arrive at a binding agreement regarding their actions. In these games, players are generally in full communication with each other and have mechanisms to assure implementation of any agreements made. The common goal of a cooperative game is to find a socially optimal outcome: one that collectively optimizes the outcomes for individual players. We will see examples of these in bargaining games (Chapter 4) and coalition games (Chapter 8). In both cooperative and non-cooperative games, variations develop as we consider the possibility of repetitive play or the amount of information that each player has about the situation and interests of the other players. In each of these chapters, we develop specific game models and the mathematical tools to analyze them. For the moment we focus on moving from the real world to the mathematical world.

1.3 EXAMPLES OF GAMES

In this section, we describe three scenarios and the information needed to begin constructing a game theoretical model for each. In particular, we write problem statements and ask many questions about assumptions that we might make and about data that we might search for. Most of the rest of this book focuses on developing the game theoretic tools needed to analyze scenarios such as these.

Tour Bus. There were 29 tourists traveling together on a guided tour of several European cities. When riding on a bus with 60 seats, the tourists moved among the various seats. The group of tourists included several older couples, a number of middle-aged couples, and five adult/child pairings. Some of the "couples" were quads in that they came on the tour together. On the first day, the tour guide advised the group that they should rotate seats — she had a pattern in mind — that would eventually move everyone from the front to the back of the bus. However, what happened in reality was that the children (and their parents) migrated to the back of the bus, as did the gentleman with the really long legs, as he appreciated the back middle seat to stretch out. Another foursome stayed together at the front, as one member claimed to get car sick riding in the back. A final group of four couples moved back and forth among middle seats.

In this scenario, we attempt to model the decision-making process of the tourists when they decide where to sit on the bus each time the group boards it. Thus our problem statement is: Can the tourists collectively find a seating arrangement which will make each tourist as happy as possible?

To construct a game theoretic model for this scenario, we must consider a series of organizing questions:

- **Who are the players?** In this scenario, the set of players may consist of the set of 29 individual tourists. Alternately, the set of players may be each family or group traveling together.

- **What are possible outcomes?** The set of outcomes is the set of all possible seating arrangements.

- **What are the rules that govern players' actions?** Each player must choose a seat on the bus. We can include the rule that the seat must be unoccupied or, alternately, the rule that also allows them to ask someone else to move. This would be a non-cooperative model. If the tourists agree to a sign-up sheet committing them to a particular seat each day, we would model the scenario as a cooperative game. The tourists make their seat choice repeatedly, on each day of the tour. They do not necessarily have much information on the first day, but as the tour progresses they learn more about each other's preferences among the seats.

- **What preferences do the players have among the outcomes?** One person wanted the front to avoid car sickness, the children prioritized the perceived anonymity of riding in the back, and the gentleman with the long legs prioritized a seat in which he could stretch his legs out. Other tourists prioritized sitting with a particular individual rather than a specific seat.

Before we proceed with the next step of the modeling process, each of these questions must be answered concretely by making assumptions, which may be supported by data. For example, this data may come from conversation with the tourists or could be revealed in a literature search of human behavior.

Black Sea Cooperation. In the 1960s, the seventeen countries of the Danube River basin in Europe rapidly expanded their use of agricultural fertilizers to increase crop production. However, since the Danube River feeds into the Black Sea, there were serious environmental consequences as pollution from fertilizer combined with untreated wastewater and other pollution from growing populations in the area. According to [53], "by the nineteen eighties much of the northwest shelf of the Black Sea was hypotoxic, a number of species and benthic ecosystems had disappeared, and economic losses — from fisheries, tourism and other sectors — were estimated at $500 million per year." The countries needed to find a way to work together to control runoff into the water system [53].

Here our problem statement is the following: Can these countries come to a mutual agreement on how each will control the run off into the water system?

- **Who are the players?** In this case, it is obvious that the 17 countries are each players, but we may ask if there are any other players, such as United Nations agencies, or other international governance bodies.

- **What are possible outcomes?** The set of outcomes consists of all of the possible agreements that the countries could reach. For example, one agreement could be that upstream countries are responsible for keeping pollution below a certain threshold and paying a penalty if they do not. Another agreement may be that each country is responsible for cleaning any water it uses.

- **What are the rules that govern players' actions?** They might fully cooperate with everyone else, or partially cooperate with only some of the other countries. Each country could have different levels of information about their own and others' costs and benefits. There may be a concurrent decision in which all seventeen countries sit down at a conference and come to an agreement. It may be either of two sequential processes: one in which the primary countries reach an agreement, then the secondary countries join the coalition, or another in which the group of countries reach an agreement on a topic, then proceed to negotiate a next topic. In all cases, it is reasonable to model this as a cooperative game with each outcome having a binding agreement.

- **What preferences do the players have among the outcomes?** It is reasonable to assume that each country prioritizes its own interests over others, but they may also prioritize the interests of some allies or neighbors.

Again, each of these questions must be answered concretely to proceed with the model.

Olympic Aerial Skiing. In 2014, NPR interviewed Olympic aerial skiers and discussed the game theory behind how they decide on their routines. Like figure skating, each of a ski jumper's attempts are scored both on their performance and on the difficulty of the trick. Unlike figure skating, ski jumpers are only allowed to perform each trick once. They must also react to others' moves, sometimes making split-second decisions to change their routines in the hopes of improving their chances of advancing to the next round [14].

This gives rise to the problem statement: Can we find an ordering of tricks for each participant in the competition which maximizes their scores? For this scenario, rather than discussing a wide range of possible answers to our organizing questions, we will instead make necessary assumptions to define a specific game.

- **Players:** The players are the individual contestants. This means that we assume that players from the same country compete aggressively even against their fellow teammates.

- **Outcomes:** An outcome is a list of scores, one for each player.

- **Rules:** To keep our model simple, we assume that each player has one hard trick and one easy trick and they must perform two tricks total. Therefore, since a player cannot repeat a trick, each must decide the order of their two tricks. They must decide this before the competition and report their decision to the judges without knowing what the other players are doing. Finally, to determine how trick ordering results in scores, we use a probability distribution for each player based on information about their abilities.

- **Preferences:** Finally, we assume that a player prioritizes outcomes in which their score is highest in the list of scores and is indifferent to the scores of the other players, including their own teammates.

The Tour Bus scenario may seem at first glance to be the least significant of the scenarios. However, it illustrates several key ideas relevant to game theoretic modeling and provides a good framework for observing the process of mathematical modeling. First, since there is no data available, assumptions need to be made about the tourist preferences as we did in modeling the Olympic Skiing scenario. Second, while the Tour Bus scenario involves collaboration, it may not be a cooperative game since there need not be a formal process of collaboration among the tourists; they may each simply chat and eventually choose places to sit. In contrast, in Olympic Skiing, we made the assumption that collaborating was completely against the rules. Both the Black Sea Cooperation scenario and the Olympic Aerial Skiing scenario hint that there is much more data underlying the scenario which should influence our assumptions.

1.4   RATIONALITY ASSUMPTION

We make the major assumption throughout our modeling processes that players act rationally. By this we mean that each player will choose actions that optimize their own utility functions. This does not necessarily mean that a player acts selfishly, but rather that any altruistic motives have already been built into their preferences among the outcomes. For modeling purposes, it is important that preferences, whether selfish or altruistic, be described by the utility functions accurately, and that players make choices to lead them toward their preferred outcomes. Otherwise, when the models seek to apply mathematical methods to the data provided by the utility functions, they will not be accurate. We assume that rationality also means that each player is acting on their best estimate of what the other players' preferences are. We will describe situations in which players know precisely what each of the other players preferences are as well as situations in which players do not know others' preferences and need to make assumptions about them.

Our assumptions about rationality give us the structure we need to construct and analyze our models mathematically. For example, we had to assume that the skiers want to win and will, therefore, choose the trick order that results in the best chance of winning in order to build the model. It is possible, of course, that perfectly reasonable models with logically precise solutions describe behavior that is rarely practiced by real human beings going about their daily lives.

A common game theory scenario, the *Ultimatum scenario*, involves two players and a sum of money given to one of them, the Proposer. The Proposer then offers to give some portion of this money to the Responder. If the Responder accepts the offer, then the Proposer keeps the remainder. If the Responder rejects the offer, then both players lose all of the money.

The mathematical analysis of a simple model of this scenario shows that the Proposer should offer the smallest unit of money possible and the Responder should accept any amount offered. Interestingly, this is not what happens in reality. Consistently, across many experimental settings, Proposers offer as much as half and Responders generally reject offers of less than a quarter [50].

Camerer [21] explains the findings in [50]. The inconsistency between what the mathematical analysis of the simple model says should happen and what actually happens can be explained by examining the motivations of the two players. The simple model is grounded in the assumption that each player is self-interested and attempting to maximize their own gain. In reality, players are often motivated by other less selfish goals. In this case, the behavior observed in the experiments reflects the players' sense of fairness. The Proposers' willingness to give more than they are expected to give demonstrates their willingness to be fair. On the other hand, the Responders' rejection of small offers demonstrates their willingness to punish unfair Proposers, even at a cost to themselves. This shows the impact of assumptions on the fidelity of a model. With different assumptions and a more sophisticated model, we can achieve higher fidelity.

Examining data or performing experiments can help us understand which assumptions are reasonable for our model and test the fidelity of a model. For example, Palacio-Huerta [79] uses many years of data from professional soccer teams and competitions to validate game theoretic constructs. He explores the relationship between the theoretically predicted behavior of kickers and goalies competing in a penalty kick situation to what happens in high-end professional leagues. To conduct this exploration, he first uses data from over 1400 penalty kicks during a five-year period to construct the payoff matrix in Table 1.1. The payoffs in the table indicate the probability of the kicker scoring a goal.

TABLE 1.1   Soccer Scoring Probabilities

	Guard Left	Guard Right
Kick Left	58.30	94.70
Kick Right	92.91	69.92

Game theorists calculate a mathematical solution for this model called the Nash equilibrium. Table 1.2 compares this solution to the observed behavior of the professional soccer players. Clearly, professional players' actions are consistent with the theory.

TABLE 1.2   Soccer Scoring: Predicted vs. Actual

	Guard Left	Guard Right	Kick Left	Kick Right
Nash equilibrium	41.99	58.01	38.54	61.46
Observed Behavior	42.31	57.69	39.98	60.02

In his book [80], Palacio-Huerta also investigates whether or not experiments conducted in controlled laboratory settings can mirror activity occurring in the real world. To do this, he conducted a series of simulated penalty kick experiments in a controlled environment. When the participants were top-quality European players playing against opponents from other teams, the strategy behavior of both kickers and goalies again closely mirrored that predicted the game theoretic solution and their own on-field performance. When the same experiment was conducted with lower-quality players, play in the controlled environment varied significantly from the predicted solution, but this same sub-optimal play was observed in the empirical data from these players' real game play. Thus, one is led to the conclusion that the ability to play optimal game theoretic strategies is a learned behavior. Participant behavior in controlled experiments will not suddenly be optimal if they had not been practiced prior to the experiment.

Not only can data from real-world observations help construct and validate game theoretic models, but game theoretic models can also help explain observed real-world behavior. In many cases, research on human behavior in the social sciences is conducted by surveys, by analysis of aggregated public data, or by experiments. Behavioral game theory investigates human behaviors, but its methods are distinctive from other fields. Unlike psychology experiments, for example, experiments in behavioral game theory attempt to be as transparent as possible to encourage players to play to known preferences. Participants have full knowledge of their available actions, the actions that the other players have, and the consequences to everyone of choosing specific actions.

A student conducted a behavioral game theory experiment at a Christian university to answer the questions: Are practicing church members more generous than non-practicing? Is there such a thing as "Catholic guilt"? To obtain answers to these questions, the student designed an experiment using the Ultimatum scenario and a Dictator scenario [92]. The Dictator scenario is even simpler than the Ultimatum scenario: The Proposer is given the money and told to distribute it between the two players; the Respondent has no option to reject what is offered. The mathematically correct action in the Dictator game is for the Proposer to keep all of the money. While the Ultimatum scenario can be seen as a test for fairness, the Dictator scenario can be interpreted as an indicator of guilt. Since there is no incentive to offer larger sums to the Respondent, doing so would indicate "guilt" on the part of the Proposer for receiving the unexpected windfall.

To conduct his experiment, the researcher recruited 43 students from introductory economics courses. These students were equally split between Catholics and non-Catholics; a few participants were non-Christian. They were uniformly distributed among the four years of college and evenly split between quantitative and non-quantitative majors. A large majority of the participants were white, reflecting the demographics of the campus. During the experiment itself, each participant played five rounds of the Ultimatum scenario and five rounds of the Dictator scenario. In each round, the participant indicated the portion of $15 that they would offer as the Proposer and what they would accept as the Responder. Their responses would then be randomly paired with another anonymous player to determine four payoff values for the participant (two each as Proposer and Responder). The two types of scenarios were played in alternating order, Ultimatum-Dictator-Ultimatum, with the participants being reminded of which type of scenario was being played and of the important differences between the two. This period of instruction was used to slow down the participant responses and to encourage them to pause to consider the type of game being played.

This student project is not asking whether any of the participants play the optimal strategy, but rather is simply using the organizational structure of game theory to explore differences in human behavior. In a similar way, Pech and Swicegood [82] used a variation of the Ultimatum scenario to explore cross-cultural trust and trustworthiness behaviors. In their experiment, the Proposer and Respondent, each either an American or Croatian college student, were given money. The amount the Proposer decided to give away was tripled by the experimenters before being given to the Respondent. The Respondent then had the opportunity to give some or all of their money back to the Proposer. The amount of money given by the Proposer was an indicator of trust (that they will get at least that amount back) and the amount of money given by the Responder was an indicator of trustworthiness (that they deserved the trust). The experiment revealed how students interacted with others of the same nationality, the other nationality, and when operating in a state of doubt about nationality.

Summarizing, experimental game theory tests whether or not game theoretic models represent real behavior by individuals, corporations, or countries. Behavioral game theory uses game theoretic models to specifically explore theories of human behavior. In the former case, we test whether players behave in mathematically predicted ways, while in the second we use mathematical theory to understand players' behaviors.

EXERCISES

1.1 **Purchasing Gasoline**. Gasoline prices change frequently and will vary from gas station to gas station, causing drivers to make decisions about where to buy their gasoline. In this problem, we build a non-game theoretic mathematical model to decide whether to drive further to buy cheaper gasoline.

    a. Precisely articulate the problem statement.

    b. Let S be the savings in the gasoline purchased from the farther station and T the cost of driving to the farther station. What assumptions have we already made by defining these variables?

    c. What factors influence the values of S and T?

    d. Build a mathematical model by describing the relationships between S and T that dictate driving farther or not.

1.2 **Buying a Home**. Should a person buy or rent a home? Construct a mathematical model to answer the question. Be sure to include the cost of housing in the community, the terms of a mortgage, rental rates, and value of the home as an asset. Give a precise problem statement, list assumptions, define variables, and explain how the model dictates choices based on the problem statement.

1.3 One can build computer simulations of various phenomena. Comment on the fidelity, cost, and flexibility of creating a model using computer simulations, as compared to collecting real-world data.

1.4 **Video Store**. Several decades ago, the owner of a small video store might have needed to decide between stocking Beta and VHS video tapes. At the time, it was unclear which system would have the superior product, but it was clear that one or the other would emerge as such. For this scenario, give a precise problem statement and discuss the four organizing questions for defining a game that models the store owner's decision-making process.

1.5 Two naval forces, R and C, must decide how to spread out their aircraft carriers around the world. Here we simplify the problem and assume two admirals must deploy their fleets between the northern and southern areas of an ocean. We'll assume Admiral R has three fleets to split among the areas and Admiral C has two fleets to disperse. For this situation, give a precise problem statement and discuss the four organizing questions for defining a game [46].

1.6 Bill is going to a see a movie at the local multiplex cinema. Bill claims that he prefers them according to the ranking (1) *Commission Possible*, (2) *Lawn of Daydreams*, (3) *Lord of the Hoops*, and (4) *Pie Rats of the Jelly Beans*. Yet, when he arrives at the multiplex cinema, Bill chooses to view *Lord of the Hoops*. Give at least two explanations for why Bill's choice would not be considered a violation of rationality.

1.7 In William Styron's book *Sophie's Choice*, Sophie is forced to choose which of her children, Eva or Jan, is saved from the gas chamber in Auschwitz. If Sophie does not choose, both children will die. From her decision to allow Jan to live, is it rational to believe that Sophie prefers that Jan lives over Eva living? Can we infer that Sophie preferred the outcomes in the following order: (1) Jan lives, (2) Eva lives, and (3) neither lives?

1.8 Have pairs of acquaintances enact the Ultimatum scenario with no incentives. Does their average action reflect what the literature suggests should happen? Interview several of the Proposers and Respondents. What justifications do they provide for their actions?

1.9 Have pairs of acquaintances enact the Ultimatum scenario with financial incentives linked to the outcomes. How do their actions change in comparison with the game with no financial incentives? What justifications do they provide for their actions?

1.10 Describe a scenario in your life that could be analyzed with a game theoretic model, identifying the key components of a game theory model in your scenario.

CHAPTER 2

Player Preferences

Player preferences among outcomes are revealed by free choice behavior of the player. Utility functions are presented as a mechanism for modeling player choices among the outcomes in a game. The properties of two common classes of utility functions, ordinal and von Neumann-Morgenstern, are developed, and multiple examples of each type are explored.

In Chapter 1, we discussed how to translate a real-world scenario to a game model by identifying the players, outcomes, rules, and preferences. We will model preferences by utility functions which assign numerical values to the possible outcomes based on the players' preferences.

In most games, possible outcomes are determined by the interaction of choices made by the players. These choices can reflect their own preferences among the outcomes as well as the preferences of the other players among those outcomes. However, if a player can freely choose among the outcomes of a game, then choice should be synonymous with preferences: to say that a player *prefers* outcome A to outcomes B and C means that the player would choose A if offered a choice among outcomes A, B, and C. To say that a player is *indifferent* between outcomes A and B means that the player would be willing to choose either outcome or even to flip a coin to determine their choice. We will construct utilities in such a way that choices and preferences align.

2.1 ORDINAL UTILITIES

Ice Cream Parlor. Scarlett goes to an ice cream parlor that usually carries three flavors: vanilla, chocolate, and strawberry. At certain times, the parlor may be out of any given flavor. Scarlett might make a variety of choices upon her arrival, including leaving with no ice cream.

In this scenario, there are four possible outcomes: Scarlett leaves with vanilla ice cream, which we will label V, leaves with chocolate ice cream, labeled C, leaves with strawberry ice cream, labeled S, or leaves with no ice cream, labeled N. Our outcome set is then $\{V, C, S, N\}$. The rules dictate that Scarlett may choose any outcome that is available.

To model Scarlett's preferences and build a utility function, we consider the choices she would make when certain outcomes are available. Supposing that all potential outcomes are available, Scarlett would choose V; if C were not an available outcome, Scarlett would choose S; if only V and N were available outcomes, Scarlett would flip a coin to determine whether to select V or N; and if only C and S were available outcomes, Scarlett would be unable to make a choice. While this behavior would be possible, most would find it bizarre: Scarlett's choice of V when presented with the outcome set $\{V, C, S, N\}$ suggests she prefers V to S, but Scarlett's choice of S when presented the outcome subset $\{V, S, N\}$ suggests she prefers S to V. To avoid such absurd possibilities, we will assume that individual behavior is governed by self-consistent internal preferences over the outcomes, which is reflected in the mathematical definition of ordinal preferences below.

Definition 2.1.1. A player i is said to have *ordinal preferences* among outcomes if there exists a utility function u_i from the set O of outcomes into the real numbers, $\mathbb{R}$, such that whenever presented with a subset $O' \subseteq O$ of outcomes, player i chooses any of the outcomes that maximize u_i over all outcomes $o \in O'$.

To ensure that ordinal preferences align with the players real-world choice behavior, we note that whenever player i prefers outcome o_j over outcome o_k, we should have $u_i(o_j) > u_i(o_k)$, and when player i is indifferent between o_j and o_k we should have $u_i(o_j) = u_i(o_k)$.

We now ask under what conditions Scarlett's outcome choice behavior can be modeled by ordinal preferences and describe three reasonable properties for a self-consistent set of choices. Since it is usually easier for a player to choose between two rather than among many outcomes, these properties will focus on pairwise choices.

First, we want Scarlett's pairwise choices to be *complete*, meaning that whenever she is presented with a pair of outcomes, Scarlett is able to make a choice. Equivalently, for each pair of outcomes, A and B, exactly one of the following conditions holds: (a) Scarlett chooses A over B, (b) Scarlett chooses B over A, or (c) Scarlett is willing to flip a coin to determine which outcome to choose (in this case we will often say Scarlett chooses either A or B). Hence, this condition excludes the following option as a possibility: (d) Scarlett chooses neither A nor B. (When we want the rules to allow a player to choose none of the options, we must include that as an outcome, as we did in the Ice Cream Parlor scenario.) For (a), we will assign utilities so that $u(A) > u(B)$. Likewise for (b) we will assign utilities so that $u(B) > u(A)$. Finally for (c), since Scarlett is willing to flip a coin to determine the outcome, we assume she is indifferent between A and B and assign $u(A) = u(B)$.

When Scarlett chooses A from the set $\{A, B\}$ we can assume that either (a) or (c) holds from above. When this happens, we say Scarlett *weakly chooses* A over B. More generally, when Scarlett chooses A from a set of outcomes O', we can assume that for every outcome $o \in O'$, Scarlett either prefers A over o or is indifferent between A and o. In this case we say Scarlett *weakly chooses* A from the set O'.

To model Scarlett's choice with ordinal utilities, we will require that her choices be *transitive*. That is, for each triple of outcomes, A, B, and C, if Scarlett weakly chooses A when presented with $\{A, B\}$, and weakly chooses B when presented with $\{B, C\}$, then Scarlett weakly chooses A when presented with $\{A, C\}$.

Finally, we want Scarlett's pairwise choices to be *generalizable*, meaning that when presented with a subset O' of the set of outcomes O, Scarlett weakly chooses an outcome A from O' if and only if for each outcome $B \in O'$ Scarlett weakly chooses A whenever presented with $\{A, B\}$.

Theorem 2.1.1 (Ordinal Preferences Characterization). A player has ordinal preferences over a finite set of outcomes if and only if the player's pairwise choices are complete, transitive, and generalizable.

Before presenting the proof of this theorem, we demonstrate how it can be used to construct a player's ordinal utility function from the player's choices among specified subsets of outcomes. We illustrate with the Ice Cream Parlor scenario. We first ask Scarlett which outcome she would choose when presented with the outcome set $\{V, C, S, N\}$. If she states she would choose V, we then define $u(V) = -1$ (the number -1 was chosen to remind us that V is Scarlett's 1st choice). We next ask Scarlett which outcome she would choose when presented with the outcome subset $\{C, S, N\}$. If she states she would flip a coin to choose between C or S, we define $u(C) = u(S) = -2$ (the number -2 was chosen to remind us that C or S is Scarlett's 2nd choice). Since there is now only one outcome with an unassigned utility, we define $u(N) = -3$. In general, we assign $u(A) = -k$ if outcome A is among the outcomes chosen at the kth step; we could say that A is *ranked* kth by the player.

Scarlett's ordinal utility function is now defined, but how do we know that it will correctly predict Scarlett's choices when presented with any of the eleven possible subsets of two or more outcomes? For example, if presented with the subset $\{V, S, N\}$, the utility function u is maximized with the outcome V. Will this be Scarlett's choice? If Scarlett's pairwise choices are generalizable, then her choice of V when presented with $\{V, C, S, N\}$ implies that V would be chosen over S and V would be chosen over N, which implies (via another application of the generalizable property) that V would be chosen when Scarlett is presented with $\{V, S, N\}$.

In summary, we have demonstrated a straightforward process which uses a player's choices among carefully chosen sets of outcomes to construct their ordinal utility function over the entire (finite) set of outcomes. The proof will show that if the player's pairwise choices are complete and transitive, we can always use such a process to construct an ordinal utility function. The proof will also show that if the player's pairwise choices are generalizable, then the constructed utility function will accurately predict the player's choices for all subsets of outcomes. To complete the proof (although it will be handled first), we will show that if a player's choices can be predicted with an ordinal utility function, then the player's pairwise choices must be complete, transitive, and generalizable. This makes it clear what is being assumed when we model a player's choices with an ordinal utility function.

Proof. ($\Rightarrow$) Suppose a player has ordinal preferences among a finite set of outcomes O. By definition, there exists a function $u : O \to \mathbb{R}$ for which the player's choices maximize u over the outcomes that are available. We will prove that the player's pairwise choices are complete, transitive, and generalizable.

Let $A, B \in O$ be outcomes. Since $u(A), u(B) \in \mathbb{R}$, exactly one of the following conditions holds: (a) $u(A) > u(B)$, (b) $u(B) > u(A)$, or (c) $u(A) = u(B)$. Since the player's choices maximize the function u, these conditions correspond to the three conditions that show the player's pairwise choices are complete.

Suppose the player weakly chooses A when presented with $\{A, B\}$, and weakly chooses B when presented with $\{B, C\}$. By the definition of u, it follows that $u(A) \geq u(B)$ and $u(B) \geq u(C)$. By the transitivity of $\geq$ over the real numbers, this implies $u(A) \geq u(C)$. Again, by the definition of u, it follows that the player will weakly choose A when presented with $\{A, C\}$. Hence, the player's pairwise choices are transitive.

Now suppose the player weakly chooses an outcome A from a subset $O' \subseteq O$. By the definition of u, it follows that $u(A) \geq u(B)$ for all outcomes $B \in O'$ Again, by the definition of u, the player weakly chooses A when presented with $\{A, B\}$ for any $B \in O'$. Conversely, suppose the player weakly chooses A whenever presented with $\{A, B\}$ for any $B \in O'$. By the definition of u, it follows that $u(A) \geq u(B)$ for all $B \in O'$. Again, by the definition of u, the player weakly chooses A when presented with the entire subset O'. Hence, the player's preferences are generalizable.

($\Leftarrow$) Suppose a player's pairwise choices for outcomes in set O are complete, transitive, and generalizable. We will show there exists a utility function $u : O \to \mathbb{R}$ for that player such that whenever that player is presented with a subset $O' \subseteq O$, the player chooses an outcome that maximizes the function u over all outcomes in O'. To do this, we first claim there must be at least one "top" outcome $o_1 \in O$ which is always weakly chosen from any set $\{o_1, B\}$ for $B \in O$. Indeed, suppose to the contrary, that for every outcome o_j there is another outcome o_{j+1} that the player chooses instead of o_j when presented with $\{o_j, o_{j+1}\}$. Because there are a finite number of outcomes, this chain of outcomes $(o_j, o_{j+1}, o_{j+2}...)$ must repeat itself eventually, say at $j + i = k$. By construction, outcome o_{k+1} is chosen over o_k. But since o_{k+1} is equal to one of the outcomes earlier in the chain than o_k, by transitivity, o_k is weakly chosen when the player is presented with o_k and o_{k+1}. This is a contradiction, so we must have at least one outcome in O which is always a weakly chosen outcome when presented in a pair of outcomes from O. Define $u(o) = -1$ for all such outcomes and remove these outcomes from consideration.

Now identify all outcomes that are always weakly chosen outcome(s) when considered as part of a pair among the remaining outcomes (the argument in the preceding paragraph ensures the existence of such outcomes). Define $u(o) = -2$ for these outcomes and remove these outcomes from consideration. Repeat this procedure to define $u(o) = -3$, $u(o) = -4$, and so forth for disjoint sets of outcomes. Because the set of outcomes is finite, this procedure exhausts all outcomes in a finite number of steps and defines our utility function. To finish the proof, we must show that with the utility function u that we have defined, when presented with a subset O' of outcomes, the player will choose any of the outcomes that maximize u over all outcomes $o \in O'$. Let A be an outcome that maximizes u in O'. We will show that A is weakly chosen. Let C be any other outcome in O'. Then we must have $u(C) \leq u(A)$. By the process we used to construct u, we know there exists a set containing C and A in which A was weakly chosen. Since the player's choices are generalizable, A is weakly chosen from the set $\{A, C\}$. Since C was arbitrary, another application of the generalizable property shows that A is chosen when the player is presented with the entire subset O'. $\qquad\square$

In summary, this theorem assures us that if a player's choice behavior satisfies three reasonable properties, then their preferences can be modeled by a utility function. If their choice behavior does not satisfy any one of the three properties, as in the example at the beginning of this section, then their preferences cannot be modeled by a utility function and we cannot (at least with the tools developed in this book) move forward in our analysis.

A player with ordinal preferences can have their choice behavior modeled by many different ordinal utility functions. In the Ice Cream Parlor scenario, the initial ordinal utility function we constructed for Scarlett was

$$u_R(V) = -1, \ u_R(C) = -2, \ u_R(S) = -2, \ u_R(N) = -3.$$

It is easy to see that

$$v_R(V) = 3, \ v_R(C) = 2, \ v_R(S) = 2, \ v_R(N) = 1$$

and

$$w_R(V) = 100, \ w_R(C) = 73, \ w_R(S) = 73, \ w_R(N) = 0$$

are ordinal utility functions predicting the same choice behavior for Scarlett among the eleven subsets of two or more outcomes, and so we will call these utility functions *ordinally equivalent*. In fact, any function u satisfying

$$u(V) > u(C) = u(S) > u(N)$$

will predict the same choice behavior for Scarlett among the eleven subsets of two or more outcomes. However, the utility function

$$u_C(C) = 100, \ u_C(S) = 100, \ u_C(V) = 68, \ u_C(N) = 0$$

would not model Scarlett's preferences because, for example, it incorrectly predicts her choice when presented with the set $\{V, C\}$.

The number of outcomes from which Scarlett might have to choose may be a very large finite number, making it difficult for her to intentionally consider all possible comparisons. For example, when Scarlett goes to her university registrar to sign up for next semester's classes, the set of outcomes available to her consists of all possible course schedules that she might have. This situation could be modeled as a game because there are many players influencing the outcome. Here the outcome Scarlett ends up with not only depends on her choices, but also on the choices of the students who have registered before her (affecting the availability of seats) and after her (affecting whether or not a course runs). But before we consider the full game, to model Scarlett's choices we must determine her preferences among all possible schedules and create a corresponding utility function. In the real world, Scarlett may only examine three or four feasible schedules, however, we will assume that Scarlett has a utility function defined on all logically possible schedules even if there are millions of such schedules and some are not possible when she registers because individual courses have filled.

2.2   VON NEUMANN-MORGENSTERN UTILITIES

A significant limitation of ordinal preferences and their associated utility functions is that they cannot describe the intensity of a player's preference for a particular outcome. That is, they cannot capture the difference between Scarlett preferring vanilla ice cream over chocolate ice cream and Scarlett so strongly preferring vanilla that she would pay for it rather than have a free serving of chocolate. Notice how we have once again translated our intuitive sense of internal preference intensity into something that is observable (a real-world choice) so we can create a utility function based on these choices. While asking players to choose among outcomes that include the receipt or payment of money would be one observable way to determine intensity of preferences, we will take an approach that does not rely on the availability of money.

We begin by introducing a new, probability-based outcome called a *lottery*. Suppose that when a second customer, Regis, enters the ice cream parlor, he encounters a college student conducting a taste test involving different flavors of ice cream. The college student offers Regis the choice of either a sample of chocolate ice cream (his second-most favorite) or an unknown sample that is either vanilla (his favorite) or strawberry (his least favorite). The second option in this example is a simple lottery.

Definition 2.2.1. Given a set of outcomes, O, a *simple lottery* is a probability distribution over this set. When $O = \{o_1, o_2, \ldots, o_m\}$, a finite set, a simple lottery can be denoted by $p_1 o_1 + p_2 o_2 + \ldots + p_m o_m$ where p_i is the probability of outcome o_i. A *compound lottery* is a probability distribution over other lotteries. Because an outcome $o \in O$ can be written as the simple lottery $1o$, we see that $O \subset \mathcal{L}$, the set of all (simple and compound) lotteries.

To reveal the strength of a player's preference for one outcome over another, we must examine not only choices between single outcomes but choices between single outcomes and lotteries. Suppose Regis prefers vanilla V over chocolate C and prefers C over strawberry S. The choice Regis makes between C and the lottery $0.5S + 0.5V$ tells us about the strength of his preference for V over C and for C over S. If he would choose either of the two possibilities, then the strength of Regis's preference for C is exactly halfway between S and V. If Regis were to choose C over $0.5S + 0.5V$, it reveals that his preference intensity for C is closer to V than to S. If Regis were willing to choose either C or the lottery $0.1S + 0.9V$, it would reveal that his preference for V over C is very small and his preference for C over S is relatively large. However, if he were instead willing to choose C or the lottery $0.9S + 0.1V$, it would reveal a strong preference for V over C and that his preferences for C over S is small. When a player is willing to choose either of two lotteries, this reveals the player is indifferent between these choices. This motivates the following generalization of the utility function concept.

Definition 2.2.2. A player i is said to have *von Neumann-Morgenstern preferences*, abbreviated *vNM preferences*, among outcomes O if there exists a function $u_i : \mathcal{L} \to \mathbb{R}$ from the set of lotteries $\mathcal{L}$ into the real numbers, $\mathbb{R}$, satisfying the following conditions:

1. The player has ordinal preferences among lotteries, that is, whenever presented with a subset $\mathcal{L}' \subseteq \mathcal{L}$ of lotteries, player i will choose any of the lotteries that maximize u_i over all lotteries $L \in \mathcal{L}'$.

2. The utility function u_i satisfies the *Expected Utility Hypothesis*, that is,

$$u_i(p_1 o_1 + p_2 o_2 + \ldots + p_m o_m) = p_1 u_i(o_1) + p_2 u_i(o_2) + \ldots + p_m u_i(o_m)$$

for any lottery $p_1 o_1 + p_2 o_2 + \ldots + p_m o_m \in \mathcal{L}$.

Obviously, allowing lotteries to be considered as outcomes increases the complexity of the space of outcomes. We will certainly want our pairwise choices over this new set of outcomes to be complete, transitive, and generalizable as an extension of the concept of ordinal preferences. There are four additional properties we want a player's choice behavior to satisfy now that the outcome space includes compound lotteries, that is, lotteries which are probability distributions over other lotteries.

First, we want Regis's pairwise choices to be *probabilistic*, meaning that he will be indifferent between two lotteries that have the same probability distribution over the original outcome set. For example, Regis should be indifferent between

$$0.3C + 0.7(0.5V + 0.5S)$$

and

$$0.3C + 0.35V + 0.35S.$$

Second, we want Regis's pairwise choices to be *monotonic*, meaning that if Regis prefers C over S, then whenever $q > p$, Regis will choose the lottery $(1 - q)S + qC$ over the lottery $(1 - p)S + pC$.

Third, we want Regis's pairwise choices to be *continuous*, meaning that if Regis prefers V over C and prefers C over S, then there exists a probability p for which Regis is indifferent between C and the lottery $(1 - p)S + pV$.

Finally, we want Regis's pairwise choices to be *substitutable*, meaning that if Regis is indifferent between the lotteries L_1 and L_2, then Regis is indifferent between the two lotteries $(1 - p)L_1 + pL_3$ and $(1 - p)L_2 + pL_3$ regardless of the value of p or the probability distribution in the lottery L_3.

The following theorem demonstrates that the Expected Utility Hypothesis not only provides an intuitive method for assigning utilities, it does so in a manner consistent with the properties we require.

Theorem 2.2.1 (vNM Preferences Characterization). A player has vNM preferences over the lotteries $\mathcal{L}$ associated with a finite set of outcomes and a vNM utility function $u : \mathcal{L} \to \mathbb{R}$ satisfying the Expected Utility Hypothesis if and only if the player's pairwise choices within $\mathcal{L}$ are complete, transitive, generalizable, probabilistic, monotonic, continuous, and substitutable [119].

Again, before we provide the proof of this theorem, we will demonstrate how to use the theorem to construct a player's utility function from the player's choices between outcomes and specific lotteries. We illustrate with Regis in the ice cream parlor. We first ask which outcome he would choose when presented with $\{V, C, S\}$. If Regis states he would choose V, we would define $u(V) = 1$. We next ask Regis which outcome he would never choose in pairwise comparisons. If Regis states he would never choose S, we then define $u(S) = 0$. Finally, we ask Regis to compare each of the remaining outcomes with lotteries involving the "worst" and "best" outcomes. In this example, Regis is asked to compare outcome C with the lotteries $(1-p)S + pV$ for different probabilities p. Assuming Regis's pairwise choices are continuous, there should be a probability p_C such that Regis would choose either outcome C or lottery $(1 - p_C)S + p_C V$. With this information, we define

$$u(C) = (1 - p_C)u(S) + p_C u(V) = (1 - p_C)(0) + p_C(1) = p_C.$$

If Regis states he is indifferent between C and $0.7S + 0.3V$, then $p_C = .3$ and we define $u(C) = 0.3$. Intuitively, Regis's preference intensity for chocolate is 30% of the distance between his preference intensity for strawberry and vanilla.

Regis's vNM utility function is now defined on outcomes and extended to lotteries using the Expected Utility Hypothesis, but how do we know that this utility function will correctly predict Regis's choices when presented with any of the infinite possible subsets of two or more outcomes and/or lotteries? For example, if presented with the subset consisting of the lotteries

$$K = 0.2S + 0.5C + 0.3V \text{ and } L = 0.6S + 0.4V,$$

the utility function u is maximized with the lottery K because

$$u(K) = 0.2u(S) + 0.5u(C) + 0.3u(V) = 0.45$$

and

$$u(L) = 0.6u(S) + 0.4u(V) = 0.4.$$

Will K be Regis's choice? If Regis's pairwise choices are substitutable, then he is indifferent between K and

$$K' = 0.2S + 0.5(0.7S + 0.3V) + 0.3V.$$

If Regis's pairwise choices are probabilistic, then he is indifferent between K' and

$$K'' = 0.55S + 0.45V.$$

If Regis's choices are monotonic, then he prefers K'' to L and so if Regis's pairwise choices are transitive, then he prefers K to L.

In summary, we have a straightforward process to construct a player's vNM utility function over a finite collection of outcomes. The proof will show that if the player's pairwise choices are complete, transitive, and continuous, then the player will be able to choose a best and worst outcome as well as determine the probabilities for which they are indifferent between the other outcomes and a lottery of the best and worst outcomes. The proof will also show that if the player's pairwise choices are transitive, probabilistic, monotonic, and substitutable, then the constructed utility function predicts the player's choices for all pairs of outcomes and/or lotteries. The proof will furthermore show that if the player's pairwise choices are generalizable, then the constructed utility function predicts the player's choices for all subsets of outcomes. Conversely (although argued first), the proof will show that if a player's choices can be predicted with a vNM utility function, then the player's pairwise choices must be complete, transitive, generalizable, probabilistic, monotonic, continuous, and substitutable. This makes it clear what is being assumed when we model a player's choices with a vNM utility function.

Proof. (⇒) First, we will assume that the player has vNM preferences and an associated utility function $u : \mathcal{L} \to \mathbb{R}$ satisfying the Expected Utility Hypothesis and will demonstrate that the player's choices satisfy all the listed properties. Suppose A and B are lotteries. Since $u(A), u(B) \in \mathbb{R}$, exactly one of the following holds: $u(A) > u(B)$, $u(B) > u(A)$, or $u(A) = u(B)$. This implies that the player either chooses A over B, chooses B over A, or is willing to flip a coin to choose between A and B, and hence the player's pairwise choices are complete. It is similarly straightforward to see that the player's pairwise choices are transitive and generalizable, as we did in the proof of Theorem 2.1.1.

Now suppose that two (possibly compound) lotteries, L_1 and L_2, simplify to the same probability distribution over the pure outcomes. This shared probability distribution can be used to define a third lottery, L_3, over the pure outcomes. By the expected utility hypothesis, $u(L_1) = u(L_3)$ and $u(L_2) = u(L_3)$ and hence $u(L_1) = u(L_2)$. Therefore, the player would choose either L_1 or L_2, and so the player's pairwise choices are probabilistic.

Suppose $q > p$ and the player chooses outcome C over outcome A. Then $u(C) > u(A)$ and

$$q\left(u(C) - u(A)\right) > p(u(C) - u(A)).$$

Adding $u(A)$ to both sides and doing some algebra, we obtain

$$(1 - q)u(A) + qu(C) > (1 - p)u(A) + pu(A).$$

By the Expected Utility Hypothesis, we obtain

$$u\left((1 - q)A + qC\right) > u\left((1 - p)A + pC\right),$$

indicating that the player would choose the first of these lotteries over the second and his preferences are therefore monotonic.

Suppose the player would choose an outcome C over an outcome B and choose outcome B over an outcome A. Then $u(C) > u(B)$ and $u(B) > u(A)$, which implies that

$$u(C) - u(A) > u(B) - u(A) > 0.$$

Let

$$p = \frac{u(B) - u(A)}{u(C) - u(A)}.$$

By the previous inequalities, $0 < p < 1$, and so p is a probability. By the Expected Utility Hypothesis,

$$
\begin{aligned}
u((1 - p)A + pC) &= (1 - p)u(A) + pu(C) \\
&= u(A) + p(u(C) - u(A)) \\
&= u(A) + \frac{u(B) - u(A)}{u(C) - u(A)}(u(C) - u(A)) \\
&= u(B).
\end{aligned}
$$

So, the player would be indifferent between the outcome B and the lottery $(1-p)A+pC$. Thus, the player's pairwise choices are continuous.

Finally, suppose the player is indifferent between the two lotteries L_1 and L_2. Then

$$u(L_1) = u(L_2).$$

Suppose p is a probability and L_3 is some lottery. By the Expected Utility Hypothesis,

$$u((1-p)L_1 + pL_3) = (1-p)u(L_1) + pu(L_3)$$
$$= (1-p)u(L_2) + pu(L_3)$$
$$= u((1-p)L_2 + pL_3).$$

So, the player is indifferent between the two lotteries

$$(1-p)L_1 + pL_3$$

and

$$(1-p)L_2 + pL_3.$$

Thus, the player's pairwise choices are substitutable.

($\Leftarrow$) Suppose the player's pairwise choices over the lotteries $\mathcal{L}$ are complete, transitive, generalizable, probabilistic, monotonic, continuous, and substitutable. By Theorem 2.1.1, we know that the player has ordinal preferences. We will further show that the player makes choices in accordance with a utility function $u : \mathcal{L} \to \mathbb{R}$ that satisfies the Expected Utility Hypothesis and thus the player has vNM preferences.

Since the player's pairwise choices over O are complete, transitive, and generalizable, by Theorem 2.1.1, there is a ranking of the pure outcomes that is consistent with the player's pairwise preferences over the pure outcomes. Let $A_1, A_2, \ldots, A_m$ be the pure outcomes listed from last to first ranked (with ties broken arbitrarily). Since the player's pairwise choices are continuous, there are probabilities $p_1, p_2, \ldots, p_m$ for which the player would be indifferent between outcome A_i and lottery $(1 - p_i)A_1 + p_i A_m$. Define $u(A_1) = 0$, $u(A_m) = 1$, and $u(A_i) = p_i$ for the other outcomes. For lotteries, define u by the Expected Utility Hypothesis. This defines the utilities, but we must still show that these utilities are consistent with the player's choices.

We digress for a moment to consider an arbitrary (possibly compound) lottery L_1 for which p_i is the probability of outcome A_i when simplified over the pure outcomes. Define the lotteries

$$L_2 = p_1 A_1 + p_2 A_2 + \cdots + p_m A_m,$$
$$L_3 = p_1((1-p_1)A_1 + p_1 A_m) + \cdots + p_m((1-p_m)A_1 + p_m A_m), \text{ and}$$
$$L_4 = (1 - p_1^2 - p_2^2 - \cdots - p_m^2)A_1 + (p_1^2 + p_2^2 + \cdots + p_m^2)A_m.$$

Since L_1 and L_2 have the same probability distribution when simplified over the pure outcomes and the player's pairwise choices are probabilistic, the player should be indifferent between L_1 and L_2. By the Expected Utility Hypothesis,

$$u((1-p_i)A_1 + p_i A_m) = (1-p_i)u(A_1) + p_i u(A_m) = (1-p_i)(0) + p_i(1) = p_i$$

for each $i = 1, 2, \ldots, m$. Since the player is indifferent between A_i and $(1-p_i)A_1 + p_i A_m$ and the player's pairwise choices are substitutable, the player should be indifferent between L_2 and L_3. Since L_3 and L_4 have the same probability distribution over the pure outcomes and the player's pairwise choices are probabilistic, the player should be indifferent between L_3 and L_4. Since the player's pairwise preferences are transitive, the player is indifferent between L_1 and L_4. By the Expected Utility Hypothesis,

$$u(L_1) = u(L_4) = p_1^2 + p_2^2 + \cdots + p_m^2.$$

We have shown, therefore, that any compound lottery that simplifies to a lottery over the pure outcomes with probability p_i for outcome A_i has utility equal to the sum of the squares of the p_i's.

We can now show that the player's pairwise choices are consistent with the assigned utilities. Suppose K and L are lotteries. Let p_i and q_i be the probabilities of original outcome A_i when lotteries K and L, respectively, are simplified over the pure outcomes. In the previous paragraph, we argued that the player should be indifferent between K and

$$(1 - p_1^2 - p_2^2 - \cdots - p_m^2)A_1 + (p_1^2 + p_2^2 + \cdots + p_m^2)A_m$$

and should be indifferent between L and

$$(1 - q_1^2 - q_2^2 - \cdots - q_m^2)A_1 + (q_1^2 + q_2^2 + \cdots + q_m^2)A_m.$$

By transitivity, the player chooses K over L if and only if the player chooses

$$(1 - p_1^2 - p_2^2 - \cdots - p_m^2)A_1 + (p_1^2 + p_2^2 + \cdots + p_m^2)A_m$$

over

$$(1 - q_1^2 - q_2^2 - \cdots - q_m^2)A_1 + (q_1^2 + q_2^2 + \cdots + q_m^2)A_m.$$

By monotonicity, the player chooses

$$(1 - p_1^2 - p_2^2 - \cdots - p_m^2)A_1 + (p_1^2 + p_2^2 + \cdots + p_m^2)A_m$$

over

$$(1 - q_1^2 - q_2^2 - \cdots - q_m^2)A_1 + (q_1^2 + q_2^2 + \cdots + q_m^2)A_m$$

if and only if

$$p_1^2 + p_2^2 + \cdots + p_m^2 > q_1^2 + q_2^2 + \cdots + q_m^2.$$

But by the result of the previous paragraph,

$$p_1^2 + p_2^2 + \cdots + p_m^2 > q_1^2 + q_2^2 + \cdots + q_m^2$$

if and only if $u(K) > u(L)$. Taken together, we have that the player chooses K over L if and only if $u(K) > u(L)$. A similar argument shows that the player chooses K or L if and only if $u(K) = u(L)$. Thus, the player's pairwise choices are consistent with the assigned utilities.

Since the player's pairwise choices are generalizable and the player's pairwise choices are consistent with the assigned utilities, the player's choices for all subsets of lotteries are consistent with the assigned utilities. ◻

In summary, this theorem assures us that when a player's choice behavior satisfies a long, but reasonable, list of properties, then their preferences can be modeled by a vNM utility function. Table 2.1 lists the three properties that characterize ordinal preferences and the seven properties that characterize vNM preferences.

We will call two vNM utility functions *vNM equivalent* if they predict the same choices for all subsets of outcomes and lotteries. It turns out that two vNM utility functions u and v are vNM equivalent if and only if one can be transformed into the other via a positive linear transformation, that is, the utility functions u and v are vNM equivalent if and only if there exists a positive number α and some number β such that $v(o) = \alpha u(o) + \beta$ for every outcome $o \in O$. For example, suppose Scarlett is playing a gambling game in which she can win any amount of money in the interval $[-10, 10]$. One vMN utility that she might use to model her preferences is

$$u_R(x) = x.$$

TABLE 2.1   Relationship between Preference Type and Properties

vNM	
Ordinal	Probabilistic
Complete	Monotonic
Transitive	Continuous
Generalizable	Substitutable

The utility function u_R is vNM equivalent to both $v_R(x) = 2x$ and $w_R(x) = x + 3$. None of these utility functions is vNM equivalent to $x_R(x) = x^2$.

When we assume a player has ordinal preferences, it is conceivable that the assumption could be checked by asking the player to state their choices when presented with every subset of two or more outcomes. When we assume a player has vNM preferences, it is impossible to check the assumption since there are an infinite number of lotteries to consider; our only recourse is the reasonableness of the seven properties used in the vNM Preferences Theorem. At times, it will be necessary to make the stronger vNM assumption because it will be important to know about the intensity of the player's preferences. For example, if we return to Scarlett in the ice cream parlor for a moment, as ordinal utility functions,

$$v_R(V) = 3, \ v_R(C) = 2, \ v_R(S) = 2, \ v_R(N) = 1$$

and

$$w_R(V) = 100, \ w_R(C) = 73, \ w_R(S) = 73, \ w_R(N) = 0$$

provide us with the exact same information. However, if v_R and w_R are interpreted as vNM utility functions, we see that they give different lotteries to indicate indifference between the lottery and the outcome S:

$$2 = v_R(S) = v_R(0.5N + 0.5V)$$

but

$$73 = w_R(S) = w_R(0.27N + 0.73V).$$

The second utility function w_R shows a preference for strawberry that is closer in intensity to the preference for vanilla than demonstrated by the first utility function v_R.

2.3   CONSTRUCTING UTILITIES

We have already suggested how we might construct a utility function to model a player's choices when there are a finite number of outcomes. Ordinal preferences can be revealed by asking the player to choose among all outcomes and assign those outcomes the highest utility, asking the player to choose among all outcomes not previously chosen and assign those outcomes the second highest utility, and so forth. vNM preferences can be obtained by asking the player to name the highest and lowest ranked outcomes o_h and o_l, assign utilities of $u(o_h) - 1$ and $u(o_l) = 0$ to these outcomes, and then for each remaining outcome o determine a probability p for which the player would be willing to choose either the outcome o or the lottery $(1 - p)o_l + po_h$ and assign $u(o) = p$.

In this section, we examine four specific scenarios to illustrate a variety of ways utility functions may be created.

Self-Interest and Other-Interest. In most scenarios, an outcome has different impacts among the several players. For example, a person may have an exciting and lucrative job offer in a smaller town for which there are no job openings for a person with her spouse's skills, and at the same time they each have more mundane job offers in a different city. A variety of outcomes could be imagined, and the couple may have different preferences over those outcomes, however, their preferences are likely to depend upon their spouse's, as well as their own, happiness.

To model Self-Interest and Other-Interest, we simplify our scenario to examine the monthly salaries of the job offers for each spouse. Suppose Scarlett and Regis receive $\$x$ thousand and $\$y$ thousand, respectively; we will denote this by (x, y). Consider the following four possible outcomes: $(7, 0)$, $(6, 6)$, $(5, 7)$, and $(1, 6)$. If Scarlett is exclusively self-interested, she would rank order these outcomes in the given order. If Scarlett is primarily interested in Regis receiving money and only secondarily interested in receiving money for herself, Scarlett would rank order the outcomes $(5, 7)$, $(6, 6)$, $(1, 6)$, and $(7, 0)$. If Scarlett had a mixture of self-interest, other-interest, and a desire for equity, she might rank order the outcomes $(6, 6)$, $(5, 7)$, $(7, 0)$, and $(1, 6)$.

In fact, this last rank order would be obtained if Scarlett considered $\$1,000$ given to Regis to be worth the same to her as her receiving $\$500$, suggesting the utility function $u(x, y) = x + 0.5y$. Of course, this is only an ordinal utility function unless, at minimum, Scarlett is indifferent between the outcome $(7, 0)$ with utility $u(7, 0) = 7$ and the lottery $L = 0.6(6, 6) + 0.4(1, 6)$ with utility

$$u(L) = 0.6u(6, 6) + 0.4u(1, 6) = 0.6(9) + 0.4(4) = 7.$$

This example demonstrates how we can incorporate both self-interest and altruistic interests into a player's utility function. Therefore, maximizing a utility function does not necessarily imply selfishness, but rather achieving the most preferred outcome based on the player's interests.

Cournot Duopoly. In many economic markets, there are only a few companies competing for customers. For example, there are two major competitors in the mobile phone operating system market and there are four major breweries in the United States. Companies, especially those that are incorporated, typically make choices so as to maximize profit.

We will focus our attention on duopolies, markets in which there are two companies competing to provide an identical (or nearly identical) good or service. This duopoly model was first proposed by Cournot in 1838 [26]. In the following paragraphs, we construct the profit function for each company and explain how this can serve as a utility function.

In this model, the two companies are the two players and we apply some basic assumptions from the field of economics. First of all, we assume that supply and demand are inversely related, that is, if a total quantity Q of a product is available in the market at a price P, we can model Q and P by the classic supply and demand function

$$Q = a - bP$$

where $a > 0$ represents the amount that could be given away for free, and $b > 0$ represents the rate at which demand falls as the price P increases. Since we've assumed that the companies are producing nearly identical products, we let Q_1 represent the amount produced by the first company and Q_2 the amount produced by the second. The total quantity of product in the market then is $Q = Q_1 + Q_2$ (since there are only two companies producing the product). The value of P then depends on the fixed parameters a, b and the total quantity Q. Each company, however, can only control the quantity it produces, so we say the strategies available to company 1 correspond to the possible amounts Q_1 they could produce, and likewise for company 2. We impose reasonable restrictions that make $Q_1, Q_2 \geq 0$ and in a range that keeps $P \geq 0$. For simplification we will also assume that the cost of production is constant and the same for each company. Let c be the cost per unit produced. The profit function for company 1 is then

$$pi_1 = \text{Revenue} - \text{Cost} = PQ_1 - cQ_1 = \left(\left(\frac{a}{b} - \frac{Q}{b} \right) - c \right) Q_1$$

and for company 2

$$pi_2 = \text{Revenue} - \text{Cost} = PQ_2 - cQ_2 = \left(\left(\frac{a}{b} - \frac{Q}{b} \right) - c \right) Q_2.$$

Notice that each company's profit depends not only on their own choices, but also on the choices of the other company. Since we have assumed that each company prioritizes its own profit, these profit functions are clearly ordinal utility functions. With certain other assumptions, which we discuss in the next section, we can also interpret them as vNM utility functions.

Job Offers. Early in his career, David was fortunate to receive multiple job offers almost simultaneously. There were differences in the salary and benefits offered, there were differences in the types of colleges and their students (e.g., engineering vs. private liberals arts vs. public comprehensive), there were differences in the living environments (urban vs. rural, quality of public schools, political climate, etc.), and there were differences in the likelihood that his wife would be able to find employment within her career.

In the Self-Interest and Other-Interest scenario, we discussed two spouses considering job offers. We limited our mathematical model to include only salaries, but in reality, couples considering job offers not only consider the salaries of jobs available to them, but also consider the work environment (large vs. small company, opportunities for advancement, etc.) and living environment (urban vs. rural, quality of life, political climate, etc.).

Major personal decisions (e.g., which college to attend, job offer to accept, person to marry, place to live) can be difficult, not only because of strategic considerations, but because the enormity of the scenario and the multifaceted aspects of the outcomes make it difficult to make even pairwise comparisons. The *analytic hierarchy process*, abbreviated *AHP* and introduced by Saaty [89], is a technique to help individuals determine their preferences in such complex scenarios. The basic idea is to make comparisons among the outcomes with respect to independent and clearly defined objectives that can then be combined knowing the relative importance of each objective. The outcomes must be identified, e.g., the jobs being offered. Personal objectives must also be determined. It is important for the person to identify all relevant objectives, understand each objective clearly, and find the objectives to be mutually independent of each other.

The hope now is that the person can more easily compare the outcomes with respect to each objective and weight the importance of each objective. If the person still has difficulties making comparisons among outcomes with respect to an objective, the objective can be separated into sub-objectives that will make comparisons easier. If the person has difficulties weighting the importance of each objective, some can be grouped together into higher-order objectives. The totality of relationships from the top-level desired utilities through one or more layers of objectives down to the outcomes is the hierarchy in AHP. The Job Offers hierarchy is shown in Figure 2.1.

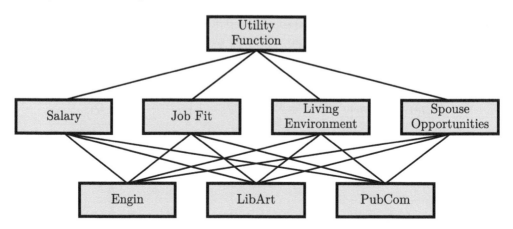

FIGURE 2.1 Job Offers Hierarchy.

While AHP simplifies the comparison of outcomes by having the decision maker consider separate objectives, it asks the decision maker to make more meaningful comparisons than those modeled by ordinal or vNM utility functions. As we have observed previously, the utility functions u and v are ordinally equivalent (i.e., represent the same ordinal preferences) if $u(x) \leq u(y)$ if and only if $v(x) \leq v(y)$ for all outcomes x and y. Such a measurement scale is called an *ordinal scale*. In addition to utility functions representing ordinal preferences, one ordinal scale used by scientists is the Mohs scale of mineral hardness (e.g., diamond is assigned 10, corundum is assigned 9, and topaz is assigned 8 because diamond scratches corundum which scratches topaz).

Similarly, the utility functions u and v are vNM equivalent (i.e., represent the same vNM preferences) if and only if there are numbers $\alpha > 0$ and β such that $v(x) = \alpha u(x) + \beta$ for all outcomes x. Observe that if $u(a) - u(b) = u(c) - u(d)$, then $v(a) - v(b) = v(c) - v(d)$, that is, equal intervals for one utility function yield equal intervals for the equivalent utility function. Such a measurement scale is called an *interval scale* since intervals are considered meaningful. Time and temperature are measured with interval scales. With vNM preferences, it is not meaningful to say that one outcome is twice as preferred to another outcome because, for example, $u(x) = x$ and $v(x) = x + 3$ are equivalent interval scales but although $u(2)$ is twice as big as $u(1)$, $v(2)$ is not twice as big as $v(1)$.

AHP asks decision makers to assign utilities for which saying "one outcome (or objective) is twice as preferred to another outcome (or objective)" is a meaningful statement. More formally, the utility functions u and v are *ratio equivalent* if there is a number $\alpha > 0$ such that $v(x) = \alpha u(x)$ for all outcomes x, and any such measurement scale is called a *ratio scale*. Many physical measurements (e.g., length, area, volume, mass) use ratio scales. In the context of the Job Offers scenario, we will describe four approaches to elicit ratio scale comparisons among boxes in the hierarchy.

The first approach is to think in percentage terms. In thinking about their contribution to his utility, David determined that salary is 10%, job fit is 40%, living environment is

30%, and spouse opportunities is 20% of the satisfaction that will be associated with the job offer decision.

A second approach is to determine the worst option and then how many times better each other option is in comparison. LibArt is the lowest salary offer, PubComp's salary offer is 1.1 times LibArt's, and Engin's salary offer is 1.3 times LibArt's. This results in the numbers $(1.3, 1, 1.1)$ for (Engin, LibArt, PubComp). We can normalize these numbers to sum to 1 by dividing each number by $1.3 + 1 + 1.1 = 3.4$, resulting in the salary numbers $(0.382, 0.294, 0.324)$ for (Engin, LibArt, PubComp). Similarly, David considered the living environment for LibArt and PubComp to be about the same and Engin to be one-third better, resulting in the numbers $(4/3, 1, 1)$, which normalize to $(0.4, 0.3, 0.3)$.

When the decision maker finds it difficult to assign numbers with any precision using the first two approaches, a third approach is to compare each pair of options. Of course, this requires more comparisons, but the additional numbers provide a cross check and bolster our confidence in the validity of the final numbers obtained. David thinks the LibArt job fit is 3 times better than the Engin job fit, the Engin job fit is 4 times better than the PubComp job fit, and the LibArt job fit is 5 times better than the PubComp job fit. Observe that if job fit could be measured on a ratio scale, it would follow that 3×4 should equal 5. Since they are not equal, we could take this as a sign that job fit cannot be measured on a ratio scale. Alternatively, we could attribute the discrepancy to the imprecision in our instruments of measurement (i.e., our intuition and reasoning). This suggests that we find numbers that best fit the data we have.

TABLE 2.2   Job Offers Scenario Job Fit Pairwise Comparisons

	Engin	LibArt	PubComp
Engin	1	1/3	4
LibArt	3	1	5
PubComp	1/4	1/5	1

Table 2.2 shows one way to record the data for Job Offers in an organized fashion, allowing us to view the data in matrix form. The number in the cell that is the intersection of row r and column c is the number of times r college is better than c college with respect to job fit. Observe the reciprocal symmetry across the main diagonal of the matrix: since LibArt is 3 times better than Engin, Engin is 1/3 times better than LibArt. If the numbers in the table reflected an underlying ratio scale, then each column would be a multiple of the first column.

TABLE 2.3   Job Offers Scenario Job Fit Pairwise Comparisons

	Engin	LibArt	PubComp	Average	Eigenvector
Engin	4/17	5/23	2/5	0.284	0.280
LibArt	12/17	15/23	1/2	0.619	0.627
PubComp	1/17	3/23	1/10	0.096	0.094

Table 2.3 shows two ways to obtain a reasonable estimate of an underlying ratio scale when the data does not conform exactly. Each column of Table 2.2 has been normalized to sum to one, and then an average of the columns is computed (because of standard rounding rules, the resulting numbers reported do not sum to 1 exactly).

When statisticians summarize multivariate data, they use principal component analysis, which in this context corresponds to the dominant eigenvector of the matrix given in Table 2.2. Plenty of software is available to calculate eigenvectors. We used wolframalpha.com [124], entering

$$\text{eigenvector of } \{\{1,1/3,4\},\{3,1,5\},\{1/4,1/5,1\}\}$$

to obtain the dominant eigenvector $(2.9876, 6.69433, 1)$, which upon normalization so that the numbers sum to one, yields the last column of Table 2.3.

The last approach to comparing options to elicit a ratio scale is employed if a person has difficulty attaching a number to how much one outcome is better than another outcome. Words are used for the comparisons instead. For example, David opined that with respect to spouse opportunities, Engin was extremely better than LibArt and moderately better than PubComp, and PubComp was strongly better than LibArt. With the translation of moderately strong to 3, strong to 5, very strong to 7, and extremely strong to 9, the matrix in Table 2.4 is obtained, and the eigenvector can be then obtained.

TABLE 2.4 Job Offers Scenario Spouse Opportunities Pairwise Comparisons

	Engin	LibArt	PubComp	Eigenvector
Engin	1	9	3	0.672
LibArt	1/9	1	1/5	0.063
PubComp	1/3	5	1	0.265

Table 2.5 summarizes the weights we have obtained. With the normalization made for

TABLE 2.5 Job Offers Scenario Comparisons

	Salary	Job Fit	Living	Spouse	Utility
Importance	10%	40%	30%	20%	
Engin	0.382	0.280	0.4	0.672	40.4
LibArt	0.294	0.627	0.3	0.063	38.3
PubComp	0.324	0.094	0.3	0.265	21.3

the comparisons among outcomes for each objective, we can interpret the decimals as the fraction of the objective obtained by the college chosen. If we weight these numbers by the relative importance if each objective, we obtain the utilities. For example, the utility for choosing the Engin job offer is

$$(10)(0.382) + (40)(0.280) + (30)(0.4) + (20)(0.672) = 40.4.$$

Similar computations can be performed for LibArt and PubComp, and the results are reported in the rightmost column of Table 2.5. The best choice is Engin because it has the highest utility. Note that many software tools are available to make using AHP intuitive for larger models.

AP Ranking. The AP ranking of college football and basketball teams combines rankings by experts. A team receives points based on the number of times they were ranked first, second, third, etc. The points range from 25 points for a first place ranking to 1 point for a last place ranking. The AP ranking is then determined by these point totals for each team [84].

Sometimes we will think of a business or an entire nation as a player in a game. The preferences for such a player can be thought of as an aggregation of the preferences of individual actors within the business or nation. The AP rankings of college teams is a simple version of how individual preferences could be combined to achieve societal preferences. Experts provide ordinal preferences which are then treated, via the point assignment process, as if they were vNM preferences (since, for example, the difference between first and second ranked teams is considered to be the same as the difference between 24th and 25th ranked teams), and each expert's preferences are given equal weight.

For contests where there is a certain amount of subjectivity in determining the ranking of contestants (e.g., ice skating, diving, and gymnastics), judges are asked to rate each contestant on a numeric scale and then the ratings are combined in some manner (usually an average, perhaps with the lowest and highest ratings excluded first). The ratings can usually be interpreted as a vNM utility because the differences between scores are thought to be meaningful. Sometimes these ratings can even be interpreted as a ratio scale because it is meaningful to say that one contestant was twice as good as another contestant. This makes it meaningful to not only combine ratings but even to weight one judge's score more heavily than another judge's score. AHP would be one way to determine weights among judges or between different aspects of judge ratings.

A more complicated example of preference aggregation is a corporation where the shareholders elect Trustees who hire and fire a Chief Executive Officer who makes day-to-day decisions which are carried out more or less faithfully by others who have been employed by the corporation. A democratic nation can be even more complicated. In the United States, legislation is passed either by simple majorities in the House and Senate with the President's approval, or by two-thirds majorities in the House and Senate over the President's disapproval. Members of the House are voted into office by pluralities of voters in the district they represent; members of the Senate are voted into office by pluralities of voters in the states they represent; and the President is voted into office by an Electoral College whose members are determined by voters in each state. Of course, the day-to-day decisions made by government officials occur with a mixture of adherence to often ambiguous interpretations of the laws passed and policies adopted by a myriad of agencies and officials.

It can sometimes seem to be a stretch to assume that a business or a country acts as a single player in accordance with a utility function. Nonetheless, sometimes this is the most natural assumption when constructing a simple model of some scenario. It is also common to treat different phenomena at very different levels. For example, we talk about and measure the temperature, volume, and pressure of a gas despite the underlying reality of a large number of molecules interacting with each other.

2.4 DETERMINING RISK

In the duopoly scenario, and in most other economic models, the utility of an outcome is equivalent to some dollar value associated with the outcome. While we can see how dollar values might capture the intensity of a player's preferences, dollar values are not necessarily vNM utilities. For example, receiving $11.00 instead of $10.00 means significantly more than receiving $1001.00 instead of $1000.00 to most people. To explore this difference, we consider the relationship between the expected utility of a lottery, as given by the Expected Utility Hypothesis, and the utility of the expected value of the lottery.

Consider the following raffle: For $25, you can purchase a $\frac{1}{400}$ chance for a $10,000 college scholarship. We can represent this lottery with our usual notation

$$\tfrac{399}{400}(\text{losing } \$25) + \tfrac{1}{400}(\text{winning } \$9,975),$$

but since the outcomes are numerical, we can calculate the expected monetary value of the raffle as

$$\tfrac{399}{400}(-\$25) + \tfrac{1}{400}(\$9,975) = \$0.$$

The expected monetary value of entering or not entering the raffle is the same, however, entering the raffle involves a small chance of a large gain offset by a large chance of a small loss, while not entering the raffle involves no chance of a gain or a loss. Entering the raffle involves *risk* while not entering the raffle does not.

Most parents of college students are willing to enter the raffle, but many college students themselves are not. For the college parents,

$$u\left(\tfrac{399}{400}(\text{losing } \$25) + \tfrac{1}{400}(\text{winning } \$9,975)\right) > u(\$0),$$

but for the students themselves,

$$u\left(\tfrac{399}{400}(\text{losing } \$25) + \tfrac{1}{400}(\text{winning } \$9,975)\right) < u(\$0).$$

For the parents, the utility of the lottery is greater than the utility of the expected value, making them risk loving in this scenario. On the other hand, the students are risk adverse since the utility of the lottery is less than the utility of the expected value. This principle holds in general, as we describe in the following definition.

Definition 2.4.1. A player is *risk adverse*, *risk neutral*, or *risk loving* with respect to money if the expected utility of every monetary lottery is less than, equal to, or greater than, respectively, the utility of the lottery's expected monetary value.

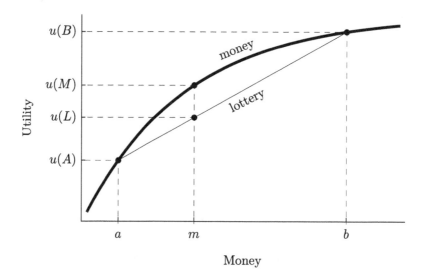

FIGURE 2.2   Utility of a Risk-Adverse Player.

Suppose A is the outcome of obtaining a units of money, B is the outcome of obtaining b units of money, and p is a probability. The lottery $L = (1-p)A + pB$ has utility

$$u(L) = u((1-p)A + pB) = (1-p)u(A) + pu(B)$$

by the Expected Utility Hypothesis. The curves in Figures 2.2 and 2.3 show a player's utility for receiving a guaranteed amount of money. For each monetary outcome $m = (1-p)a + pb$ between a and b, we can graph both the utility to the player of receiving that amount of

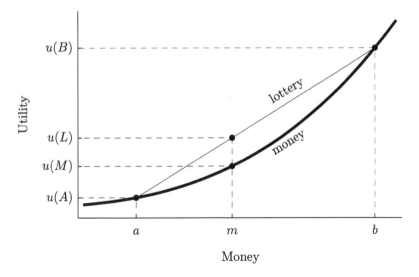

FIGURE 2.3 Utility of a Risk-Loving Player.

money, $u(M)$, as well as the expected value of a lottery between receiving a units and b units of money, $u(L)$. The points corresponding to the expected value of the lottery form the line segment between $(a, u(A))$ and $(b, u(B))$. When the curve corresponding to $u(M)$ is above this line segment, the player values the guaranteed risk-free money over the risky lottery and is therefore risk adverse. When this curve is below the line segment, the player values the risky lottery more than the guaranteed risk-free money and is therefore risk loving. A player is risk neutral if the $u(M)$ curve is the line segment $u(L)$. When a player is risk neutral, the dollar values themselves may serve as a vNM utility function.

Observe that these concepts of risk could also be applied to any single type of object received that can be measured on an interval scale, not just money. These include points received by a student towards their final course grade, the amount of sulfur dioxide pollution in the atmosphere of a city, or the number of progeny for an organism.

EXERCISES

2.1 For the Ice Cream Parlor scenario, we constructed for Scarlett the utility function $u(V) = -1$, $u(C) = u(S) = -2$, and $u(N) = -3$. List the eleven subsets of two or more outcomes, and for each subset, state Scarlett's choice if her choices are consistent with the constructed utility function.

2.2 Make a list of four or five weekend leisure activities (or use a list provided by your instructor). Let an outcome be engaging in one of these activities for free. Determine your ordinal utilities among these outcomes. Determine one other person's ordinal utilities among these four or five outcomes. By doing all of the necessary pairwise comparisons, check to see whether your utilities and the other person's utilities truly are ordinal. If both are ordinal utility functions, are they the same utility function?

2.3 Consider the last time you registered for classes and the thousands of schedules that were available to you. Clearly, you did not consider every one of them. Create an ordinal utility function that assigns a utility to every possible schedule and reflects your preferences over them.

2.4 Suppose there are m outcomes to a game.

a. We can construct an ordinal utility function for any player as we did for Scarlett in the Ice Cream Parlor scenario: First present all possible outcomes, assign a utility to the player's choice, remove that outcome from the list, present the remaining outcomes, assign a smaller utility to the player's choice at this step, and continue until all outcomes have been chosen and utilities have been assigned. What is the maximum number of subsets that we will present to the player to make a choice?

b. To verify that Scarlett has ordinal preferences, we presented 11 subsets of two or more outcomes. How many subsets need we present to the player facing m outcomes to verify that they have ordinal preferences?

c. Make a table to compare the previous two formulas with $m = 3, 4, 5, 6$.

d. What do you observe about the relative growths of the two formulas?

2.5 In the text, we assigned negative numbers to Scarlett's ordinal preferences so that $u(o) = -k$ if outcome o was ranked kth best by Scarlett. Later, we emphasized the fact that the specific values of the numbers did not matter, but rather the ordering of them. Why might one pick a particular set of numbers to represent ordinal preferences over another set of numbers?

2.6 Andreoni and Miller [3] asked undergraduate students to divide tokens between a randomly chosen and anonymous other student and themselves. Tokens were converted into points at the specified rate and the points were converted into money at $0.10 per point. For example, a student might divide 50 tokens by keeping 40 at one point each and passing 10 to the other student at three points each. The divider would receive $40 \times 1 \times \$0.10 = \$4.00$ and the other student would receive $10 \times 3 \times \$0.10 = \$3.00$.

a. How would you make the following divisions presented in the experiment described above?

i. Divide 60 tokens: keep ___ at two points each, and pass ___ to the other student at one point each.

ii. Divide 100 tokens: keep ___ at one point each, and pass ___ to the other student at one point each.

iii. Divide 75 tokens: keep ___ at one point each, and pass ___ to the other student at two points each.

iv. Divide 40 tokens: keep ___ at one point each, and pass ___ to the other student at three points each.

b. Andreoni and Miller found that roughly half of the students maximized their own monetary payoff most of the time; that is, they kept all available tokens and passed nothing to the other student. The remaining students had one of two forms of altruism. Roughly two-fifths of the students maximized the total monetary payoff most of the time; that is, in division (i), they would keep all of the tokens, but in divisions (iii) and (iv), they would pass all of the tokens to the other student. Roughly three-fifths of the students gave equal monetary payoffs to themselves and the other student; that is, in division (i) they would keep 20 and pass 40 so that each would receive $4.00, in division (ii) they would keep 50 and pass 50 so that each would receive $5.00, in division (iii) they would keep 50 and pass 25 so that each would receive $5.00, and in division (iv) they would keep 30 and pass 10 so that each would receive $3.00. How did your choices compare with what was experimentally observed? How would you compare the experimental results with your *ad hoc* observations about the altruism of people?

2.7 Here is an argument, commonly referred to as the "money pump," for why it is rational to have ordinal preferences: Suppose Firstus is money-loving, but does not have ordinal preferences among the outcomes A, B, and C. In particular, suppose Firstus prefers A to B, prefers B to C, but does not prefer A to C. Since Firstus prefers A to B and is money-loving, she should be willing to pay some amount, say $1, to obtain A instead of B. Since Firstus prefers B to C and is money-loving, she should be willing to pay some amount, say $1, to obtain B instead of C. Since Firstus does not prefer A to C and is money-loving, she should be willing obtain C, with any amount of money, instead of A alone. Finally, suppose Secondus has the ability to choose which of outcomes A, B, or C occurs.

a. If Secondus says to Firstus that outcome C will occur unless Firstus pays him $1 to obtain outcome B, explain why Firstus will pay Secondus $1.

b. If Secondus says to Firstus that outcome B will occur unless Firstus pays him $1 to obtain outcome A, explain why Firstus will pay Secondus $1.

c. If Secondus says to Firstus that outcome A will occur unless Firstus accepts $1 and outcome C, explain why Firstus would accept the $1.

d. If Secondus makes the offers suggested in parts (a), (b), and (c) in that order, explain why Firstus is $1 poorer and expecting Secondus to select outcome C.

e. If Secondus repeats the process suggested in part (d), explain why Firstus can be made very poor. Further, explain why this is an absurdity.

f. Evaluate whether the above argument shows that it is rational to have ordinal preferences.

2.8 Peter Fishburn, at a 1987 Operations Research Society of America meeting [38], proposed the following six games of chance (A, B, C, D, E, and F). In each game of chance, a six-sided die is rolled. Depending upon which face lands face up (1, 2, 3, 4, 5, or 6), the player will receive the amount of money given in Table 2.6.

TABLE 2.6   Payoffs for Dice Games

		Die Face					
		1	2	3	4	5	6
	A	$100	$200	$300	$400	$500	$600
	B	$600	$100	$200	$300	$400	$500
Game	C	$500	$600	$100	$200	$300	$400
	D	$400	$500	$600	$100	$200	$300
	E	$300	$400	$500	$600	$100	$200
	F	$200	$300	$400	$500	$600	$100

Fishburn stated that he would prefer A to B because in five of the six cases (all except when a 1 is rolled), he would win more money. For the same reason, he would prefer B to C, prefer C to D, prefer D to E, prefer E to F, and prefer F to A. Hence, a renowned game theorist admitted that in this situation, he did not have ordinal preferences. He stated that those were his preferences even though he knew of the "money pump" scenario described in the previous exercise. How does this information change or not change your beliefs about whether players must always have ordinal preferences?

2.9 Suppose a player's pairwise choices are transitive.

    a. Prove that if the player chooses either outcome if presented with $\{A, B\}$, and the player chooses either outcome if presented with $\{B, C\}$, then the player chooses either outcome if presented with $\{A, C\}$.

    b. Prove that if the player chooses A over B and chooses B over C, then the player chooses A over C.

*2.10 This exercise explores how the Ordinal Preferences Theorem need not generalize to an infinite set of outcomes. Lex will receive a portion a of an apple pie where $0 \leq a \leq 1$ and v pints of vanilla ice cream where $0 \leq v \leq 1$; we denote such an outcome by the ordered pair (a, v). Lex prefers more than less of each, but apple pie is of primary importance while vanilla ice cream is of secondary importance. This means that Lex will choose the outcome (a_1, v_1) over the outcome (a_2, v_2) if and only if (i) $a_1 > a_2$, or (ii) $a_1 = a_2$ and $v_1 > v_2$.

    a. Prove that Lex's pairwise choices are complete and transitive.

    b. Given a subset of outcomes, describe what Lex must choose to make his pairwise choices generalizable.

    c. Exhibit a subset of outcomes for which Lex cannot make a choice if his pairwise choices are generalizable.

    d. Prove that Lex's preferences cannot be modeled by a utility function. Hint: Suppose, to the contrary, that Lex's preferences can be modeled by the utility function u. Given any $0 \leq a_1 < a_2 \leq 1$, what is true about $u(a_1, 0)$ in comparison with $u(a_2, 0)$? Show that this implies that $\{u(a, 0) : 0 \leq a \leq 1\}$ is an uncountable set. Given any $0 \leq a \leq 1$, what is true about $u(a, 0)$ in comparison with $u(a, 1)$? Show that this implies that $\{u(a, 0) : 0 \leq a \leq 1\}$ is a countable set.

2.11 Consider the simple Self-Interest and Other-Interest scenario with the four outcomes $(7, 0)$, $(6, 6)$, $(5, 7)$, and $(1, 6)$, where (x, y) now represents you receiving $\$x$ thousand and an acquaintance receiving $\$y$ thousand. For most people, their preferences will differ depending upon who the acquaintance is.

    a. Construct a vNM utility function to represent your preferences if the acquaintance is a randomly chosen person in the world whom you are not likely to ever meet. Explicitly state the two lotteries used to determine the utilities of the middle ranked outcomes.

    b. Construct a second vNM utility function to represent your preferences if the acquaintance is your spouse, significant other, or a close family member. Explicitly state the two lotteries used to determine the utilities of the middle ranked outcomes.

2.12 Return to the list of four or five weekend leisure activities you created in an earlier exercise (or use a list provided by your instructor). Let an outcome be engaging in one of these activities for free. Determine your vNM utilities among these outcomes. Determine one other person's vNM utilities among these four or five outcomes. Check whether the Expected Utility Hypothesis is satisfied for two lotteries; that is, show that you are indifferent between engaging in one of the middle ranked activities and an appropriate lottery between your highest and lowest ranked activities. Record the vNM utilities found and any difficulties you or others had in coming up with their vNM utilities.

2.13 Suppose

$$u(A) = 0, \ u(B) = 20, \ u(C) = 90, \ u(D) = 100.$$

a. Suppose u is an ordinal utility function. List the outcomes in order of highest to lowest preference.

b. Suppose u is a vNM utility function. Explain qualitatively what the utility values tell us about preferences.

c. Give the lotteries between A and D for which the player would be indifferent to each of the other two outcomes.

d. Suppose u is a vNM utility function and L is the lottery in which a fair six-sided die is rolled, outcome A is chosen if a 1 or 2 is rolled, outcome B is chosen if a 3, 4, or 5 is rolled, and outcome C is chosen if a 6 is rolled. Find $u(L)$.

e. Suppose u and v are equivalent vNM utility functions, with $v(B) = 0$ and $v(D) = 400$. Find $v(A)$ and $v(C)$.

f. If $u(0.5A + 0.5C) = 20$, is u a vNM utility function? Using the Expected Utility Hypothesis, explain why or why not.

g. If $u(0.8A + 0.2D) = 20$ and $u(0.4A + 0.6D) = 60$, is u a vNM utility function? Using the Expected Utility Hypothesis, explain why or why not.

2.14 Suppose a player has vNM preferences; would choose outcome D if presented with $\{A, B, C, D\}$; is indifferent between outcome B and the lottery $0.3A + 0.7C$; and is indifferent between outcome C and the lottery $0.5B + 0.5D$. Find a vNM utility function that models this player's preferences.

2.15 Suppose that a collection of possible outcomes for you is to receive one of $0, $200, $400, $600, and $800.

a. Determine your utility function, u, for your preferences among this collection of outcomes.
The following parts assume that your utility function satisfies $u(\$0) < u(\$200) < u(\$400) < u(\$600) < u(\$800)$.

b. Find a value of p for which you would be indifferent between choosing a lottery of the form $(1 - p) \cdot \$200 + p \cdot \$600$ or receiving $400. After determining p, find $u((1 - p) \cdot \$200 + p \cdot \$600)$ and compare with $u(\$400)$ as determined in part (a).

c. Find a value of p for which you would be indifferent between choosing a lottery of the form $(1 - p) \cdot \$200 + p \cdot \$800$ or receiving $600. After determining p, find $u((1 - p) \cdot \$200 + p \cdot \$800)$ and compare with $u(\$600)$ as determined in part (a).

d. Based on your answers to parts (b) and (c), do you have vNM preferences among the monetary outcomes above?

2.16 Suppose instead of receiving different fixed amounts of money, you are offered four different lotteries with amounts of money based on which color ball you draw from an urn. Ellsburg [33] describes players offered a chance to draw balls from an urn. The urn contains 30 red balls and 60 other balls that are either black or yellow. You do not know how many black balls or yellow balls there are, but that the total number of black and yellow balls is 60. The balls are well mixed so that each ball is as likely to be drawn as any other.

a. Create an ordinal utility function u for the following four prize lotteries. Do this before reading about the rest of the problem.

   A You receive $100 if you draw a red ball; otherwise, you receive nothing.

   B You receive $100 if you draw a black ball; otherwise, you receive nothing.

   C You receive $100 if you draw a red or yellow ball; otherwise, you receive nothing.

   D You receive $100 if you draw a black or yellow ball; otherwise, you receive nothing.

b. The probability of drawing a red ball is $30/90 = 1/3$. Let the unknown probabilities of drawing a black ball and drawing a yellow ball be denoted b and y, respectively. Assume now that u is a vNM utility function. Show that $u(A) > u(B)$ if and only if $u(C) > u(D)$.

c. Are parts (a) and (b) consistent?

d. Ellsberg [33] reports that most people choose lottery A over lottery B and choose lottery D over lottery C. Ellsberg argues that this shows that most people do not have vNM preferences. Do you agree? Is this consistent with the preferences you created in part (a)?

2.17 Complete the proof of the vNM Preferences Theorem by showing if a player has vNM preferences, then the player's pairwise choices are transitive and generalizable.

2.18 Suppose that the set of outcomes is finite. Prove that the vNM utility functions u and v are vNM equivalent if and only if one can be transformed into the other via a positive linear transformation, that is, there exists a positive number α and some number β such that $v(o) = \alpha u(o) + \beta$ for every outcome o.

*2.19 Investigate the following problem originally posed by Allais [1].

a. Create an ordinal utility function u for the following four outcomes, and include an explanation. Do this before reading about the rest of the problem.

   A You receive $1 million with complete certainty.

   B You receive $5 million with probability 0.10, you receive $1 million with probability 0.89, and you receive nothing with probability 0.01.

   C You receive $1 million with probability 0.11 and you receive nothing with probability 0.89.

   D You receive $5 million with probability 0.10 and you receive nothing with probability 0.90.

b. Allais found that most people ranked A over B, because $1 million for sure is better than having a 1% chance ending up with nothing even if a bigger prize is possible. Allais also found that most people ranked D over C, because the probabilities of winning anything are almost the same but the prize in D is so much bigger than the prize in lottery C. Was your ranking consistent with Allais's empirical findings?

c. Use the Expected Utility Hypothesis to show that a player should choose A over B if and only if the player should choose C over D.

d. Allais argued that his empirical findings and the result of part (c) show that most people's preferences do not follow the Expected Utility Hypothesis. Do you agree with his conclusions? Does this experiment support that conclusion?

2.20 Rework the Cournot Duopoly scenario by creating a utility function for Firm 1 when there is a third firm in the market.

2.21 Find the utility function for a Cournot Duopoly firm if the consumer demand function is $Q = a/(1 + bP)$ for some $a > 0$ and $b > 0$ rather than $Q = a - bP$.

2.22 Choose a category of objects to rank (e.g., songs, college football teams, Presidential candidates, potential spouses) and three or more of those objects. Choose three or more criteria against which the objects can be ranked. Use the Analytical Hierarchy Process to produce a ratio scale ranking of the chosen objects. Lay out the information so that it is easy to understand what calculations have been performed. At least one set of comparisons should use a pairwise comparison approach to determine the ratio scale.

2.23 Students are often assigned letter grades A, B, C, D, or F in the courses that they take. These letter grades are often assigned the numbers 4, 3, 2, 1, and 0, respectively, so that an average of all grades can be computed. Describe the implications for course grades to be an ordinal, interval, or ratio scale. Conclude, with justification, what type of scale course grades are.

2.24 Evaluation forms often ask people to state whether they strongly disagree, disagree, are neutral, agree, or strongly agree with a statement. People's answers are coded as the numbers 1, 2, 3, 4, and 5. The coded responses of several people are then averaged. A typical summary might be that people, on average, agreed with the statement, because the average of the coded responses was 3.9. Discuss whether this is a correct conclusion to draw after averaging. Are there implicit assumptions about the respondents' interpretations of the type of scale used?

2.25 What are the advantages and disadvantages of measuring preferences with a ratio scale? Interval scale? Ordinal scale?

2.26 Biologists talk about "survival of the fittest" with respect to organisms, phenotypes, and genes [28]. How might you construct utilities for organisms, phenotypes, or genes? That is, how could you measure success of an organism?

*2.27 **Merit Raise**. Faculty work at universities is usually partitioned into three major areas: research, teaching, and service, and faculty engage in each of these activities at different levels and with different levels of productivity. This makes it difficult for administrators to assign merit raises as it requires comparing work in different categories and at different levels. How does one compare a year of great teaching to one with multiple publications? How does one compare faculty work in different disciplines? Use the Analytical Hierarchy Process to develop a method that an administrator might use to rank faculty for merit raises.

2.28 Suppose Evan's utility for receiving $\$x$ is given by $u(x) = \sqrt{x}$ for $x \geq 0$.

a. Consider the lottery in which Evan receives $10,000 with a 10% chance and otherwise receives nothing. What is Evan's utility for this lottery? What is the expected value of the lottery? What is Evan's utility for the expected value of the lottery? Does this indicate Evan is risk adverse, risk neutral, or risk loving?

b. Draw a graph of Evan's utility for amounts from \$0 to \$10,000. Does this indicate Evan is risk adverse, risk neutral, or risk loving?

c. How much money would need to be offered to Evan for him to take the offer rather than the lottery described earlier?

d. Consider the lottery in which Evan receives \$4 with probability $1/2$, receives $\$16 = \$4^2$ with probability $1/4 = 1/2^2$, receives $\$64 = \$4^3$ with probability $1/8 = 1/2^3$, ..., receives $\$4^k$ with probability $1/2^k$, and so on. How much money must Evan be offered for him to prefer it over the lottery? Discuss the reasonableness of this result.

2.29 Suppose Fern's utility for receiving an x point bonus to her course grade is given by $u(x) = x^2$ for $0 \leq x \leq 20$.

a. Consider the lottery in which Fern receives 20 points with a 30% chance and otherwise receives nothing. What is Fern's utility for this lottery? What is the expected value of the lottery? What is Fern's utility for the expected value of the lottery? Does this indicate Fern is risk adverse, risk neutral, or risk loving?

b. Draw a graph of Fern's utility for amounts from 0 to 20 points. Does this indicate Fern is risk adverse, risk neutral, or risk loving?

c. How many points would need to be offered to Fern for her to take the offer rather than the lottery described earlier?

2.30 When it comes to large but unlikely winnings (the multimillion-dollar jackpot) and small but definite losses (the purchase price of a ticket), the popularity of state lotteries indicates that many people are risk loving in such situations. When it comes to large but unlikely losses (the destruction of a home or the early death of a spouse and parent) and small but definite losses (the premium payment), the popularity of home and life insurance indicates that many people are risk adverse in such situations. Draw the graph of a utility function for a person who is risk loving for gains and risk adverse for losses. Note that this should be a single graph with the horizontal axis including both positive (for gains) and negative (for losses) values.

2.31 Consider the power functions $u(x) = x^p$. For which value(s) of p is u the utility function for a risk-adverse player? For which value(s) of p is u the utility function for a risk-neutral player? For which value(s) of p is u the utility function for a risk-loving player?

CHAPTER 3

Simultaneous Play

Strategic games model situations in which the players can be thought to act simultaneously, but with complete information about each other's possible actions and preferences. We introduce the basic solution concepts, solution techniques, and key theorems. Then, we build models demonstrating a range of options for players and strategy choices.

We begin with a formal definition of a strategic game. This definition specializes Definition 1.2.1 of a game and identifies the elements of the models that we build.

Definition 3.0.1. A *strategic game* consists of the following:

1. A set $N = \{1, 2, \ldots, n\}$ of at least two players.

2. A set O of outcomes that can occur when the game is played.

3. A set S_i of strategies available to player i for each $i \in N$, which generates the set of strategy profiles $S = S_1 \times S_2 \times \cdots \times S_n$.

4. Each player simultaneously, without knowledge of any other player's choice, chooses a strategy, and the resulting strategy profile determines the outcome $o \in O$.

5. Utility functions $u_i : O \to \mathbb{R}$ that specify the preferences among outcomes for each player $i \in N$.

Although a strategic game has players make their choices once and simultaneously, it is possible for a single strategy assigned to a player to encompass many actions. For example, we can imagine replacing the myriad decisions made by someone playing chess with a single set of instructions dictating actions; in this case we could call this set of instructions a *strategy*. Regardless of whether a strategy dictates one or many actions, the combined strategies of all players, the strategy profile $s = (s_1, s_2, \ldots, s_n)$, determines the outcome o. This then determines the players utility $u_i(o)$. Since strategies lead to outcomes and ultimately to utilities, we will sometimes abuse notation by writing $u_i(s)$ for $u_i(o)$.

The definition of a strategic game allows for finite, infinite, and continuous sets of strategies for each player, and does not specify whether or not the utility functions are ordinal or von Neumann-Morgenstern. Strategic games are non-cooperative since there are no binding agreements for how players must act.

3.1 STRATEGIC GAMES

Prisoner's Dilemma is a canonical example of a strategic game because, as we will see shortly, it typifies many scenarios that confront decision makers. Further, being a simple scenario, it can be used to illustrate many of the fundamental concepts of game theory, and it also clearly demonstrates a fundamental dilemma in our (human) decision-making processes.

Prisoner's Dilemma. Two suspects are taken into custody by the police for a major crime. They are put into separate rooms for interrogation. Evidence against them is slim, so the police need one or both to confess to the crime. Therefore, each of the two suspects is informed that if they confess to the crime at hand and implicate the other as the principal culprit, then the confessor will get the lightest possible sentence and the principal culprit will get the heaviest possible sentence. They both will be convicted of a misdemeanor crime if neither confesses. If both confess, they each get moderate sentences as the police cannot identify the principal culprit.

We model this scenario as a strategic game in which the two suspects, each confined in a separate interrogation room, are the players. We will often refer to our two players in strategic games as Rose and Colin. (This convention helps later to emphasize the distinction between row and column players and was popularized by Phil Straffin in his book *Game Theory and Strategy* [110].) They each have two strategies available to them which we name Quiet and Confess. Table 3.1 lists each of the strategy profiles in the form (Rose, Colin) and the resulting outcome.

TABLE 3.1   Prisoner's Dilemma Outcomes

Strategy Profile	Outcome
(Quiet, Quiet)	Each suspect receives a three-year sentence.
(Quiet, Confess)	Rose receives a six-year sentence and Colin receives a one-year sentence.
(Confess, Quiet)	Rose receives a one-year sentence and Colin receives a six-year sentence.
(Confess, Confess)	Each suspect receives a five-year sentence.

We assume that each suspect is primarily concerned about their own sentence and wants to minimize it. Table 3.2 provides *payoffs* (a common synonym for utilities) for each player. Here we use the utility function 6 minus the number of years in prison; this is consistent with the player preferences. Based on our assumptions, these payoffs are ordinal. For these payoffs to also be vNM, we would need to assume that the suspects are risk neutral in the number of years to be served in prison.

TABLE 3.2   Prisoner's Dilemma Payoffs

Strategy Profile	Rose's Payoff	Colin's Payoff
(Quiet, Quiet)	3	3
(Quiet, Confess)	0	5
(Confess, Quiet)	5	0
(Confess, Confess)	1	1

Tables 3.1 and 3.2 complete the construction of the model by identifying the strategies, outcomes, and payoffs. We will refer to this model of the Prisoner's Dilemma scenario as the Prisoner's Dilemma strategic game.

We are now ready to look for a solution that maximizes the payoffs to the players. By observing that $5 > 3$, we see that Confess is the best response strategy for Rose if she knows that Colin will choose Quiet. Further, we can observe that Confess is also a best response for Rose if she knows Colin will choose Confess. We formalize the definition of a best response strategy below.

Definition 3.1.1. Define $s_{-i} = (s_1, \ldots, s_{i-1}, s_{i+1}, \ldots, s_n)$ to be the strategy profile s with player i's strategy removed. Define $u_i(t_i, s_{-i})$ to be the value of the utility function to player i of the strategy profile with s_i removed and replaced by t_i. Player i's strategy s_i is a *best response* to the profile s_{-i} of other player strategies if

$$u_i(s_i, s_{-i}) \geq u_i(t_i, s_{-i})$$

for all other strategies $t_i \in S_i$.

Because Rose's best response is the same regardless of Colin's choice of strategy, we say that the strategy Confess dominates the strategy Quiet for Rose. This notion is generalized in the following definition.

Definition 3.1.2. Player i's strategy s_i *strongly dominates* player i's strategy t_i if

$$u_i(s_i, s_{-i}) > u_i(t_i, s_{-i})$$

for all strategy profiles $s_{-i} \in S_{-i} = S_1 \times S_2 \times \cdots \times S_{i-1} \times S_{i+1} \times \cdots \times S_n$ available to the remaining players. Player i's strategy s_i *dominates* player i's strategy t_i if

$$u_i(s_i, s_{-i}) \geq u_i(t_i, s_{-i})$$

for all strategy profiles $s_{-i} \in S_{-i}$, and strict inequality holds for at least one strategy profile $s_{-i} \in S_{-i}$. A strategy is *strongly dominant* [resp., *dominant*] for player i if it strongly dominates [resp., dominates] all other strategies for player i.

Because the players' choices are modeled by their payoffs, when strategy s_i strongly dominates strategy t_i, player i should select strategy s_i over strategy t_i. When strategy s_i dominates strategy t_i, player i should select strategy s_i over strategy t_i unless the strategy selections by the other players result in player i obtaining the same utility for either.

We have already noted that Confess strongly dominates Quiet for Rose. By symmetry, Confess also strongly dominates Quiet for Colin. Thus, both players select Confess resulting in payoffs of 1 for each player, which we denote with the *payoff pair* $(1, 1)$. Further, knowing that this was the thinking of the other suspect, neither regrets or second guesses their own decision to confess. Such a regret-free strategy profile is known as a Nash equilibrium.

Definition 3.1.3. A strategy profile s is a *Nash equilibrium* if

$$u_j(s) \geq u_j(t_j, s_{-j})$$

for all players $j \in N$ and all strategies $t_j \in S_j$ available to that player.

That is, s is a Nash equilibrium if, given what the other players have chosen to do, s_{-j}, each player j cannot unilaterally improve their payoff by replacing their current strategy, s_j, with a new strategy, t_j, and, hence, has no regrets about their strategy selection.

Alternately, a player might choose a strategy by looking at the worst thing that can happen with each strategy choice, and choosing the strategy that makes the worst case as "least bad" as possible. More formally, a player would choose a strategy that maximizes their minimum payoff with respect to the strategy choices of the other players. For example, if Rose chooses Quiet, the worst that can happen is a payoff of 0 when Colin chooses Confess. On the other hand, Rose's worst payoff if she chooses Confess is 1. This suggests that in the scenario, Rose should confess if she is strategically risk adverse. We formalize this notion in the following definition.

Definition 3.1.4. Player i's strategy s_i is *prudential* if

$$u_i(s_i, s_{-i}) = \max_{t_i \in S_i} \left\{ \min_{s_{-i} \in S_{-i}} \{u_i(t_i, s_{-i})\} \right\}.$$

The value of $u_i(s_i, s_{-i})$ is called the *security level* for player i.

In the case of Prisoner's Dilemma, the prudential strategies are the same as the strongly dominant strategies, and the pair of prudential strategies coincides with the Nash equilibrium. Some of this is true for all strategic games: a dominant strategy (if it exists) is always prudential, a profile of dominant strategies is always a Nash equilibrium, and any Nash equilibrium must contain strongly dominant strategies if they exist. On the other hand, the concurrence of these three solution concepts for the Prisoner's Dilemma strategic game is a special case: a profile of prudential strategies need not be a Nash equilibrium, and a Nash equilibrium can contain dominated strategies and be bereft of prudential strategies.

Confess is the unique dominant and unique prudential strategy for each player, and although (Confess, Confess) is the unique Nash equilibrium, the two players in the Prisoner's Dilemma strategy game would be better off if they both chose Quiet, resulting in the payoff pair $(3, 3)$ instead of the payoff pair $(1, 1)$. This tells us that these three solution methods do not always yield the best overall payoff for each player.

Definition 3.1.5. A strategy profile s, and its associated outcome o, are *efficient* if there does not exist a strategy profile $t \in S$ such that

$$u_j(t) \geq u_j(s)$$

for all players j, with at least one of the inequalities being strict.

Notice that the concept of efficient is defined in the negative: a strategy profile is efficient if we cannot find another strategy profile that at least maintains the utility for all players, while strictly improving the utility for at least one player. Generally, because we do not want to ignore an outcome that is as good or better for each player, we want the resolution of a strategic game to be efficient.

For the Prisoner's Dilemma strategic game, the strategy profile (Confess, Confess) is not efficient because both players obtain larger utilities with (Quiet, Quiet). Each of the other three strategy profiles are efficient because it is impossible to make a change without reducing at least one of the player's payoffs.

Since utilities model individual choices, players optimize their personal utility without regard to the utilities of the other players (or stated somewhat differently, whatever interest a player has in the welfare of others has already been modeled in the player's utility function). As in the Prisoner's Dilemma strategic game, an efficient strategy profile may not individually maximize player utilities, and so players may not have an incentive to choose a strategy that is part of an efficient strategy profile. There is no dilemma when Nash equilibria are also efficient.

In the Prisoner's Dilemma strategic game, there is a tension between (1) choosing the dominant strategy (Confess), which will always yield a higher payoff regardless of the other player's choice, and (2) knowing that a better outcome might be possible if both players choose their dominated strategy (Quiet). This tension is what puts the dilemma into the

Prisoner's Dilemma: each player selecting the logical, rational strategy does not lead to an efficient outcome!

One way to resolve this dilemma would be for the players to enter into a binding agreement to stay quiet. In our original scenario, who would enforce adherence to the agreement? Certainly not the police nor the courts, but if the suspects were part of a crime family or gang, private agreements might be enforced. Of course, if we introduce binding agreements, then we no longer have a strategic game.

There are many real-world prisoner's dilemma scenarios and they may, and frequently do, involve more than two players. Consider the following scenario:

> **Office Coffee.** In many offices, there is a shared coffee pot, and the employees voluntarily contribute to a pool of money to replenish the supplies. Each employee who drinks the coffee must decide whether or not to contribute to the pool.

In the model for this scenario, player strategies are Contribute or Not Contribute. Clearly, Not Contribute is the strongly dominant strategy since everyone prefers free coffee, no one wants to be the only person contributing to the pool. Thus, everyone selects Not Contribute and there are no funds to replace the coffee. There are obvious issues with this strategy profile, since it eventually leads to the outcome of no coffee for anyone!

Generally, a scenario is said to be a *prisoner's dilemma scenario* if it can be modeled as a strategic game in which there is a single strategy for each player that strongly dominates all of that player's other strategies, but all players would receive a higher payoff if they would together choose a specific dominated, rather than the dominant, strategy. Since the mutual benefit result requires all players to cooperate by choosing the dominated strategy, it is often called the *Cooperate* strategy. In the original prisoner's dilemma scenario this was Quiet and in the Office Coffee scenario this is Contribute. Since there is always an incentive for any individual player to switch their choice to the dominant strategy, the dominant strategy is often called the *Defect* strategy. Confess and Not Contribute are the Defect strategies in our scenarios.

The traditional choice of the words "Cooperate" and "Defect" can deliver unwanted connotations of "good" and "bad." It would have been equally valid to choose the words "Collude" and "Resist" which would reverse the connotations for most people.

If the prisoner's dilemma scenario involves exactly two players and each player has exactly two strategies, then their payoffs can be described generally as in Table 3.3. Here

TABLE 3.3 General Prisoner's Dilemma Payoffs

Strategy Profile	Player 1's Payoff	Player 2's Payoff
(Cooperate, Cooperate)	R_1	R_2
(Cooperate, Defect)	S_1	T_2
(Defect, Cooperate)	T_1	S_2
(Defect, Defect)	P_1	P_2

the "sucker" payoff to player i for cooperating when the other player is defecting is S_i, the "punishment" payoff when both players defect is P_i, the "reward" payoff when both cooperate is R_i, and the "temptation" payoff to entice player i to defect is T_i. In order for these payoffs to have the intended interpretation, they must satisfy

$$S_1 < P_1 < R_1 < T_1 \text{ and } S_2 < P_2 < R_2 < T_2.$$

The parameters have player subscripts to emphasize that the necessary relationships among the payoffs are by player: there need not be any relationship between the payoffs of the two players.

Given this ordering of the payoffs, the reader can verify that Defect is the strongly dominant and prudential strategy for each player, the strategy profile (Defect, Defect) is the unique Nash equilibrium, each player would be better off if the strategy profile (Cooperate, Cooperate) were chosen instead of the strategy profile (Defect, Defect), and (Defect, Defect) is the only strategy profile that is not efficient.

Note that all of this is true whether the payoffs are ordinal or vNM utilities. The restriction $S + T < 2R$ is often added to ensure that the two players could not achieve even greater payoffs by publicly flipping a coin and choosing (Cooperate, Defect) if heads and (Defect, Cooperate) if tails. Of course, in this circumstance, the payoffs must be vNM utilities to justify our use of the Expected Utility Hypothesis.

There is another way to view the results of the previous paragraph. We modeled the original Prisoner's Dilemma scenario with payoffs being 6 minus the number of years in prison, resulting in Table 3.2. It would be reasonable to question this modeling assumption. Do the prisoners really know the exact number of years of prison each would get in each situation? Are they risk neutral in the number of years of prison? A "no" answer to either question would necessitate a revision in the numerical values of the payoffs. Nonetheless, the rank order of the four possible outcomes would not change, and the results of the previous paragraph show that the game theoretic predictions would also be unchanged. This shows flexibility in our model of the original scenario.

The preceding discussion suggests that the strategic game modeling process is fairly straightforward, particularly in simple scenarios. The following two scenarios demonstrate some potential complications.

Normandy Breakout. In the late summer of 1944, during World War II, the American Third Army broke out of the Normandy beachhead headed south, creating a gap between the American First Army and itself. It also created an opportunity for the American and German generals to make strategic decisions. The American generals had three options as to how to deploy their reserves: use them to reinforce the gap, move them quickly east to support the Third Army flank, or hold off on this decision. The German generals had two options: attack the gap or retreat to the east [48].

This brief description suggests the players (Americans and Germans) and strategies (the options for the Americans abbreviated by Reinforce, Move, and Hold, and the options for the Germans abbreviated Attack and Retreat) but does not give us much else to work with. However, we can turn to military historians to fill in more of the details. Given the options available to each country, Table 3.4 describes the possible outcomes, based on the model created by Haywood in his now famous paper on military strategy and game theory [48].

TABLE 3.4 Normandy Breakout Outcomes

Strategy Profile	Outcome
(Reinforce, Attack)	Americans hold the gap and Germans do not retreat.
(Reinforce, Retreat)	Germans retreat unimpeded.
(Hold, Attack)	Germans are likely surrounded and captured.
(Hold, Retreat)	Germans retreat under light pressure.
(Move, Attack)	Germans capture the gap, cutting off the Third Army.
(Move, Retreat)	Germans conduct a fighting retreat.

We will analyze this as a strategic game based on historical information from Haywood's paper and the techniques of this chapter.

It's safe to assume that surrounding and capturing the German army would be most highly valued for the Americans. Since the Americans needed to expand their hold in France, if they couldn't completely defeat the Germans, they at least wanted to force the German army to retreat so that the Americans can break out of Normandy. Using this type of analysis for each outcome, we can construct a *payoff matrix* as displayed in Table 3.5.

TABLE 3.5   Normandy Breakout Payoff Matrix

(American, German)	Attack	Retreat
Reinforce	(2, $\boxed{5}$)	(3, 4)
Hold	($\boxed{6}$, 1)	(4, $\boxed{3}$)
Move	(1, $\boxed{6}$)	($\boxed{5}$, 2)

Instead of listing the strategy profiles, we list the strategies for the Americans in the rows and the German strategies in the columns. The payoff pairs correspond to the strategy profile of that row and column and list the American payoff first. The best responses by the Americans and the Germans to each other's strategy choices are boxed.

An analysis of this strategic game reveals that Hold strongly dominates Reinforce for the Americans, and no other strategy is dominated. Analysis also reveals that there are no Nash equilibria, and that all strategy pairs are efficient. Hold is the prudential strategy for the Americans, with a security level of 4, and Retreat is prudential for the Germans, with a security level of 2. The point being made, from a game theoretic perspective, is that our solution methods (dominant, prudential, Nash, efficient) do not always agree on what is the "best" solution.

From a modeling perspective, this analysis suggests that the Americans should hold their reserves (their prudential strategy and strategy that dominates reinforcing the gap) and that the Germans should retreat to the east (perhaps predicting the likely American strategy choice or thinking prudentially). These were, in fact, the decisions made by the generals in the field, suggesting that the model has reasonable fidelity. In reality, however, Hitler overrode his generals' decision and ordered an attack on the gap, resulting in the Americans surrounding and capturing the German army.

Trio. Throughout the course of human history, civilizations have merged and separated to form new countries or redefine boundaries on existing countries. This can happen as the result of conflict or peaceful negotiations among leaders. The European Union, for example, created an economic alliance among 28 independent countries that now share a currency but maintain independent governments [34]. While the factors that contribute to countries deciding to merge or remain independent are complicated and cannot be completely explained by mathematics, we assume here that a driving factor that leads to countries merging is the economic benefit, and a compelling reason to remain separated is to maintain independent decision making. Here we consider the fictitious region of Trio, made up of three independent nations Alpha, Beta and Gamma whose leaders are making the decision to merge for economic benefit or remain independent.

We will model this situation as a game in which we treat each nation as a single player, acknowledging that the manner in which nations come to consensus may be too complicated to include in our model. Thus, the players in this game are the three nations Alpha, Beta and Gamma, each with two strategy choices (Merge or Remain). We will assume the nations' leaders meet to peacefully decide on the economic merger. Further we assume that the best outcome economically is for all three groups to merge, but they will each have to give up some independence. The next best option economically is for all three to choose independence, since the mutually agreed upon independence would allow for some level of economic cooperation among the independent nations. If two nations elect to remain independent and the third to merge, then the three nations continue to operate separately without any economic agreements benefiting any of them. If two groups elect to merge and third to remain independent, we assume that the merger cannot happen, but there is now tension causing some economic hardship, particularly to the detriment of the group that chose independence.

The construction of our model rests heavily on how we create the utility functions for each player. There are many ways to assign utilities. Depending on the historical, cultural, and political context, independent research can inform our choices. Since this is a fictitious example, we will consider only the following case: Alpha strongly prioritizes independence over economic benefits achieved through merging. Beta is indifferent between independence and merging for economic benefit, and Gamma strongly prefers the economic gain over the benefits of independent decision making. All three would prefer to avoid the situation that involves economic hardship.

Table 3.6 lists the eight strategy profiles, the corresponding outcomes, and ordinal utilities consistent with the preferences just described. For example, Gamma most prefers (Merge, Merge, Merge) because they value economic gain over independence, and so we assign the highest payoff of 8. The next highest outcomes for Gamma are the tense situations in which Alpha and Beta lose the most since Gamma has some economic gain. However, Gamma is indifferent between these two outcomes, and so we assign the average of 7 and 6 to both of these. They least prefer a tense situation in which they lose the most, so $u_\gamma((\text{Merge, Merge, Remain})) = 1$.

TABLE 3.6   Partition of Trio Outcomes and Payoffs

Strategy Profile s	Outcome o	$u_\alpha(s)$	$u_\beta(s)$	$u_\gamma(s)$
(Merge, Merge, Merge)	Most economic gain Some independence lost	4	7.5	8
(Merge, Merge, Remain)	Economic instability Gamma loses most	2.5	2.5	1
(Merge, Remain, Merge)	Economic instability Beta loses most	2.5	1	6.5
(Remain, Merge, Merge)	Economic instability Alpha loses most	1	2.5	6.5
(Merge, Remain, Remain)	No economic gain Independence retained Against Alpha's choice	5	5.5	2.5
(Remain, Merge, Remain)	No economic gain Independence retained Against Beta's choice	6.5	4	2.5
(Remain, Remain, Merge)	No economic gain Independence retained Against Gamma's choice	6.5	5.5	4
(Remain, Remain, Remain)	Some economic gain Independence retained	8	7.5	5

If both Beta and Gamma choose the strategy Merge, then Alpha prefers the strategy Merge for a payoff of 4 rather than Remain for a payoff of 1. That is, Merge is a best response by Alpha to Beta and Gamma both choosing Merge, which is noted in Table 3.6 by boxing 4 in the first row of Alpha's column u_α and not boxing 1 in the fourth row of Alpha's column u_α. Throughout the table, payoffs that correspond to best response strategies are boxed. A strategy profile in which all three strategies are best responses is a Nash equilibrium. Thus, there are two Nash equilibria: all three groups choose Merge or all three groups choose Remain.

Since Merge is the unique best response for Alpha when both Beta and Gamma choose Merge, but Remain is the unique best response for Alpha when both Beta and Gamma choose Remain, Alpha does not have a dominant strategy. Similar reasoning reveals that neither Beta nor Gamma has a dominant strategy.

Somewhat surprisingly, each player's prudential strategy is Merge. If Alpha chooses Merge, its worst payoff is 2.5, which occurs when exactly one of Beta or Gamma also selects Merge. If Alpha chooses Remain, its worst payoff is 1, which occurs when both Beta and Gamma select Merge. Since $2.5 > 1$, Alpha's prudential strategy is Merge. Beta and Gamma may also obtain payoffs of 1 if they were to select Remain, so each has Merge as their prudential strategy with minimum possible payoffs of 2.5 and 4, respectively.

Not only does (Merge, Merge, Merge) represent the profile of prudential strategies, but it is also efficient because any outcome change will result in Gamma receiving a smaller payoff. The Nash equilibrium (Remain, Remain, Remain) is also efficient because any outcome change will result in Alpha receiving a smaller payoff. None of the remaining strategy profiles is efficient because either (Merge, Merge, Merge) or (Remain, Remain, Remain) is at least as good for each player and strictly better for at least one player.

In summary, no strategy is dominant or dominated, (Merge, Merge, Merge) is the profile of prudential strategies, while (Merge, Merge, Merge) and (Remain, Remain, Remain) are the Nash equilibria and the efficient strategy profiles. Given the utilities assigned, our solution concepts predict the three groups will mutually choose to Merge or mutually choose to Remain. It turns out that these two possibilities are the only ones predicted by our solution concepts for a wide range of utility functions. That is, even if we modify our utility functions somewhat, our model predicts the same end behavior. This suggests that the model we have built has flexibility.

The model is also flexible in the sense that it can apply to many real-world scenarios where independent groups are debating whether or not to give up independence for economic or political gain. In recent history, we've seen the formation of the European Union, followed by the Brexit movement in the United Kingdom to leave the EU. We've also seen unions formed for ideological reasons or even forced political alliances and dissolutions. With some adjustments to the model, game theory can help predict behavior in these political situations by quantifying countries' preferences for independence or economic and political gain.

We have now illustrated the computation and interpretation of four solution concepts for the strategic games used to model four scenarios. The primary solution used by game theorists is the Nash equilibrium. It generalizes the use of utility functions as a model for the choice behavior of a single decision maker to the choice behavior of multiple decision makers in the presence of complete information. Dominance is also a reasonable generalization; however, it is rarely helpful since the strategic game model constructed often excludes dominated strategy choices from the beginning. Prudential strategies ignore the payoffs to the other players, and so can be thought of as one way of dealing with not knowing the other player's preferences. Efficiency can be thought of as a normative goal rather than a prediction about what will happen.

3.2   FINAL JEOPARDY

As we observed in Section 3.1, the phrase "Prisoner's Dilemma" has been used to describe many real-world scenarios; however, not all of these scenarios actually fit the mathematical definition. This can only be revealed by constructing and analyzing a model of the scenario.

Final Jeopardy. The popular TV game show Jeopardy!^{TM}, in which three contestants compete against each other for cash prizes, would seem to be an ideal scenario for multiple game theoretic models to be applicable. In fact, fan websites such as J! Archives often use game theoretic terminology, including Prisoner's Dilemma. In some cases, this terminology may not be accurate [55] and [35].

We examine a situation in Jeopardy! which fans have identified as a Prisoner's Dilemma. In the Final Jeopardy round, each contestant makes a wager as to whether they can answer a specific question correctly. When making the wager, contestants know the category of the question, but not the question itself, and the amount of money each of the other contestants has available. Each player's wager can be between 0 and their current winnings. Depending on whether the contestant answers the question correctly, they win or lose the amount of money wagered. The contestant with the most money after this final round of play wins the game. The winner keeps all of their winnings, and the other two contestants lose essentially all of their money. If there is a tie at the end of the round, a simple, essentially random, tie-breaker rule is applied to identify the winner.

The so-called Prisoner's Dilemma situation occurs when two contestants are tied for the lead, and the third contestant has less than half of the money of either of the first two contestants. For simplicity we will assume that it is contestants 1 and 2 who are tied with the most money.

In this situation, aficionados of Jeopardy! often refer to "Jeek's Rule," which asserts that while they could wager any amount up to their current winnings, contestants 1 and 2 should either wager nothing or everything. We discuss the reasonableness of this rule and then make it an assumption when we define our strategic game.

Let E be the amount of money contestants 1 and 2 have each won at the time Final Jeopardy begins. Let w_i denote the wager of contestant i and suppose that contestant 1's wager satisfies $0 < w_1 < E$. There are four cases to consider:

Case 1: Both contestants answer the question correctly. In this case, if $w_1 < w_2$, contestant 1 regrets not wagering E in order to win. If $w_1 \geq w_2$, then contestant 1 regrets not wagering E to maximize their winnings.

Case 2: Contestant 1 answers the question correctly and contestant 2 does not. Here contestant 1 regrets not wagering E in order to maximize their winnings.

Case 3: Contestant 1 answers the question incorrectly and contestant 2 answers correctly. Then contestant 1 is indifferent about their bet unless $w_2 = 0$, in which case they regret not wagering $w_1 = 0$.

Case 4: Both contestants answer the question incorrectly. Here, if $w_1 \geq w_2$, contestant 1 regrets not wagering $w_1 = 0$ in order to win. If $w_1 < w_2$, then contestant 1 regrets not wagering $w_1 = 0$ to maximize their winnings.

In all four cases, contestant 1 prefers to wager either everything or nothing over wagering w_1. By symmetry, contestant 2 also will wager everything or nothing.

Observe that this argument is based upon each contestant's view of the situation after the wagers and answers are known. At the time the two contestants are deciding what to wager, they can only guess how likely each contestant is to answer correctly and what their opponents will wager. With some rare set of guesses, a contestant may maximize their expected utility with a wager strictly between 0 and E. Nonetheless, we will accept Jeek's Rule, for to do otherwise would immediately mean that this scenario is not a Prisoner's Dilemma.

Given that contestants 1 and 2 will either wager nothing or everything, contestant 3's best response is to wager all but one dollar of his holdings and hope that the other two contestants both wager everything and answer the question wrong. Because of contestant 3's weak position, we will assume this is a two-player game and we can continue to explore whether this is a Prisoner's Dilemma.

For our first model, we assume both contestants 1 and 2, whom we will name Rose and Colin, will answer the question correctly. The resulting strategic game looks like the one displayed in Table 3.7. Since a contestant only wins money and only advances to the next game if they win this game, we also assume that winning the game is much more important than the amount of money won. Therefore, the payoffs in Table 3.7 represent the chances of the player winning the game.

TABLE 3.7   Final Jeopardy Payoff Matrix when Both Players Answer Correctly

(Rose, Colin)	Everything	Nothing
Everything	(0.5, 0.5)	(1.0, 0.0)
Nothing	(0.0, 1.0)	(0.5, 0.5)

As we see from Table 3.7, wagering everything is a strongly dominant strategy for both players and the strategy profile (Everything, Everything) is a Nash equilibrium. But this strategic game is not a Prisoner's Dilemma since this strategy profile is also efficient.

If we change our assumption that both players can always answer the question correctly, then the question of whether this is a Prisoner's Dilemma is less straightforward. So now we let p_i be the probability that contestant i is able to answer the question correctly. Assuming that this probability can be reasonably known to both contestants (perhaps based on their performance to this point), we construct the Generic Final Jeopardy strategic game in Table 3.8. For example, we calculate the probability of Rose winning when both contestants choose the strategy Everything by summing the three following probabilities.

1. The probability that Rose wins by getting the answer correct while Colin gets it wrong is $p_1(1 - p_2)$.

2. The probability that Rose wins given that they both answer correctly is $0.5p_1p_2$.

3. The probability that Rose wins if they both guess wrong is 0, since contestant 3 wins in this case.

TABLE 3.8   Generic Final Jeopardy Payoff Matrix

(Rose, Colin)	Everything	Nothing
Everything	$(p_1(1 - 0.5p_2),\ p_2(1 - 0.5p_1))$	$(\ p_1,\ 1 - p_1)$
Nothing	$(\qquad 1 - p_2,\qquad\qquad p_2)$	$(0.5,\quad 0.5)$

For the strategy Everything to be strongly dominant for Rose, we need

$$p_1(1 - .5p_2) > 1 - p_2$$

and

$$p_1 > .5.$$

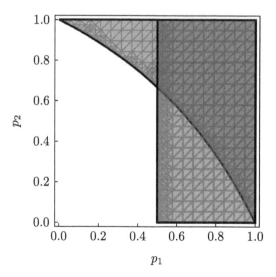

FIGURE 3.1 Where Everything Is a Dominant Strategy for Rose.

The first inequality corresponds to the region in Figure 3.1 above the curve $p_2 = (1 - p_1)/(1 - 0.5p_1)$. The second inequality corresponds to the rectangular region to the right of the vertical line $p_1 = 0.5$. The doubly shaded region in the upper right, excluding the boundary, is where choosing Everything is the strongly dominant strategy for Rose.

For the strategy Nothing to be strongly dominant for Rose, we need

$$p_1(1 - .5p_2) < 1 - p_2$$

and

$$p_1 < .5.$$

with at least one inequality being strict.

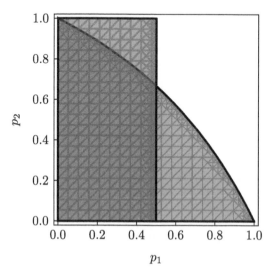

FIGURE 3.2 Where Nothing Is a Strongly Dominant Strategy for Rose.

The doubly shaded region in the lower left of Figure 3.2, again excluding the boundary, illustrates when choosing Nothing is the strongly dominant strategy for Rose.

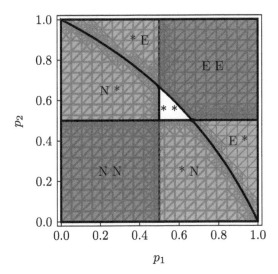

FIGURE 3.3 Where Each Strategy Is Strongly Dominant for Each Player.

By symmetry of the strategic game payoffs, we can construct Figure 3.3. Each region, excluding boundaries, is labeled with two letters. The first letter denotes which strategy is strongly dominant for Rose in that region, and the second letter denotes which strategy is strongly dominant for Colin. An asterisk (*) indicates that neither strategy is dominant for that player. Note in particular the unshaded region in the middle; there are no dominant strategies for either player in this region.

Let's return now to the question of whether or not there is a Prisoner's Dilemma within the model. A Prisoner's Dilemma requires each player to have a strongly dominant strategy and for there to be a benefit to each player by each choosing the dominated rather than the strongly dominant strategy. Figure 3.3 displays two regions where both players have strongly dominant strategies: lower left and upper right. In the lower left, corresponding to $p_1 < 0.5$ and $p_2 < 0.5$, the strategy Nothing is strongly dominant for both players. Unfortunately, it then follows that

$$p_1(1 - 0.5p_2) < 0.5 \text{ and } p_2(1 - 0.5p_1) < 0.5;$$

hence, the strategy pair (Everything, Everything) is not better than (Nothing, Nothing). Thus there is not a Prisoner's Dilemma in this region.

The upper right EE region corresponds to $0.5 < p_1$, $0.5 < p_2$, $p_1(1 - 0.5p_2) > 1 - p_2$, and $p_2(1 - 0.5p_1) > 1 - p_1$ (the last two inequalities are equivalent to each other). For there to be a Prisoner's Dilemma here, (Nothing, Nothing) needs to provide a larger payoff to each player than (Everything, Everything), which corresponds to $p_1(1 - 0.5p_2) < 0.5$ and $p_2(1 - 0.5p_1) < 0.5$. Incorporating these last two inequalities gives the small region in Figure 3.4 where we have a true Prisoner's Dilemma strategic game (under the assumptions that Jeek's Rule holds and contestant 3 is not considered a player but has a role in reducing the other contestant's probability of winning when each wagers everything).

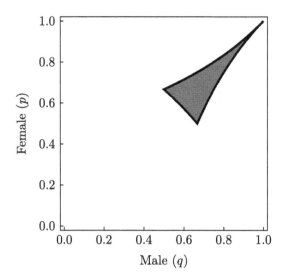

FIGURE 3.4 Where the Generic Jeopardy Strategic Game Is a Prisoner's Dilemma.

Now that we have completed an analysis of Generic Final Jeopardy, we should interpret it in the context of Jeopardy! Recall that in every Prisoner's Dilemma strategic game there are two strategies: Cooperate and Defect. In Generic Final Jeopardy, the strategy Everything is strongly dominant for large values of p_i, so it is equivalent to Defect. The strategy Nothing must therefore be equivalent to Cooperate and we have (Nothing, Nothing) as the cooperative, efficient solution.

As we interpret this solution in the context of an actual Jeopardy game, our assessment of it shows that we have not accounted for the fact that the players not only want to win, but also to win as much money as possible. Therefore, we might want to build into our model a player's preference for wagering everything rather than nothing if they have a high probability of getting the answer correct and are not risk adverse. However, if both players are willing to wager everything, we need to re-visit our assumption to remove contestant 3 as a player in our model. Specifically, since a contestant's goal is to win the game, Rose and Colin wagering everything allows contestant 3 a chance at winning the game.

Observe that when $p_1 = p_2 = 1$, we obtain Table 3.7, which we have already seen is not a Prisoner's Dilemma. This suggests that we might wish to consider boundary conditions more carefully than we have so far.

We close with the observation that for each probability pair (p_1, p_2), we are playing a different strategic game, and in each strategic game the values of p_1 and p_2 are common knowledge. Our work in this section solves the problem for each specific value of p_1 and p_2. In Chapter 6, we will return to the problem of modeling a scenario when we have incomplete information.

3.3   MIXED STRATEGIES

Mating within species can be modeled as a strategic game with each male and female making choices about the conditions in which they will mate.

Mating Behavior. In the animal kingdom, a measure of success is often the number offspring an animal can produce. For this reason, animals engage in a variety of mating behaviors to attract partners. Male peacocks, for example, display flashy plumage in the hopes of drawing more mating partners, despite the risk that they become more of a target for predators. Seahorses are the only animal species in which the males carry the young, and before a female will deposit her eggs into a male seahorse, the two engage in extensive courtship behaviors, including a daily dance. Biologists believe that courtship behaviors for many species can be a way for an animal to determine the suitability of a mate. Dawkins uses game theory to analyze how males and females in the animal kingdom decide to mate. Dawkins' model applies to many species where the female carries most of the cost of reproduction so she has a vested interest in selecting a mate who will help her care for the young and might require an extensive courtship before selecting a mate. Males of such species might agree to such courtship if it increases their ability to produce offspring and might even agree to help protect or care for offspring after birth if it increases their chance of survival [28] [88] [61].

In this section, we will investigate Dawkins' famous game theory model of mating behaviors. We assume as he did that a female of a species might adopt one of two behaviors during the mating season. She can either insist on some courtship before mating or accept mating without courtship. A male of the species can either agree to stay with one female after mating or leave to mate with other females. If a female who insists on courtship mates with a male who stays, the result is offspring raised by both parents. A female who insists on courtship will not mate with a male who chooses to leave. If a female who does not insist on courtship mates with a male who stays, the offspring is still raised by both parents, but without the courtship costs. If a female mates with a male who leaves, she will raise her offspring alone [28].

We will model the scenario as a strategic game. The players will be named Female and Male, and abbreviated by F and M. Female's strategies are Insist on courtship and Accept mating without courtship, and Male's strategies are a willingness to Stay for courtship and child raising and Leave if courtship or child raising are needed. We assume that outcomes can be separated into three components: courtship, having offspring, and raising offspring. We assume that the benefit associated with having offspring is $h = 9$, the cost associated with courtship is $c = 2$, and the cost associated with raising offspring is $2r = 12$ (and that cost can be shared). Finally, we assume that payoffs are additive in the benefits and costs associated with each component. This information is summarized in Table 3.9. Note that by describing payoffs as the difference of benefits and costs, we are assuming that these payoffs are vNM utilities.

TABLE 3.9 Mating Behavior Outcomes and Payoffs

Strategy Profile $s = (s_F, s_M)$	Outcome o Court	Offspring	Raising	Female Payoff $u_F(s)$		Male Payoff $u_{M(s)}$	
(Insist, Stay)	yes	yes	both	$h - c - r =$	1	$h - c - r = 1$	
(Insist, Leave)	no	no	none		0		0
(Accept, Stay)	no	yes	both	$h - r$	= 3	$h - r$	= 3
(Accept, Leave)	no	yes	female	$h - 2r$	= -3	h	= 9

Table 3.10 rearranges the strategies and payoffs into a payoff matrix. Female's best response to Stay is Accept and her best response to Leave is Insist. Male's best response to Insist is Stay and his best response to Accept is Leave. The payoffs corresponding to best responses are boxed in Table 3.10. There is no dominant or dominated strategy for either

TABLE 3.10   Mating Behavior Payoff
Matrix

(Female, Male)	Stay	Leave
Insist	(1, $\boxed{1}$)	($\boxed{0}$, 0)
Accept	($\boxed{3}$, 3)	(-3, $\boxed{9}$)

player. Further, since no pair of strategies are mutually best responses there is no Nash equilibrium.

If Female chooses Insist [respectively, Accept], her smallest possible payoff is 0 [resp. –3]. Hence Female's prudential strategy is Insist, with a security level of 0. Similarly, if Male chooses Stay [resp. Leave], his smallest possible payoff is 1 [resp. 0]. Hence his prudential strategy is Stay, with a security level of 1.

However, we see that this profile of prudential strategies (Insist, Stay) is not efficient, as the strategy pair (Accept, Stay) yields a higher payoff for both the Female and Male. (Accept, Stay) and (Accept, Leave) are the two efficient strategy profiles.

Let's explore further Female's response to Male's selection of the Stay strategy. In an initial encounter, she is likely to select her prudential strategy of Insist, but if she discovers in repeated encounters that the male continues to choose Stay, then she becomes increasingly likely to select Accept, her best response to Stay. The language of the previous sentence ("likely" and "increasingly likely") suggest that we should consider our payoffs from a von Neumann-Morgenstern perspective and allow the possibility of strategies consisting of lotteries over the pure strategies given above.

Therefore, let p denote the probability that Female will select Accept and let q denote the probability that Male will select Leave. The expected payoffs to Female choosing a pure strategy versus Male choosing a mixed strategy are

$$u_{\text{Female}}(\text{Insist},\ (1-q)\text{Stay} + q\text{Leave}) = (1)(1-q) +\ \ (0)q$$
$$u_{\text{Female}}(\text{Accept}, (1-q)\text{Stay} + q\text{Leave}) = (3)(1-q) + (-3)q.$$

These payoffs are plotted in the left half of Figure 3.5. The dashed line gives the Female's expected payoff as a function of q when she chooses Accept. The solid line gives the Female's expected payoff as a function of q when she chooses Insist. Corresponding to boxing the best response payoffs in the payoff matrix, we thicken the graphs of the best response payoffs in the payoff graphs: for sufficiently low q, Accept is Female's best response to Male's mixed strategy, and for sufficiently high q, Insist is Female's best response to Male's mixed strategy.

Female is indifferent between her strategy choices when the two payoffs are equal. This happens if and only if

$$u_{\text{Female}}(\text{Insist}, (1-q)\text{Stay} + q\text{Leave}) = u_{\text{Female}}(\text{Accept}, (1-q)\text{Stay} + q\text{Leave})$$
$$(1)(1-q) + (0)q = (3)(1-q) + (-3)q$$
$$q = 2/5.$$

For $q = 2/5$, all of Female's mixed strategies yield her the same payoff 3/5, and so all of her mixed strategies are best responses. This best response analysis is summarized by

$$BR_{\text{F}}(q) = \begin{cases} p = 1 & \text{if } q < \frac{2}{5} \\ p \in [0,1] & \text{if } q = \frac{2}{5} \\ p = 0 & \text{if } q > \frac{2}{5} \end{cases}$$

and the solid line in Figure 3.6.

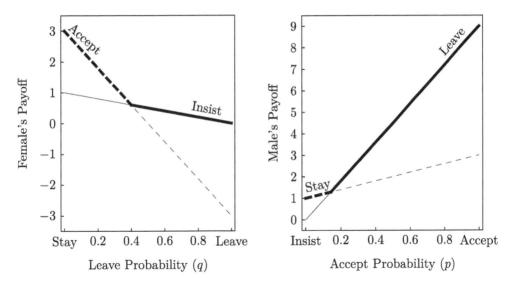

FIGURE 3.5 Player Payoff vs. Opponent's Mixed Strategy.

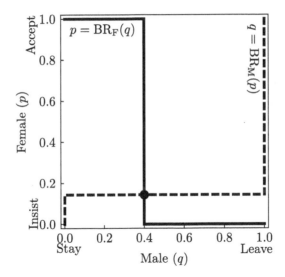

FIGURE 3.6 Mating Behavior Best Response Functions.

A similar analysis can be done from Male's perspective. His pure strategy versus Female's mixed strategies payoffs are

$$u_{\text{Male}}((1-p)\text{Insist} + p\text{Accept}, \text{Stay}) = (1)(1-p) + (3)p$$
$$u_{\text{Male}}((1-p)\text{Insist} + p\text{Accept}, \text{Leave}) = (0)(1-p) + (9)p$$

which are graphed in the right half of Figure 3.5 with the payoffs to his best responses thickened. His payoffs are equal if and only if $(1)(1-p) + (3)p = (0)(1-p) + (9)p$ if and only if $p = 1/7$. Male's best responses are

$$BR_{\mathrm{M}}(p) = \begin{cases} q = 0 & \text{if } p < \frac{1}{7} \\ q \in [0,1] & \text{if } p = \frac{1}{7} \\ q = 1 & \text{if } p > \frac{1}{7} \end{cases}$$

which is graphed by the dashed line in Figure 3.6.

We can view Figure 3.6 as an extension of the boxes in the payoff matrix with the solid line corresponding to Female boxes and the dashed line corresponding to Male boxes. Nash equilibria correspond to places the two best response graphs intersect. The unique Nash equilibrium is

$$\left(\frac{6}{7}\text{Insist} + \frac{1}{7}\text{Accept}, \frac{3}{5}\text{Stay} + \frac{2}{5}\text{Leave} \right)$$

in which the first lottery is the female's strategy and the second is the male's strategy.

We can interpret this Nash equilibrium in several ways. Focusing on the female, we may say that in 6/7 of her mating opportunities she insists on courtship. We might also say that in any given mating opportunity, she randomly decides to insist on courtship with probability 6/7. Thinking more generally, we might also interpret the strategy to say that 6/7 of all females of this species will insist on courtship when mating while 1/7 will not.

The Mating Game did not have any Nash equilibrium when we considered only pure strategies, but we were able to find one using the lotteries that become viable strategy options when we considered the payoffs as vNM utilities. The following theorem, from Nash's 1950 paper [76], establishes the fact that we will always have a Nash equilibrium with vNM utility functions.

Theorem 3.3.1 (Nash Equilibrium Existence). Every strategic game with VNM utility functions, and in which each player has a finite number of strategies, contains at least one Nash equilibrium in mixed strategies.

Nash's proof of this extraordinary theorem generalizes the methodology that we used in the Mating strategic game to an arbitrary finite number of players and strategies. His proof relies heavily on the Brouwer Fixed Point Theorem from topology and is beyond the scope of this book.

> **Attrition Behavior**. In many species, there is a direct competition between the males to win mating privileges. Yet these competitions rarely result in significant injuries to the animals involved. We attempt to model these interactions in a way that explains the outcomes observed in nature.

This question was first investigated by Smith ([70], [101]), and many subsequent authors have developed various presentations of their work to focus on different perspectives of the question. Our development is similar to [28] and [110]. Once the model is developed, we use it to discuss how to find Nash equilibria in strategic games in larger matrix forms.

To begin, we recognize that in this scenario, our players are two males of the species that arrive at a moment of competition. In that particular moment, each must choose whether to engage in a serious fight or to retreat from the scene. If both decide to fight, then both risk serious injury. The winner of a fight (i.e., the male without injury), or the male that remains on the scene, obtains the privilege of mating.

The strategies that each male might use arise from considering how they might behave over multiple encounters with other males. For example, a male might be a Hawk, engaging in a fight in every opportunity and stopping only upon winning or injury. Alternatively, he might be a Dove in the anthropomorphic sense and therefore flee from every engagement after a bit of posturing. But there are other options as well. A male might be a Bully, who bluffs at fighting; a Bully wins against a Dove after some posturing, but loses against a Hawk immediately. A Retaliator starts to posture like a Dove, will continue to posture if the other male postures, but will fight if confronted by a male who shows signs of fighting (such as a Hawk or Bully). A Bourgeois behaves like a Hawk in his own territory and a Dove when away from his territory. After making some assumptions about the utility of mating, injury, fighting, and posturing, Table 3.11 is the resulting payoff matrix for this strategic game.

TABLE 3.11   Attrition Behavior Payoff Matrix

(Male 1, Male 2)	Hawk	Dove	Bully	Retaliator	Bourgeois
Hawk	(−30, −30)	(40, 0)	(40, 0)	(−30, −30)	(5, −15)
Dove	(0, 40)	(12, 12)	(−8, 32)	(12, 12)	(6, 26)
Bully	(0, 40)	(32, −8)	(12, 12)	(−8, 32)	(16, 16)
Retaliator	(−30, −30)	(12, 12)	(32, −8)	(12, 12)	(−9, −9)
Bourgeois	(−15, 5)	(26, 6)	(16, 16)	(−9, −9)	(20, 20)

We arrived at these payoffs by making the reasonable assumption that mating, status quo, posturing, fighting, and injury are in descending order of male bird preferences. Quite arbitrarily, let us assign 40, 0, −8, −20, and −60 points to mating, status quo, posturing, fighting, and injury, respectively. Further, we will assume that a player's payoff is a sum of the individual points obtained. Finally, we assume that either male is equally likely to win if they both fight or they both posture, and a male is equally likely to be within or outside of his territory. Thus, if two Hawks interact, both fight at a cost of 20 points, one wins the fight and obtains 40 points, and the other loses with an injury at a cost of 60 points. Since there is a 50-50 chance of winning or losing the fight, the average payoff to a Hawk in a Hawk-Hawk interaction is $-20 + (1/2)(40 - 60) = -30$ points. This information is given in the upper left cell of Table 3.11. If two Doves interact, one wins after posturing and obtains $40 - 8 = 32$ points, and one loses after posturing and obtains $0 - 8 = -8$ points. Since there is a 50-50 chance of winning or losing the fight, the average payoff to a Dove in a Dove-Dove interaction is $(1/2)(32 - 8) = 12$ points. This information is given in the Dove-Dove cell of Table 3.11. If a Hawk interacts with a Dove, the Dove may start to posture but flees upon the recognition that the Hawk will fight, resulting in the Hawk mating for 40 points while the Dove avoids a fight and the time wasted by posturing for an extended time yielding 0 points. This information is given in the Hawk-Dove and Dove-Hawk cells of Table 3.11.

Similar calculations can be done to complete the other cells in the table. For example, if a Bourgeois male, who acts like a Hawk half the time and acts like a Dove the other half the time, meets a Dove, then the payoff profile can be calculated as

$$u(\text{Bourgeois}, \text{Dove}) = \tfrac{1}{2}u(\text{Hawk}, \text{Dove}) + \frac{1}{2}u(\text{Dove}, \text{Dove})$$

$$= \frac{1}{2}(40, 0) + \frac{1}{2}(12, 12) = (26, 6).$$

If two Bourgeois males meet, they are equally likely to be on the territory of the first or second male, so the payoff profile is given by

$$u(\text{Bourgeois}, \text{Bourgeois}) = \frac{1}{2}u(\text{Hawk}, \text{Dove}) + \frac{1}{2}u(\text{Dove}, \text{Hawk})$$

$$= \frac{1}{2}(40, 0) + \frac{1}{2}(0, 40) = (20, 20).$$

The reader can easily box the payoffs corresponding to best response strategies and thereby find the six pure-strategy Nash equilibria: (Hawk, Dove), (Dove, Hawk), (Hawk, Bully), (Bully, Hawk), (Retaliator, Retaliator), and (Bourgeois, Bourgeois). Calculating mixed-strategy Nash equilibria in two-player strategic games having more than two strategies per player can become quite involved mathematically. However, the following theorem provides one approach to doing these calculations. A full development of this approach can be found in [78]. This, and other approaches, can also be found in [8].

Theorem 3.3.2 (Finding Nash Equilibria). Suppose $G = (N, S, O, u)$ is a strategic game with $N = \{1, 2\}$, $S_1 = \{r_1, r_2, \ldots, r_l\}$, and $S_2 = \{c_1, c_2, \ldots, c_m\}$. The mixed-strategy profile

$$(s_1, s_2) = (x_1 r_1 + x_2 r_2 + \cdots + x_l r_l, \; y_1 c_1 + y_2 c_2 + \cdots + y_m c_m)$$

is a Nash equilibrium for the strategic game G if and only if

1. $u_1(r_i, s_2) = u_1(s_1, s_2)$ for each strategy r_i with $x_i > 0$,

2. $u_2(s_1, c_j) = u_2(s_1, s_2)$ for each strategy c_j with $y_j > 0$,

3. $u_1(r_i, s_2) \leq u_1(s_1, s_2)$ for each strategy r_i with $x_i = 0$,

4. $u_2(s_1, c_j) \leq u_2(s_1, s_2)$ for each strategy c_j with $y_j = 0$.

The first two conditions of the theorem say that each player could unilaterally deviate from their mixed strategy to any of the pure strategies that could be chosen by their mixed strategy without changing their expected payoff, and the last two conditions say that neither player would benefit by unilaterally deviating to a pure strategy that could not be chosen by their mixed strategy. The last two conditions are special cases of the Nash equilibrium definition, and if one of the first two conditions were false, then the corresponding player would have a pure-strategy response that would give her a higher payoff.

Following the path suggested by the four conditions of the theorem, we select which x_i and y_j we wish to be positive, solve the system of linear equations formed by the first two conditions and the constraints

$$\sum_{i=1}^{l} x_i = \sum_{j=1}^{m} y_j = 1,$$

and then test the solutions against the inequality constraints in the last two conditions, $x_i \geq 0$ for all i, and $y_j \geq 0$ for all j. The difficulty arises not from solving the system of equations, but rather from testing all possible sets of system of equations that arise from the problem, one for each pair of nonempty subsets of the strategies of the two players. For example, in the Attrition Behavior strategic game, there are $(2^5 - 1)^2 = 961$ ways to select which of the x_i and y_j will be positive.

Most of the time, practitioners use a specialized software package to do these computations. The authors suggest *Gambit* [71], a free downloadable program. Gambit lists twenty Nash equilibria for the Attrition Behavior strategic game, including the six pure-strategy Nash equilibria identified earlier by hand. Four of the others are

$$s^1 = \left(\frac{1}{2} \text{Dove} + \frac{1}{2} \text{Retaliator}, \text{Retaliator} \right),$$

$$s^2 = \left(\text{Retaliator}, \frac{1}{2} \text{Dove} + \frac{1}{2} \text{Retaliator} \right),$$

$$s^3 = \left(\frac{1}{2} \text{Dove} + \frac{1}{2} \text{Retaliator}, \frac{1}{2} \text{Dove} + \frac{1}{2} \text{Retaliator} \right), \text{ and}$$

$$s^4 = \left(\frac{14}{29}\text{Hawk} + \frac{15}{29}\text{Bully}, \frac{14}{29}\text{Hawk} + \frac{15}{29}\text{Bully}\right).$$

The list Gambit provides are only the vertices of polygonal regions of Nash equilibria. For example, all of the strategy profiles bounded by s^1, s^2, s^3, and (Retaliator, Retaliator), that is,

$$((1 - \lambda)\text{Dove} + \lambda\text{Retaliator}, (1 - \mu)\text{Dove} + \mu\text{Retaliator})$$

for all $\frac{1}{2} \leq \lambda \leq 1$ and $\frac{1}{2} \leq \mu \leq 1$, are Nash equilibria.

While we may want to leave the long computational task of finding Nash equilibria to computer software, it is straight-forward to check whether a given strategy profile is a Nash equilibrium by directly applying Theorem 3.3.2. Let $s = s^4$ so that $s_1 = s_2 = (14/29)(\text{Hawk}) + (15/29)(\text{Bully})$, and calculate

$$
\begin{aligned}
u_1(\text{Hawk}, s_2) &= \tfrac{14}{29}(-30) + \tfrac{15}{29}(40) = \tfrac{180}{29} \\
u_1(\text{Dove}, s_2) &= \tfrac{14}{29}(0) + \tfrac{15}{29}(-8) = -\tfrac{120}{29} \\
u_1(\text{Bully}, s_2) &= \tfrac{14}{29}(0) + \tfrac{15}{29}(12) = \tfrac{180}{29} \\
u_1(\text{Retaliator}, s_2) &= \tfrac{14}{29}(-30) + \tfrac{15}{29}(32) = \tfrac{60}{29} \\
u_1(\text{Bourgeois}, s_2) &= \tfrac{14}{29}(-15) + \tfrac{15}{29}(16) = \tfrac{30}{29}
\end{aligned}
$$

and observe that $u_1(\text{Hawk}, s_2) = u_1(\text{Bully}, s_2)$ which are at least as great as the other payoffs, verifying conditions 1 and 3. By the symmetry of the payoffs, conditions 2 and 4 immediately follow.

The abundance of Nash equilibria to the Attrition Behavior strategic game leads one to several questions in the modeling process. For example, are we able to provide a meaningful interpretation of the solution in the context of the model? While (Hawk, Dove) is a pure-strategy Nash equilibrium, it does not reflect the reality that the two players do not perceive themselves as either the first or second player. A more accurate interpretation is that the first player is always me, and the second player is always the other. In fact, if we examine the payoff matrix we notice that it is symmetrical. We call a strategic game in which each of the two players have the same strategies available that results in a symmetric payoff matrix a *symmetric strategic game*. This symmetry leads each of the two players to select the Hawk strategy. However, the (Hawk, Hawk) strategy profile is not a Nash equilibrium.

We can reasonably argue that we are not really modeling the one-time behavior of two males of the species but rather repeated interactions of one of two types. These types are either the repeated interactions between two particular males of the species or the random interactions between pairs of males from a larger population of males. Again, the symmetric roles of the two males causes (Hawk, Dove) to be a problematic solution. This symmetry suggests that in repeated interactions, a male should be selecting Hawk half the time and Dove the other half, yielding the strategy profile $(1/2)(\text{Hawk}) + (1/2)(\text{Dove}), (1/2)(\text{Hawk}) + (1/2)(\text{Dove}))$, which is not a Nash equilibrium.

Since the roles of the players are symmetrical, it would seem reasonable that they would each be playing the same strategy. This type of solution works when modeling two individuals meeting in a single strategic game, or when modeling repeated interactions between two individuals, or when modeling population dynamics. Thus we are inclined to think of (Retaliator, Retaliator), (Bourgeois, Bourgeois), s^3, or s^4 as good solutions. We must now investigate whether or not these have reasonable interpretations. Suppose a small number of Retaliators invaded a population of Bourgeois. A Bourgeois would primarily interact with other Bourgeois, each time receiving a payoff of 20, but would occasionally interact with a Retaliator, each time receiving a payoff of -9. A Retaliator would primarily interact with Bourgeois, each time receiving a payoff of -9, but would occasionally interact with another

Retaliator, each time receiving a payoff of 12. On average, each Bourgeois would receive nearly 20 points while each Retaliator would receive close to -9 points. Thus, the invasion of Retaliators would die off in the population of Bourgeois. Since potential invasions by Hawks, Doves, Bullies, or Retaliators will die off because their payoffs in interactions with a Bourgeois (5, 6, 16, and -9) are strictly less than the payoff to a Bourgeois interacting with another Bourgeois (20), we will say that Bourgeois is evolutionarily stable.

Definition 3.3.1. A pure or mixed strategy s^* in a symmetric strategic game is an *evolutionarily stable strategy*, commonly abbreviated ESS, if for every pure or mixed strategy $s \neq s^*$

1. $u_1(s^*, s^*) \geq u_1(s, s^*)$, and

2. if $u_1(s^*, s^*) = u_1(s, s^*)$, then $u_1(s^*, s) > u_1(s, s)$.

We observed that $u_1(\text{Bourgeois}, \text{Bourgeois}) > u_1(s, \text{Bourgeois})$ for all pure (and therefore for all mixed) strategies $s \neq$ Bourgeois; hence, the second condition in the definition is vacuously satisfied. Condition 1 is equivalent to saying that (s^*, s^*) is a Nash equilibrium, and is clearly necessary because if $u_1(s^*, s^*) < u_1(s, s^*)$ for some strategy $s \neq s^*$, then s could profitably invade a population of s^*. Condition 2 guards against what biologists call "genetic drift."

Suppose a small number of Doves invaded a population of Retaliators. All interactions would give each male a payoff of 12. The Doves could increase in proportion to the Retaliators, and if more than half of the population become Doves, then Bullies could profitably invade. It is because of this possibility that Retaliator, which violates condition 2 with $s^* =$ Retaliator and $s =$ Dove, is not deemed to be evolutionarily stable.

We can verify that $s^* = (14/29)(\text{Hawk}) + (15/29)(\text{Bully})$ is evolutionarily stable. We earlier showed that

$$u_1(\text{Hawk}, s_2) = u_1(\text{Bully}, s_2) > u_1(\text{Dove}, s_2), u_1(\text{Retaliators}, s_2), u_1(\text{Bourgeois}, s_2),$$

verifying condition 1. If $u_1(s^*, s^*) = u_1(s, s^*)$, then $s = (1 - \lambda)\text{Hawk} + \lambda\text{Bully}$ for some $0 \leq \lambda \leq 1$, and so

$$
\begin{aligned}
u_1(s^*, s) - u_1(s, s) &= \left(\tfrac{14}{29} - (1 - \lambda)\right)(1 - \lambda)u_1(\text{Hawk}, \text{Hawk}) \\
&\quad + \left(\tfrac{14}{29} - (1 - \lambda)\right)\lambda u_1(\text{Hawk}, \text{Bully}) \\
&\quad + \left(\tfrac{15}{29} - \lambda\right)(1 - \lambda)u_1(\text{Bully}, \text{Hawk}) \\
&\quad + \left(\tfrac{15}{29} - \lambda\right)\lambda u_1(\text{Bully}, \text{Bully}) \\
&= \tfrac{2}{29}(15 - 29\lambda)^2 > 0
\end{aligned}
$$

if $\lambda \neq 15/29$, verifying condition 2.

It turns out that Retaliator and $(14/29)(\text{Hawk}) + (15/29)(\text{Bully})$ are the only evolutionarily stable strategies. Either could be the endpoint of an evolutionary process. A population of Retaliators with an expected payoff of 20 are generally cooperative with mating decided by simple posturing. A population almost equally split between Hawks and Bullies with an expected payoff of $180/29 \approx 6.2$ are often attacking each other resulting in injuries in about one-quarter of the interactions.

Our definition of an evolutionarily stable strategy is static but is motivated by dynamic considerations: the time evolution of a population of players each using a particular strategy in their interactions with others in the population. This dynamic approach with be considered in Section 7.3.

As the examples in this section show, game theory has applications to the field of biology. Known as evolutionary game theory, this subfield uses game theory to explain evolutionary behavior among species. It is important to note that the application of evolutionarily stable solutions extends well beyond biological scenarios. For further investigation into this area, see [17].

3.4 NON-MATRIX MODELS

Not all scenarios are easily modeled in a matrix format, but the conceptual framework for the modeling and the analysis still applies.

In Chapter 2, we constructed the utility functions for two firms interacting as a Cournot duopoly. We complete that investigation by determining the Nash equilibrium for the two firms. That is, we will find the pair of production levels (Q_1, Q_2) which are simultaneously best responses to each other. We begin by recalling that the utility (profit) functions for the two firms are given by

$$u_1(Q_1, Q_2) = \pi_1(Q_1, Q_2) = (\frac{a}{b} - \frac{Q_1 + Q_2}{b})Q_1 - cQ_1$$

and

$$u_2(Q_1, Q_2) = \pi_2(Q_1, Q_2) = (\frac{a}{b} - \frac{Q_1 + Q_2}{b})Q_2 - cQ_2.$$

Each firm's goal is to maximize its profit, which suggests that we calculate the two partial derivatives to find the best response each firm has to the other's action. Thus, considering Firm 1,

$$\frac{\partial u_1}{\partial Q_1} = -\frac{2}{b}Q_1 - (\frac{Q_2}{b} - \frac{a}{b} + c),$$

and, when we set this equal to zero to find the critical point, we obtain

$$Q_1 = -\frac{1}{2}(Q_2 - a + bc).$$

Since

$$\frac{\partial^2 u_1}{\partial Q_1^2} = -\frac{2}{b} < 0,$$

the critical point, as long as it is positive, is a maximum; otherwise Firm 1 should produce nothing. We have found Firm 1's best response function; given a production level Q_2 for Firm 2, Firm 1 is able to calculate a production level Q_1 which will maximize its profit.

By symmetry, Firm 2's best response function is

$$Q_2 = -\frac{1}{2}(Q_1 - a + bc).$$

Summing the two best response functions, we obtain

$$Q = -\frac{1}{2}Q + a - bc$$

where $Q = Q_1 + Q_2$. By the symmetry of the two best response functions and solving for Q in the above equation, it is clear that

$$Q_1 = Q_2 = \frac{1}{2}Q = \frac{a - bc}{3}.$$

Thus, if $a - bc > 0$, then each firm will produce $(a - bc)/3$ units of the good and obtain a profit of $u_i((a - bc)/3, (a - bc)/3) = (a - bc)^2/(9b)$. If $a - bc \leq 0$, then each firm will produce nothing and obtain no profit. The condition for positive production can also be written $a/b > c$, which means that the price at which consumers will demand nothing must be higher than the cost per unit—quite a reasonable result.

Ascending Bid Auction. Auctions are used to sell works of art, used cars, livestock, assets of a bankrupt company, items from the estate of a deceased person, bands of frequencies to be used by cell phone companies, rights for mineral exploration on public lands, and more. In an oft-used method, an auctioneer displays an object for sale and takes bids from the crowd of bidders. The bids start at some number and continue to increase until no one wants to call out a higher bid. The last one to bid is given the object for the amount of their last bid.

We will model this scenario from the perspective of the n potential bidders, who will be taken to be the players. We will assume that player i values the object at the commonly known positive integer v_i. A complicated sequence of actions can occur. Whether a player enters a bid at any particular point in time and how large that bid might be can depend on the entire history of bidding that has already taken place. In order to model this scenario as a strategic game, we will assume that player i selects a nonnegative integer b_i that can be thought of as the highest bid they are willing to make.

Suppose player 1 is willing to bid up to \$30 for an object, player 2 is willing to bid up to \$20 for the same object, and all other players' potential bids are less than \$20. Player 1 will win the object by bidding any amount over \$20 but less than \$30, and in reality should only bid slightly higher than \$20. Generally speaking, the player i who chooses the largest potential bid b_i will win the object, paying an amount less than or equal to b_i, but only slightly more than the second largest potential bid. To simplify this process in our model, we assume that the player with the highest potential bid will obtain the object, but they will pay exactly the amount of the second highest potential bid. Finally, we assume the payoff for the winning player i is the difference between their commonly known valuation v_i of the object and the amount that they paid, ties are broken randomly, and players with lower bids obtain a 0 payoff.

To begin our analysis of a strategy b_i for player i, let $c_i = \max\{b_j : j \neq i\}$ be the largest bid by the players other than player i. If $b_i > c_i$, then player i is the winner and obtains a payoff of $v_i - c_i$. If $b_i = c_i$, then player i is one of $m \geq 2$ winners and obtains a payoff of $(1/m)(v_i - c_i)$. If $b_i < c_i$, then player i is a loser and obtains a payoff of 0. Player i's payoff bidding $b_i \leq v_i$ is nonnegative for all values of c_i and 0 for $c_i > v_i$, but player i's payoff bidding $b_i > v_i$ is negative for $c_i = v_i$. Thus, player i's payoff bidding any $b_i \leq v_i$ are prudential.

A bid by player i larger than their valuation $b_i > v_i$ is dominated by bidding the player's valuation $b_i^* = v_i$ because (1) if $c_i > b_i$, then player i's payoff is 0 with either bid; (2) if $b_i \geq c_i > v_i$, then player i's payoff is negative with the bid b_i and 0 with the bid v_i; and (3) if $v_i \geq c_i$, then player i's payoff is $v_i - c_i$ with either bid. A bid by player i smaller than their valuation $b_i < v_i$ is dominated by bidding the player's valuation $b_i^* = v_i$ because (1) if $c_i > v_i$, then player i's payoff is 0 with either bid; (2) if $v_i \geq c_i > b_i$, then player i's payoff is 0 with the bid b_i and $v_i - c_i \geq 0$ with the bid v_i; and (3) if $b_i \geq c_i$, then player i's payoff is $v_i - c_i$ with either bid. Therefore, player i has a unique dominant strategy: bid their valuation $b_i = v_i$.

The Nash equilibrium concept is less helpful at prediction. Choose any player i and have that player bid an amount no less than the highest valuation and have all other players bid no more than v_i. Then player i's payoff is nonnegative and the only bid change that could change their payoff is to bid low enough to lose, resulting in a 0 payoff, and each other player's payoff is 0 and the only bid change that could change their payoff is to bid high enough to win, resulting in a nonpositive payoff. Thus, we have described a Nash equilibrium, and in many of these equilibria some or all of the players are using dominated strategies and at least one player is not using a prudential strategy.

To finish the analysis, we note there are two types of efficient strategy profiles. First, if each player bids 0, then player i obtains a payoff of $(1/n)v_i > 0$. This is efficient because for any other strategy profile either (1) at least one player loses and thereby obtains a payoff of 0, or (2) all players bid the same positive amount which reduces player i's payoff to $(1/n)(v_i - b_i)$. The second type of efficient strategy profile is for one player i to bid 1 while the other players bid 0, because for any other strategy profile player i will obtain a smaller payoff. As is typical with our interpretation of efficient strategy profiles, the players may jointly desire an outcome associated with an efficient strategy profile, it seems that it could only happen with some sort of binding agreement because there is a strong incentive for each player to raise their bid slightly.

Escalation. Sometimes people, companies, animals, or nations engage in behaviors to win arguments, obtain contracts, capture territory, or achieve political gains. The behaviors are costly but are worthwhile if the goal is achieved.

Scenarios like this commonly occur, both in simple and complex forms. For a simple example, a person is placed on hold during a telephone call to a remote service center. A sufficiently long wait can lead to a satisfactory resolution to the caller's problem. If the caller hangs up too soon, they have lost all of the time invested in the call without resolving the problem. For a more complex example, when two nations are in conflict over some issue, each may have a propensity to invest more and more resources — financial, people, prestige — into resolving the issue in the their favor. The nation in whose favor the issue is ultimately resolved has incurred the costs of its investments, but can claim the issue. The other nation has incurred the cost of its resources, but has nothing to show for this cost. Regardless of the complexity of the real-world scenario, the mathematics are the same, illustrating the flexibility of the model.

To build a strategic game model for an Escalation scenario, we assume there are n players vying for an object known to be worth $v > 0$ to each player. Player i selects a nonnegative real number b_i that measures the time, money, or other resources invested in order to acquire the object. We will call b_i the amount that player i bids for the object because of the standard auction-like rule that the player with the highest bid wins the object, and if there is a tie among m players, then each player with the highest bid obtains the object with probability $1/m$. What makes this "auction" different is that each player must pay their bid, whether or not they win the object, because the bid corresponds to the investment of resources to obtain the object. Assuming that it is meaningful to take the difference of the object's worth and the investment cost (e.g., both have been expressed in monetary terms), the payoff to player i is

$$u_i(b) = \begin{cases} -b_i & \text{if } b_i < c_i \\ v/m - b_i & \text{if } b_i = c_i \\ v - b_i & \text{if } b_i > c_i \end{cases}$$

where $c_i = \max\{b_j : j \neq i\}$ is the highest bid by players other than i and m is the number of players with the highest bid.

We first observe that if $b_i = 0$, then $u_i(b) \geq 0$, and if $b_i > 0$, then $u_i(b) < 0$ when $c_i > b_i$. Thus, $b_i = 0$ is player i's unique prudential strategy with the security level 0. The strategy profile of all 0 bids is efficient because the sum of payoffs here is equal to the value of the object, v, and the sum of payoffs can never be greater than v. If b is any other strategy profile, then the highest bidders could increase their payoffs without changing any other payoffs by slightly decreasing their bids. Therefore, the only efficient strategy profile is the one in which each player selects their prudential strategy 0. Unfortunately, each player has an incentive to increase their bid slightly.

We now consider dominance. If $b_i > v$, then $u_i(b) < 0$. Hence, the bid 0 strongly dominates any bid greater than the object's worth. If $b_i = v$, then $u_i(b) = 0$ when $c_i < v$ and $u_i(b) < 0$ when $c_i \geq v$. Hence, bidding 0 dominates bidding the object's worth. Finally, suppose $0 \leq b_i < b_i' < v$. If $c_i > b_i'$, then $u_i(b) = -b_i > -b_i' = u_i(b_i', b_{-i})$, and if $b_i < c_i < b_i'$, then $u_i(b) = -b_i \leq 0 < v - b_i' = u_i(b_i', b_{-i})$. Hence, no bid less than the object's worth dominates any other bid less than the object's worth. In particular, there is no dominant strategy.

We now look for Nash equilibria. Observe first that since bids larger than v are strongly dominated, only bids between 0 and v, inclusive, can be part of a Nash equilibrium. Next observe that player i's best response against a strategy profile b is 0 if $c_i \geq v$ and does not exist if $c_i < v$ because a bid of $c_i + \epsilon$ for sufficiently small $\epsilon > 0$ results in a positive payoff that increases as ϵ approaches 0 but a bid of c_i or lower yields smaller payoffs. These two observations together imply that in a Nash equilibrium, $c_i = 0$ and $b_i = 0$ for each player, which is impossible. Thus, there is no pure-strategy Nash equilibrium.

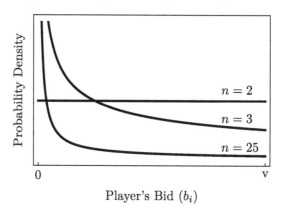

FIGURE 3.7 Escalation Mixed Strategies.

A symmetric mixed-strategy Nash equilibrium exists. Let σ_j be the strategy in which player j selects their bid randomly in accordance with the probability density and cumulative distribution functions

$$g(b_j) = \frac{1}{n-1} \left(\frac{b_j}{v} \right)^{-(n-2)/(n-1)} \quad \text{and} \quad G(b_j) = \left(\frac{b_j}{v} \right)^{1/(n-1)}$$

where b_j is assumed to satisfy $0 \leq b_j \leq v$. Figure 3.7 shows graphs of the probability densities g for three different numbers of players. To show that this is a Nash equilibrium, suppose that all players other than i adopt the mixed strategies σ_j. The payoff to player i selecting the pure strategy b_i satisfying $0 \leq b_i \leq v$ is

$$u_i(b_i, \sigma_{-i}) = vG(b_i)^{n-1} - b_i = 0.$$

Thus, any randomization of the strategies in the closed interval $[0, v]$ will be a best response for player i. In particular, σ_i is a best response. Baye, Kovenock and de Vries [9] show that this is the only Nash equilibrium when $n = 2$ but that there are additional non-symmetric Nash equilibria when $n > 2$. Gneezy and Smorodinsky [43] had undergraduate economics students play several rounds of a version of this Escalation game auction for money. They found that their subjects consistently overbid in comparison with the symmetric Nash equilibrium although less so in later rounds.

In the Ascending Bid Auction strategic game, we limited bids to non-negative integers, while in the Escalation strategic game, we allowed bids to be non-negative real numbers. Although it seems reasonable to use a discrete set for strategies representing real-world monetary bids and a continuous set of strategies representing amounts of effort in a competition, the choice of a discrete or continuous strategy set is more often based on which approach makes the mathematics tractable. It turns out that either approach will generally lead to similar results.

> **Senators Voting.** One often hears United States Senators accused of being "for something before they were against it" and their votes on bills that appeared before the Senate are used as evidence of this. Do these changes in voting decisions reflect true changes in their position, or do they reflect some kind of strategic voting pattern?

We begin to build our model of voting patterns by observing that most of the accusations of changes of position occur on bills which have amendments attached to them, so Senators are accused of voting for the amendment, and then not voting for an un-amended version of the bill.

Let's assume that there are n Senators, with $n \geq 3$ and odd (to eliminate the possibility of ties). They have a two-stage voting process in which they first vote between a bill b and an amended version of the bill, labeled a. In the second stage, they will be choosing between the bill that they passed in the first stage and the status quo, q. During each stage, the winning alternative will have received at least $(n + 1)/2$ votes.

A (pure) strategy for a Senator can be represented by an ordered list of three letters, (x, y, z), that abbreviate: vote for x (chosen from a or b) in the first stage, vote for y (chosen from a or q) in the second stage if a passes in the first stage, and vote for z (chosen from b or q) in the second stage if b passes in the first stage. Thus each Senator has eight strategies.

The smallest reasonable set of outcomes contains the three final statuses of the legislation: the bill passes, B, the amended bill passes, A, or neither pass, Q. Of course, some Senators may see a difference between the status quo when the amendment passes in the first stage and the status quo when the amendment fails in the first stage. Some Senators may also draw a distinction based upon how they or others voted in addition to the final status of the legislation. In the most expansive observable view of outcomes, each profile of votes could correspond to an outcome, resulting in 4^n different outcomes. Even more expansive would be the set of strategy profiles, resulting in 8^n different outcomes. The modeling decision being made when we choose a particular set of outcomes is to place or not place limits upon the possible preferences expressible by the Senators.

We first use the most simple set of outcomes, $\{B, A, Q\}$, and we consider an example with $n = 5$ Senators. The Senators' ordinal utility functions and three different strategy choices are given in Table 3.12.

The *sincere strategy* has a Senator always voting for the more favored alternative. For example, Senator 2 will initially vote for the amended bill over the original bill since $u_2(A) > u_2(B)$, will vote for the amended bill over the status quo since $u_2(A) > u_2(Q)$, and will vote for the original bill over the status quo since $u_2(B) > u_2(Q)$. If all of the Senators use

TABLE 3.12   Model 1 Senators' Ordinal Utility Functions and Strategies

Senator	Ordinal Utility	Sincere Strategy	Equilibrium B Strategy	Equilibrium Q Strategy
1	$u_1(B) > u_1(Q) > u_1(A)$	(b, q, b)	(b, q, b)	(b, q, b)
2	$u_2(A) > u_2(B) > u_2(Q)$	(a, a, b)	(b, a, b)	(a, a, b)
3	$u_3(A) > u_3(B) > u_3(Q)$	(a, a, b)	(b, a, b)	(a, a, b)
4	$u_4(Q) > u_4(B) > u_4(A)$	(b, q, q)	(a, q, q)	(a, q, q)
5	$u_5(Q) > u_5(B) > u_5(A)$	(b, q, q)	(a, q, q)	(a, q, q)

their sincere strategies, then the bill wins in the first stage by a vote of 3–2, and the bill goes on to win in the second stage by a vote of 3–2. The profile of sincere strategies is not a Nash equilibrium because either Senator 4 or Senator 5 could change their strategy to (a, q, q) which would result in the amendment winning in the first stage by a 3–2 vote and then the status quo winning in the second stage by a 3–2 vote, yielding a better outcome for Senators 4 and 5.

The equilibrium B strategies listed in Table 3.12 for Senators 2–5 are not sincere because each votes in the first stage against their preferences. In the first stage, the bill wins by a 3–2 vote. In the second stage, the bill wins by a 3–2 vote. Since B is Senator 1's most preferred outcome, their strategy is a best response. Since Senators 2 and 3 cannot unilaterally change the outcome from B to their only more preferred outcome A, their strategies are best responses. Finally, since no unilateral change by Senator 4 or 5 results in an outcome change, their strategies are best responses. Thus, the equilibrium B strategy profile is a Nash equilibrium.

The equilibrium Q strategies listed in Table 3.12 are not sincere for Senators 4 and 5 because they vote for the amendment in stage 1 despite preferring the bill. In the first stage, the amendment wins by a 4–1 vote. In the second stage, the status quo wins by a 3–2 vote. The only unilateral strategy changes that would result in a different outcome would be for Senators 1, 4, or 5 voting for the amended bill in the second stage, but since this would change the outcome from Q to A and each of these Senators prefers Q over A, every player's strategy is a best response; hence, the equilibrium Q strategy profile is a Nash equilibrium.

Observe that each strategy listed in Table 3.12 is sincere in the second stage; it is only in the first stage that insincerity sometimes appears. Also the use of insincere strategies in equilibrium Q were beneficial to Senators 4 and 5 because had they both adopted their sincere strategy, then the outcome would have been B rather than Q and both Senators prefer Q over B.

In the general case (an odd number of Senators with strict preference orders on the outcome set $\{A, B, Q\}$), there will always be a Nash equilibrium. One can be found by first having each Senator select their second stage actions sincerely. From this much of the strategies, it can be determined which of A versus Q would win and which of B versus Q would win. The four possibilities are given in the first two columns of Table 3.13. We

TABLE 3.13   First Stage Votes in Model 1 Nash Equilibrium Strategy Profile

Second Stage Winners		Vote in First Stage for Senator i	
A vs. Q	B vs. Q	a	b
A	B	if $u_i(A) > u_i(B)$	if $u_i(B) > u_i(A)$
A	Q	if $u_i(A) > u_i(Q)$	if $u_i(Q) > u_i(A)$
Q	B	if $u_i(Q) > u_i(B)$	if $u_i(B) > u_i(Q)$
Q	Q	if $u_i(A) > u_i(B)$	if $u_i(B) > u_i(A)$

complete each Senator's strategy by inserting the first stage vote that best aligns with their preferences and the potential second stage outcomes as given in the last two columns of Table 3.13. With all Senators adopting the strategies just described, no Senator has an incentive to unilaterally change their strategy, and so a Nash equilibrium has been constructed.

The above strategy profile construction was used to obtain the equilibrium B strategies in Table 3.12. For this example, the outcomes of the potential second stage votes correspond to the third row of Table 3.13, and so insincerity can arise depending upon where A is in the Senator's ordinal ranking. This can also happen if the potential second stage votes correspond to the second row of Table 3.13; however, the first and fourth rows correspond to the profile of sincere strategies.

Groseclose and Milyo [44] obtained different results when they analyzed Senator voting patterns with a different model. They prove that within their model whenever there is a Nash equilibrium, the Senators use sincere strategies and pure-strategy Nash equilibria usually but not always exist. One difference between their model and ours is that their Senators have preferences over their first stage votes as well as the final disposition of the legislation; however, they limit what preferences are possible. We build such a model now.

For this second model, we will use the same strategies as in the first model but expand the set of outcomes to be $\{aA, bA, aB, bB, aQ, bQ\}$ where the lower-case letters correspond to the personal first stage vote of a Senator and upper-case letters correspond, as before, to the final disposition of the legislation. Player preferences will be restricted so that there is no ambiguity about the ranking of first stage votes or the ranking of legislative dispositions. Specifically, we assume that the ordinal utility function u for a player must satisfy the following two conditions:

1. One of the following holds:

    (a) $u(aA) > u(bA)$, $u(aB) > u(bB)$, and $u(aQ) > u(bQ)$, or

    (b) $u(bA) > u(aA)$, $u(bB) > u(aB)$, and $u(bQ) > u(aQ)$.

2. If $\{X, Y, Z\} = \{A, B, Q\}$, then $u(aX) > u(aY) > u(aZ)$ if and only if $u(bX) > u(bY) > u(bZ)$.

If condition 1(a) holds, we can unambiguously state that, assuming the legislative disposition does not change, the player prefers a first stage vote of a to a first stage vote of b, which we will denote as $a \succ b$. If $u(aB) > u(aA) > u(aQ)$, then condition 2 implies that $u(bB) > u(bA) > u(bQ)$, and we can unambiguously state that, assuming the first stage vote does not change, the player prefers legislative outcome B to legislative outcome A, prefers legislative outcome A to legislative outcome Q, and prefers legislative outcome B to legislative outcome Q, which we will denote as $B \succ A$, $A \succ Q$, and $B \succ Q$.

TABLE 3.14　Model 2 Senators' Ordinal Utility Functions and Strategies

Ordinal Utility	Sincere Strategy	Equilibrium Strategy
$u_1(bB) > u_1(aB) > u_1(bQ) > u_1(aQ) > u_1(bA) > u_1(aA)$	(b, q, b)	(b, q, b)
$u_2(aA) > u_2(bA) > u_2(aB) > u_2(bB) > u_2(aQ) > u_2(bQ)$	(a, a, b)	(a, a, b)
$u_3(aA) > u_3(bA) > u_3(aB) > u_3(bB) > u_3(aQ) > u_3(bQ)$	(a, a, b)	(a, a, b)
$u_4(bQ) > u_4(aQ) > u_4(bB) > u_4(aB) > u_4(bA) > u_4(aA)$	(b, q, q)	(b, q, q)
$u_5(bQ) > u_5(aQ) > u_5(bB) > u_5(aB) > u_5(bA) > u_5(aA)$	(b, q, q)	(a, q, q)

Table 3.14 displays an example set of preferences. Once again the sincere strategy profile, which results in the B legislative disposition, is not a Nash equilibrium because Senator 4 or 5 can unilaterally change their strategy to (a, q, q) changing the legislative disposition

to Q and Senators 4 and 5 each prefer aQ to bB. For the equilibrium strategy profile, the legislative disposition is Q, Senator 5 is insincere in the first stage and the profile is a Nash equilibrium (in particular, Senator 5 would not want to change to their sincere strategy because aQ is preferred to bB).

Again a Nash equilibrium in pure strategies always exists, and the construction is similar to the one used for our first model. To construct the Nash equilibrium, we assign sincere second stage votes for each Senator. Next we determine the second stage winners based on these partial strategy assignments. Finally we assign first stage votes in accordance with Table 3.15. This construction ensures that no Senator has an incentive to change their strategy.

TABLE 3.15 First Stage Votes in Model 2 Nash Equilibrium Strategy Profile

Second Stage Winners		Vote in First Stage for Senator i	
A vs. Q	B vs. Q	a	b
A	B	if $u_i(aA) > u_i(bB)$	if $u_i(bB) > u_i(aA)$
A	Q	if $u_i(aA) > u_i(bQ)$	if $u_i(bQ) > u_i(aA)$
Q	B	if $u_i(aQ) > u_i(bB)$	if $u_i(bB) > u_i(aQ)$
Q	Q	if $u_i(aQ) > u_i(bQ)$	if $u_i(bQ) > u_i(aQ)$

Summarizing, both of our models have pure-strategy Nash equilibria for all possible preferences, but sometimes the sincere strategy profile is not a Nash equilibrium. In contrast, in the model developed by Groseclose and Milyo, Nash equilibria in pure strategies do not always exist but when they do they use sincere strategies. The remaining difference between our second model and their model is that they collapse the two stages of voting into a single stage with the comment that they are interested in situations when "a vote for a [in the first stage of voting] is really a vote for q" [44]. In their model, players have preferences over their first stage vote and the legislative outcome in the first stage. This effectively means in the context of our second model, Groseclose and Milyo's model has players who are indifferent among the final legislative outcomes! It is in this limited model that they obtain their nice results.

Even though their model is limited, Groseclose and Milyo [44] claim strong fidelity and report a meta-analysis of data from the United States Congress that demonstrates empirically that Senators and Congresspersons almost always vote sincerely, consistent with their theoretical result. Interestingly, in a subsequent paper, [45], Groseclose and Milyo further demonstrate that this result is very much contingent on the method of voting used in the Senate and House of Representatives. If representatives were required to vote sequentially and with no possibility of changing their votes, then the opportunity and incentives for strategic voting increases significantly.

While Groseclose and Milyo argue that the empirical evidence showing the rarity of insincere voting is consistent with their model in which insincere voting never occurs in equilibrium, we could argue that our models showing that insincere voting can sometimes occur could be more in alignment with the data if sincere voting is an equilibrium most of the time. In fact, in either of our models, the only situations in which sincere voting may not be a Nash equilibrium are when a Senator can change the first stage outcome, which can only occur if there are $(n + 1)/2$ Senators who favor A over B and $(n - 1)/2$ Senators who favor B over A or vice versa. If the preference between A or B is determined with a fair coin flip for each Senator, then the probability of the sincere strategy profile not being a Nash equilibrium is no greater than $2n!/(((n - 1)/2)!((n + 1)/2)!2^n)$. For $n = 101$ (corresponding to the U.S. Senate and Vice President), this upper bound is 0.16, and for $n = 435$ (corresponding to the U.S. House of Representatives), this upper bound is

0.08. Since the nonexistence of Nash equilibria in the Groseclose and Milyo model roughly corresponds to when sincere strategy profiles in our two models are not Nash equilibria, all three models support the empirical observations that non-equilibrium voting profiles and non-sincere votes are rare in the United States Senate and House of Representatives.

3.5 LIMITATIONS

We've described several desirable characteristics of strategies and strategy profiles and suggested that players who act rationally and have well-defined utility functions should select strategies aligning with these characteristics. That is, the models we have created should reflect or predict actual human behavior correctly. However, it is reasonable to wonder if humans consciously or subconsciously employ actual game theory strategies when they make decisions in the scenarios we model. Certainly some of the techniques we have described are complex or time consuming and empirically we don't often hear of people seeking out Nash equilibria to make decisions.

However, in the modeling process, there is still value in the final checking and revising stages which are designed to improve the match between the model and the scenario. When we are modeling past events such as the Normandy Breakout, the goal of the model is to provide some insight into why the participants acted as they did. We can compare the model outcomes to the actual historic outcome and check the fidelity of the model. When modeling events that will occur only once, the goal of modeling is to develop some frame of reference from which to guess the outcome of the event. If a scenario is repeated several times, we can check our model predictions against actual outcomes and make adjustments to improve fidelity. As a model improves, its predictive powers do as well.

Here we consider two scenarios that can be modeled and analyzed as strategic games using the techniques and theory of this chapter, but for which we also have empirical data about the outcomes that actually occurred over a period of time. First we consider the problem of how a group of players are able, or not, to select from among several Nash equilibria. Second, we challenge our assumption that people are able, in real situations, to do the analysis necessary to arrive at optimal decisions. Both of the strategic games that we examine are easy to solve analytically; the concerns and issues are with how people interpret and interact with them.

Continental Divide. In an experiment, a group of people were asked to select a number between 1 and 14. Each person received money depending on their choice of number and the median choice of the group as given in Table 3.16 ([117] and reproduced in [21]). When modeled as a strategic game, this scenario gives rise to two very different Nash equilibria, suggesting a "Continental Divide" between the two.

In the strategic game model, the players are the people in the group, the strategies are the choices of numbers between 1 and 14, and assuming risk-neutrality, the payoffs are the amounts given in the table.

There are two Nash equilibria, one with each player selecting 3 and another in which each player selects 12. The interesting problem arising in practice occurs when the same group plays the strategic game multiple times. The players, without communication, should move to the equilibrium play of 12, which has a larger payoff. However, experiments have shown that plays of the strategic game which begin with an initial median greater than 7 tend to end at the higher equilibrium, while strategic games which begin with an initial median of seven or less end up at the equilibrium play of 3 even though the players are aware of the possibility of a higher payoff.

TABLE 3.16   Continental Divide Payment [117]

Choice	Median Choice of the Group													
	1	2	3	4	5	6	7	8	9	10	11	12	13	14
1	45	49	52	55	56	55	46	-59	-88	-105	-117	-127	-135	-142
2	48	53	58	62	65	66	61	-27	-52	-67	-77	-86	-92	-98
3	48	54	60	66	70	74	72	1	-20	-32	-41	-48	-53	-58
4	43	51	58	65	71	77	80	26	8	-2	-9	-14	-19	-22
5	35	44	52	60	69	77	83	46	32	25	19	15	12	10
6	23	33	42	52	62	72	82	62	53	47	43	41	39	38
7	7	18	28	40	51	64	78	75	69	66	64	63	62	62
8	-13	-1	11	23	37	51	69	83	81	80	80	80	81	82
9	-37	-24	-11	3	18	35	57	88	89	91	92	94	96	98
10	-65	-51	-37	-21	-4	15	40	89	94	98	101	104	107	110
11	-97	-82	-66	-49	-31	-9	20	85	94	100	105	110	114	119
12	-133	-117	-100	-82	-61	-37	-5	78	91	99	106	112	118	123
13	-173	-156	-137	-118	-96	-69	-33	67	83	94	103	110	117	123
14	-217	-198	-179	-158	-134	-105	-65	52	72	85	95	104	112	120

Game theory can explain why players do not move to the larger equilibrium if the medians in previous strategic games were close to the smaller equilibrium, but the sensitivity to initial conditions is difficult to explain using game theory. Player choices are subject to whimsical variations such as the selection of lucky numbers which can perhaps be better explained by psychology than mathematics. Other examples of behavior that is difficult to explain mathematically include the dominance of the QWERTY keyboard over the Dvorak keyboard, the dominance of the PC over MACs, and swings in clothing fashions.

Predicting Scaled Averages. In this scenario, people are asked to select a number between 0 to 100. These numbers are then averaged and adjusted by a multiplier $p < 1$ to determine a target number. The player whose guess is closest to this target is declared the winner and receives a cash prize [74].

This scenario might be modeled as a strategic game where the players are the people in the group. The strategy choices are integers between 0 and 100. The payoff for the winner is 1 and 0 for everyone else.

This strategic game can be used to examine how deeply players are reasoning about the game. Nagel [74] among others have conducted experiments using variations of this strategic game. Using $p = .70$, most players can guess that the average might be about 50 (if it is assumed that each player chooses each number with equal probability), so that the target is $(0.7)(50) = 35$. But players who are thinking a bit more deeply realize that the other players are thinking this way, so they guess that the average will be around 35, so they pick near to the expected target of $(0.7)(35) = 24.5$. But truly careful thinkers take this one iteration further and choose a number close to a target of just over $(0.7)(24) = 17$.

In fact, this reasoning about iteratively better strategies demonstrates that the only Nash equilibrium in this strategic game is for all players to select 0. But the problem here is not in playing the Nash equilibrium, but rather in guessing how carefully and deeply the other players are reasoning about their strategies in the strategic game. The winner wants to be only one step ahead of the group! In practice it is difficult for individuals to do all of this reasoning in a single step, before the initial play of the strategic game. But experiments have demonstrated that given repetitive play of the strategic game, players move towards the Nash equilibrium strategy of selecting 0 as they learn about the strategic game.

These two examples call into question the notion of individuals' capability of implementing a Nash equilibrium strategy, even if one is known to exist. Our discussion of these inherent limits of the modeling process leads again to the field of experimental game theory to validate or refute solutions in theoretical game theory. Three sources of detailed discussion of the field of experimental game theory are [21], [39], and [47].

EXERCISES

3.1 **Stag Hunt**. Each of several hunters can choose between working with others to catch a stag, which is swift and large, or working independently to catch a much smaller hare. It takes everyone to catch the stag, which they will all share. A caught hare only belongs to the hunter who catches it. Table 3.17 gives the outcomes and payoffs for the canonical two-player version of this strategic game.

TABLE 3.17   Stag Hunt

Strategy Profile s	Outcome o	$u_1(s)$	$u_2(s)$
(Stag, Stag)	Catch and share the stag	2	2
(Stag, Hare)	Hunter 2 catches a hare	0	1
(Hare, Stag)	Hunter 1 catches a hare	1	0
(Hare, Hare)	Each hunter catches a hare	1	1

a. For the payoffs to be ordinal utilities, describe what must be true about a hunter's preferences among the outcomes.

b. For the payoffs to be vNM utilities, describe what must be true about a hunter's preferences among the outcomes.

c. Determine all, if any, dominated strategies.

d. Determine all, if any, prudential strategies and the corresponding security level.

e. Determine all, if any, Nash equilibria.

f. Determine all, if any, efficient strategy profiles.

g. Discuss what you think is likely to happen. Does it depend on whether there is communication between the hunters before each makes a decision?

3.2 **Battle of the Sexes**. A couple wishes to go out to a concert. Rose prefers Bach and Colin prefers Stravinsky, but each is willing to attend the less preferred concert if the two of them are together. Table 3.18 gives the payoffs for this canonical strategic game. Although the given payoffs are consistent with the described scenario, all that

TABLE 3.18   Battle of the Sexes

(Rose, Colin)	Bach	Stravinsky
Bach	(10, 8)	(2, 2)
Stravinsky	(0, 0)	(8, 10)

matters for this game to be considered a Battle of the Sexes canonical game is that the off-diagonal payoffs are smaller than the main diagonal payoffs.

a. For the payoffs to be ordinal utilities, describe what must be true about Rose's preferences among the outcomes.

b. For the payoffs to be vNM utilities, describe what must be true about Rose's preferences among the outcomes.

c. Determine all, if any, dominated strategies.

d. Determine all, if any, prudential strategies and the corresponding security levels.

e. Determine all, if any, Nash equilibria.

f. Determine all, if any, efficient strategy profiles.

g. Discuss what you think is likely to happen. Does it depend on whether there is communication between Rose and Colin before each makes a decision?

3.3 **Chicken**. Chicken is played by two people who drive towards each other in automobiles. The one who turns away first is the "chicken" and loses face. Of course, both are seriously hurt if neither turns away. Table 3.19 gives the payoffs for the canonical version of this scenario.

TABLE 3.19   Chicken

(Rose, Colin)	Turn	Straight
Turn	(2, 2)	(−2, 4)
Straight	(4, −2)	(−4, −4)

a. For the payoffs to be ordinal utilities, describe what must be true about Rose's preferences among the outcomes.

b. For the payoffs to be vNM utilities, describe what must be true about Rose's preferences among the outcomes.

c. Determine all, if any, dominated strategies.

d. Determine all, if any, prudential strategies and the corresponding security level.

e. Determine all, if any, Nash equilibria.

f. Determine all, if any, efficient strategy profiles.

g. Discuss what you think is likely to happen. Does it depend on whether there is communication between the players before each makes a decision?

3.4 Write a new scenario for each of the following canonical strategic games. Give the corresponding payoff matrix and explain how your scenario matches the characteristics of the canonical strategic game.

a. Prisoner's Dilemma

b. Stag Hunt

c. Battle of the Sexes

d. Chicken

3.5 In our initial model for the Prisoner's Dilemma scenario, we assumed the specific numbers in Table 3.2 for the players' payoffs. Later, we generalized the payoffs by

TABLE 3.20   Generic Strategic Game

(Rose, Colin)	A	B
A	(R_{11}, C_{11})	(R_{12}, C_{12})
B	(R_{21}, C_{21})	(R_{22}, C_{22})

replacing the specific numbers with the parameters in Table 3.3 where the parameters satisfied $S < P < R < T$ and perhaps $S + T < 2R$. Identify the strategies and write similar inequalities relating the parameters in the payoff matrix given in Table 3.20 that generate each of the following canonical strategic games.

a. Stag Hunt

b. Battle of the Sexes

c. Chicken

3.6 **Basketball Show-down.** In 2014, Hopkins High School and Shakopee High School met in the Minnesota state basketball tournament. They found themselves tied with three minutes left and, significantly, there was no shot clock. Hopkins was very, very good on offense and Shakopee was very, very good on defense. Hopkins had the ball. What is the optimal way for each team to play the rest of the game? What might the two teams do if the game were to go into overtime? [25]

3.7 **Virginia Governor's Race.** In the 2017 Virginia Governor's race, the two leading candidates, Ralph Northam and Ed Gillespie, each refused to share their full income tax returns when the *Richmond Times Dispatch* requested them. It's possible that each candidate believed that their returns could give their opponent information that could be used against them, and they didn't want to take the risk of being the only one to release theirs. But hiding returns could also lead to the voters distrusting them, so perhaps the best solution is for them to both reveal their returns, gaining voter trust and eliminating any advantage of one over the other by having both returns available to the public. Model this scenario as a prisoner's dilemma and explain why neither decided to release his returns [91].

3.8 Suppose that in the Normandy Breakout scenario, the Germans could use a skillful retreat to inflict a significant defeat on any rapidly pursuing reserves. Table 3.21 indicates possible changes to the payoff matrix. Why are the changes to the payoffs reasonable? Apply the four solution methods to the resulting strategic game. What would you predict would happen in the new model? How is this similar to or different from what was predicted by the original model?

TABLE 3.21   Normandy Breakout Payoff Matrix

(American, German)	Attack	Retreat
Reinforce	(3, 4)	(4, 3)
Hold	(6, 1)	(5, 2)
Move	(1, 6)	(2, 5)

3.9 In our Trio scenario we assumed only that nations would prefer to Merge or Remain Independent, but we didn't account for the fact that these preferences could be based on who their potential partners are. Construct a new set of utility functions for the

Partition of Trio scenario that assumes that the Alphas would prefer to be in a nation with the Betas rather than in a nation with the Gammas. Using the new utility functions, apply the four solution methods and interpret the results.

3.10 Argue that if A is a dominant strategy for Rose, then A is a prudential strategy for Rose.

3.11 Argue that if A is a dominant strategy for Rose and B is a dominant strategy for Colin in a two-player strategic game, then (A, B) is a Nash equilibrium.

3.12 Argue that if A is a strongly dominant strategy for Rose, then Rose uses A in any Nash equilibrium.

3.13 If for some player, strategy A dominates strategy B, then the player is always at least as well off choosing A instead of choosing strategy B and, under some circumstances, will be better off choosing A instead of choosing B. This sounds like a good justification for never choosing a dominated strategy. As a counter-argument, exhibit an example of a strategic game whose unique pure-strategy Nash equilibrium is made up of dominated strategies and is also the unique efficient pure-strategy profile. Be sure to explain why your example has these properties.

3.14 Eliminating a dominated strategy for player i may result in a reduced strategic game which now contains a dominated strategy for player j. This process can be iterated to possibly greatly simplify a strategic game. Apply this process to the strategic games in Table 3.22 and Table 3.23.

TABLE 3.22

(Rose, Colin)	Left	Middle	Right
Top	(3, 5)	(1, 4)	(3, 3)
Middle	(4, 1)	(2, 3)	(1, 3)
Bottom	(3, 2)	(0, 4)	(1, 3)

TABLE 3.23

(Rose, Colin)	Left	Middle	Right
Top	(3, 4)	(2, 1)	(1, 3)
Middle	(4, 2)	(3, 3)	(2, 1)
Bottom	(5, 3)	(1, 6)	(0, 2)

3.15 Argue that if iterated elimination of dominated strategies results in a single strategy choice remaining for each player, then this strategy profile is a Nash equilibrium. Hint: use a proof by induction on the total number of strategies.

*3.16 Describe a scenario (different from one described in the text) in which either there are more than two players and/or some player has more than two strategies. Model the scenario as a strategic game (e.g., provide an outcome table or matrix and a payoff table or matrix). Find the prudential strategies, dominated strategies, Nash equilibria, and efficient strategy pairs. Interpret these solutions in the context of the original scenario.

3.17 Suppose the strategic game has $n = 2$ players and is *zero sum*: the vNM utility functions u_i satisfy $u_1(r, c) + u_2(r, c) = 0$ for all strategy pairs (r, c).

a. Prove that in a zero-sum strategic game all strategy pairs are efficient.

b. Prove that in a zero-sum strategic game if the strategy pairs (r_1, c_1) and (r_2, c_2) are Nash equilibria, then the payoffs $u_1(r_1, c_1) = u_1(r_1, c_2) = u_1(r_2, c_1) = u_1(r_2, c_2)$, and the strategy pairs (r_1, c_2) and (r_2, c_1) are Nash equilibria.

3.18 Suppose that only contestants 1 and 2 have made it into Final Jeopardy (because contestant 3 ended Double Jeopardy with a negative score) and they each have exactly E dollars. Furthermore, suppose that contestant 1 believes she has a 3/4 probability of answering correctly and that contestant 2 will either (1) wager E but answer incorrectly, or (2) wager 0 but answer correctly, each with probability 1/2. Show that contestant 1's expected winnings are not maximized with a wager of 0 or E.

3.19 Recall Table 3.8, repeated as Table 3.24, defining the Generic Final Jeopardy strategic games. In the text, we calculated the (Everything, Everything) entries. Give calculations that result in the entries for the other three strategy pairs.

TABLE 3.24   Generic Final Jeopardy Payoff Matrix

(Rose, Colin)	Everything		Nothing
Everything	$(p_1(1 - 0.5p_2),$	$p_2(1 - 0.5p_1))$	$(p_1, 1 - p_1)$
Nothing	$($ $1 - p_2,$	$p_2)$	$(0.5, 0.5)$

3.20 In the Generic Final Jeopardy strategic games, argue why the strategy profile (Nothing, Nothing) is a Nash equilibrium when $p_1 < 0.5$ and $p_2 < 0.5$.

*3.21 In the Final Jeopardy scenario, develop and analyze a model for the case when contestants 1 and 2 are tied for the lead, but contestant 3 has more than half of the money that contestants 1 and 2 have.

3.22 For the Battle of the Sexes strategic game, a previous exercise showed that there are two pure-strategy Nash equilibria.

a. Draw a graph of Rose's best response expected payoffs versus Colin's mixed-strategy choice.

b. Draw a graph of Colin's best response expected payoffs versus Rose's mixed-strategy choice.

c. Draw a graph each player's best response function on the same pair of axes (Rose's and Colin's mixed-strategy choices).

d. Use the previous graphs to argue that there are three Nash equilibria in mixed strategies. What are the expected payoffs?

e. Provide an interpretation of the completely mixed-strategy Nash equilibrium.

3.23 Consider the Stag Hunt strategic game defined in Table 3.17.

a. Draw a graph of Hunter 1's best response expected payoffs versus Hunter 2's mixed-strategy choice.

b. Draw a graph of Hunter 2's best response expected payoffs versus Hunter 1's mixed-strategy choice.

c. Draw a graph each player's best response function on the same pair of axes (the two hunter's mixed-strategy choices).

d. Use the previous graphs to argue that there are three Nash equilibria in mixed strategies. What are the expected payoffs?

e. Provide an interpretation of the completely mixed-strategy Nash equilibrium.

3.24 Consider the Chicken strategic game defined in Table 3.19.

a. Draw a graph of Rose's best response expected payoffs versus Colin's mixed-strategy choice.

b. Draw a graph of Colin's best response expected payoffs versus Rose's mixed-strategy choice.

c. Draw a graph each player's best response function on the same pair of axes (Rose's and Colin's mixed-strategy choices).

d. Use the previous graphs to argue that there are three Nash equilibria in mixed strategies. What are the expected payoffs?

e. Provide an interpretation of the completely mixed-strategy Nash equilibrium.

3.25 Explain why the General Prisoner's Dilemma strategic game has no completely mixed Nash equilibrium and provide an interpretation.

3.26 For the strategic game displayed in Table 3.25, construct each player's best response function and determine their Nash equilibrium in mixed strategies.

TABLE 3.25

(Rose, Colin)	Left	Right
Top	$(-1, 3)$	$(2, 6)$
Bottom	$(-3, 0)$	$(3, 1)$

3.27 For the strategic game displayed in Table 3.26, construct each player's best response function and determine the Nash equilibrium in mixed strategies.

TABLE 3.26

(Rose, Colin)	Left	Right
Top	$(-25, 5)$	$(5, 0)$
Bottom	$(0, 3)$	$(3, 15)$

3.28 When we analyzed the Normandy Breakout strategic game with payoff matrix given in Table 3.5, we stated that there was no Nash equilibrium. To be more precise, there is no pure-strategy Nash equilibrium. By Theorem 3.3.1, there must be a Nash equilibrium in mixed strategies. Find it and provide an interpretation.

3.29 Identify which of the following strategic games are symmetric: Prisoner's Dilemma, Office Coffee, General Prisoner's Dilemma, Normandy Breakout, Trio, Generic Final Jeopardy, Mating Behavior, Attrition Behavior, Cournot Duopoly, Ascending Bid

Auction, Escalation, and Senators Voting. For some, you may need to specify parameters that make the games symmetric.

3.30 Consider the Attrition Behavior strategic game whose payoff matrix is reproduced in Table 3.27. Some of the payoffs given in Table 3.27 were justified in the text with

TABLE 3.27   Attrition Behavior Payoff Matrix

(Male 1, Male 2)	Hawk	Dove	Bully	Retaliator	Bourgeois
Hawk	(−30, −30)	(40, 0)	(40, 0)	(−30, −30)	(5, −15)
Dove	(0, 40)	(12, 12)	(8, 32)	(12, 12)	(6, 26)
Bully	(0, 40)	(32, −8)	(12, 12)	(−8, 32)	(16, 16)
Retaliator	(−30, −30)	(12, 12)	(32, −8)	(12, 12)	(−9, −9)
Bourgeois	(−15, 5)	(26, 6)	(16, 16)	(−9, −9)	(20, 20)

descriptions of the interactions and how the points were awarded. Provide similar justifications for the remaining payoffs.

a. Bully versus Bully

b. Retaliator versus Retaliator

c. Bully versus Hawk

d. Bully versus Dove

e. Retaliator versus Hawk

f. Retaliator versus Dove

g. Retaliator versus Bully

h. Bourgeois versus Hawk

i. Bourgeois versus Bully

j. Bourgeois versus Retaliator

3.31 Consider the Attrition Behavior strategic game whose payoff matrix is given in Table 3.27.

a. Box the payoffs corresponding to best response strategies and determine the pure-strategy Nash equilibria.

b. Verify that $((1 − \lambda)\text{Dove} + \lambda\text{Retaliator}, (1 − \mu)\text{Dove} + \mu\text{Retaliator})$ is a Nash equilibrium for all $(1/2) \leq \lambda \leq 1$ and $(1/2) \leq \mu \leq 1$.

c. Verify that $((1/2)(\text{Hawk}) + (1/2)(\text{Dove}), (1/2)(\text{Hawk}) + (1/2)(\text{Dove}))$ is not a Nash equilibrium.

3.32 Consider the Attrition Behavior strategic game whose payoff matrix is given in Table 3.27.

a. Find all dominated pure strategies.

b. Find all efficient pure-strategy profiles.

    c. Find the pure prudential strategies and security levels.

    d. Show that if Male 1 chooses the mixed-strategy $(1/2)$Dove$+(1/2)$Bully, then Male 2 must receive an expected payoff of at least 0. Also observe that if Male 2 chooses the pure strategy Hawk, then Male 1 cannot obtain an expected payoff of more than 0. We therefore say that $(1/2)$Dove$+(1/2)$Bully is a *mixed prudential strategy* for Male 1.

3.33 Use Gambit to find the Nash equilibria for the symmetric strategic game whose payoff matrix is given by Table 3.28. Describe the strategy profiles in terms of the pure strategies. Explain why this strategic game has no evolutionarily stable strategy.

TABLE 3.28

(Rose, Colin)	High	Medium	Low
High	(8, 8)	(8, 8)	(2, 9)
Medium	(8, 8)	(8, 8)	(4, 3)
Low	(9, 2)	(3, 4)	(0, 0)

3.34 Use Gambit to find all of the Nash equilibria in mixed strategies for the strategic game in Table 3.29. Describe the strategy profiles in terms of the pure strategies.

TABLE 3.29

(Rose, Colin)	Left	Middle	Right
Top	(2, 9)	(1, 4)	(7, 6)
Middle	(9, 2)	(8, 3)	(3, 7)
Bottom	(5, 5)	(4, 8)	(6, 1)

3.35 Use Gambit to find all of the Nash equilibria in mixed strategies for the strategic game in Table 3.30. Describe the strategy profiles in terms of the pure strategies.

TABLE 3.30

(Rose, Colin)	Left	Middle	Right
Top	(2, 6)	(1, 4)	(7, 9)
Middle	(9, 2)	(8, 8)	(3, 7)
Bottom	(5, 5)	(4, 3)	(6, 1)

3.36 In the *Simplified Attrition Behavior* strategic game displayed in Table 3.31, we use the same point assignment as in Table 3.11. Find the Nash equilibria of this strategic game. Determine the evolutionarily stable strategies, if any.

TABLE 3.31   Simplified Attrition Behavior

(Male 1, Male 2)	Hawk	Dove
Hawk	(−30, −30)	(40, 0)
Dove	(0, 40)	(12, 12)

3.37 Modify the payoffs in the Simplified Attrition Behavior strategic game displayed in Table 3.31 by increasing the value of mating from 40 points to 120 points. How does this affect the equilibrium solution?

*3.38 Use Gambit to analyze the Attrition strategic game when different points are assigned to the value of mating. Describe your observations of the resulting equilibria.

*3.39 Find or collect data concerning offensive and defensive play calling on 3rd and short situations in the National Football League (NFL). Construct a strategic game to model this scenario. Does your strategic game match any of the canonical strategic games already described? Analyze your strategic model and comment on its cost, flexibility, and fidelity.

3.40 Find the Nash equilibrium for a duopoly firm if the consumer demand function is $Q = a/(1 + bP)$ for some $a > 0$ and $b > 0$ rather than $Q = a - bP$.

3.41 How does the Cournot Duopoly model with the demand function given in the text change when considering a Cournot Oligopoly with n players instead of 2?

*3.42 The strategic game model we constructed for the Escalation scenario is sometimes called an *all pay auction* because all bidders pay the amount they bid even though only the highest bidder wins the object. If bidders are monitoring each other during the bidding process, then the highest bidder need only pay slightly more than the second highest bidder. Model this variation as a strategic game with only two bidders. One modeling choice is whether to allow only nonnegative integer bids or to allow nonnegative real number bids, and a second modeling choice is whether the winner will pay the amount of the other bidders bid or some fixed amount more. Find the Nash equilibria and provide an interpretation.

*3.43 The strategic game model we constructed for the Ascending Bid Auction scenario is sometimes called a *second price auction* because the winner pays the second highest bid or a *Vickery auction* in honor of the person who first studied these auctions academically [118]. In a *first price auction*, bidders submit sealed bids, the winner is the highest bidder (ties broken randomly), and the winner pays the amount they bid. Model a first price auction as a strategic game (the main modeling choice is whether to allow only nonnegative integer bids or to allow nonnegative real number bids). Find the Nash equilibria and provide an interpretation.

*3.44 Modify the Senators Voting scenario in some way and update an appropriate model to reflect the changes. For example, suppose there are two different amendments offered or allow more sophisticated player preferences. Describe the modified scenario and the modified strategic game. Find the Nash equilibria and provide an interpretation.

CHAPTER 4

Bilateral Agreements

When two players can communicate and enter into a binding agreement, they can achieve an efficient solution. We first examine two pure bargaining scenarios and their resolution by the egalitarian, Raiffa, and Nash solution methods. Next, we show how any strategic game can be considered a bargaining game when binding agreements are permitted. Finally, we characterize the three solution methods axiomatically, modeling fairness with formal properties and thereby providing one way of selecting the appropriate solution in a given scenario.

In a strategic game, players announce strategy choices simultaneously with no prior knowledge of each other's choices. Although communication may be permitted, no binding agreements are allowed. In bargaining games, players must negotiate and reach a binding agreement as to which payoff pair and resulting outcome they collectively choose.

Definition 4.0.1. A *bilateral bargaining game* consists of the following:

1. A set $N = \{1, 2\}$ of two players.

2. A set O of outcomes that includes a disagreement outcome $o_d \in O$.

3. A rule stating the game ends when both players agree upon an outcome $o \in O$, or a pre-specified time limit is reached resulting in the disagreement outcome, o_d.

4. Utility functions $u_i : O \to \mathbb{R}$ that specify the preferences for each player $i \in N$.

The *feasible payoff pairs* are $F = \{(u_1(o), u_2(o)) : o \in O\}$, and the *disagreement payoff pair* is $d = (d_1, d_2) = (u_1(o_d), u_2(o_d))$.

The players can bargain over the feasible payoff pairs in order to avoid the disagreement payoff pair. Once a payoff pair has been determined, a corresponding outcome can be chosen.

While we limit ourselves to two players in this chapter, the model and its solutions can be readily generalized to any number of players as long as there are no advantages for proper subsets of players to enter into an agreement.

4.1 NEGOTIATIONS

There are many scenarios in which negotiations occur between two parties. Friends negotiate where to go to dinner, consumers negotiate prices with vendors, and countries negotiate bilateral trade agreements. We begin with the simple example of sharing a prize at the county fair.

County Fair Prize. A vendor at the county fair has attracted a crowd. He offers two people in the crowd, Ann and Beth, a prize of a half-gallon of chocolate ice cream and $4 if they can agree upon a fair split. If no agreement can be reached within three minutes, each will receive a consolation prize of an uncirculated Sacagawea dollar coin. While the crowd observes, Ann suggests each take one-half of the ice cream and one-half of the money. Beth notes that it is hot out, she is at the fair alone, and although chocolate is her favorite, she is not hungry enough to eat one-half of the ice cream before it would melt. So, she would prefer less of the ice cream but more of the money. Ann is at the fair with her family and so could readily eat any portion of the half-gallon of ice cream; however, chocolate is not the family favorite. So, Ann does not concede that she should receive less money. As for the consolation prize, Ann just sees it as $1 to spend while Beth knows that she could sell it to a coin dealer for $2.

We model this scenario as a bilateral bargaining game in which Ann and Beth are the players. The set of outcomes consists of (1) all possible splits of the 64 ounces of chocolate ice cream and $4 between Ann and Beth, and (2) the disagreement outcome consisting of Ann and Beth each receiving an uncirculated Sacagawea dollar coin. Clearly, the game ends when Ann and Beth agree upon an outcome, or three minutes have passed resulting in the disagreement outcome.

As for preferences, we assume that Ann and Beth are self-interested and risk neutral with respect to money. The difference in desirability of the chocolate flavor will be captured by assuming Ann and Beth value the ice cream at $0.05 per ounce and $0.10 per ounce, respectively. To capture Beth's limited capacity for ice cream, we assume that she has zero additional utility for more than 16 ounces of ice cream. These assumptions yield the utility functions for ice cream given in Figure 4.1.

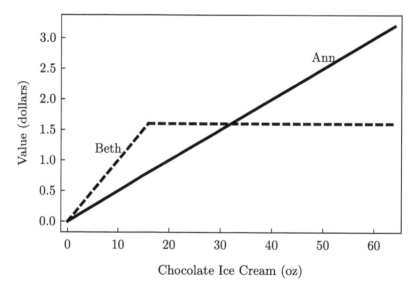

FIGURE 4.1 Ann's and Beth's Utility Functions for Ice Cream.

Implicit in our assumption that the players can ascribe a monetary value to the ice cream is that their utility for ice cream and money is just the sum of their monetary valuations. Hence, the utility to Ann and Beth, respectively, when receiving x ounces of chocolate ice cream and m dollars is assumed to be given by the following formulas:

$$u_A(x, m) = 0.05x + m$$

$$u_B(x, m) = \begin{cases} 0.10x + m, & \text{if } x \leq 16 \\ 1.6 + m, & \text{if } x > 16 \end{cases}$$

We complete the County Fair Prize bargaining game model by specifying the disagreement outcome o_d as each player receiving a Sacagawea dollar coin, $u_A(o_d) = 1$, and $u_B(o_d) = 2$.

It is useful to consider all feasible payoff pairs. Since Ann and Beth value the consolation prize at \$1 and \$2, respectively, the payoff pair for the disagreement outcome is $(1, 2)$. If Beth receives x ounces of chocolate ice cream and m dollars, then Ann receives $64 - x$ ounces of chocolate ice cream and $4 - m$ dollars, and the corresponding payoff pair is $(u_A(64 - x, 4 - m), u_B(x, m))$. The set of feasible payoff pairs is $F = \{d\} \cup F_1 \cup F_2$ where

$$d = (1, 2)$$
$$F_1 = \{(0.05(64 - x) + (4 - m), 0.10x + m) : \quad 0 \leq x \leq 16, 0 \leq m \leq 4\}$$
$$F_2 = \{(0.05(64 - x) + (4 - m), 1.6 + m) : \quad 16 \leq x \leq 64, 0 \leq m \leq 4\}.$$

The set of payoff pairs F is shown in Figure 4.2. As can be seen from its definition, F_1

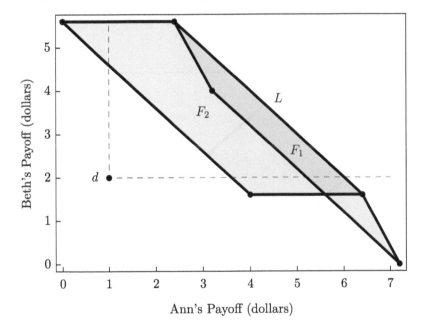

FIGURE 4.2 County Fair Feasible Payoff Pairs.

is a quadrilateral with vertices equal to

$$(u_A(64 - x, 4 - m), u_B(x, m))$$

for

$$(x, m) \in \{(0, 0), (16, 0), (16, 4), (0, 4)\}.$$

These vertices are (counter-clockwise from the lower right)

$$(7.2, 0.0), (6.4, 1.6), (2.4, 5.6), \text{ and } (3.2, 4.0).$$

Similarly, F_2 is a quadrilateral with vertices equal to

$$(u_A(64 - x, 4 - m), u_B(x, m))$$

for

$$(x, m) \in \{(16, 0), (16, 4), (64, 4), (64, 0)\}.$$

These vertices are (counter-clockwise from the lower right)

$$(6.4, 1.6), (2.4, 5.6), (0.0, 5.6), \text{ and } (4.0, 1.6).$$

The upper right line segment L for both quadrilaterals $(6.4, 1.6)$ to $(2.4, 5.6)$ can be described parametrically by

$$L = \{(1 - s)(2.4, 5.6) + s(6.4, 1.6) : 0 \le s \le 1\}$$

with s being the fraction of the distance down L, or algebraically by

$$\{(x, y) : x = 8 - y \text{ for } 1.6 \le y \le 5.6\},$$

and these payoff pairs correspond to Beth obtaining 16 ounces of chocolate ice cream and $4s = y - 1.6$ dollars.

The bilateral bargaining game now consists of Ann and Beth agreeing on a payoff pair in F, which might include settling for d. Note that a feasible payoff pair may sometimes be obtained by splitting the ice cream and money in more than one way. Once a payoff pair has been chosen, any of the corresponding outcomes can be randomly chosen to provide Ann and Beth with the agreed upon payoffs.

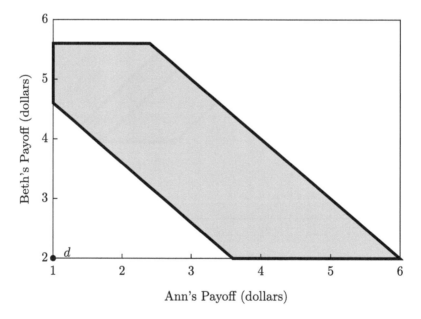

FIGURE 4.3 County Fair Feasible and Rational Payoff Pairs.

The vertical dashed line in Figure 4.2 indicates payoff pairs in which Ann receives the same utility as she would receive if no agreement is made. Hence, Ann should agree only to a payoff pair on or to the right of the vertical dashed line. Similarly, Beth should only agree to a payoff pair on or above the horizontal dashed line. Figure 4.3 zooms into the set of payoff pairs to which Ann and Beth could agree. We call these payoff pairs *rational* since they are at least as good as the disagreement payoff pair for each player.

The game theory literature has suggested many solution methods for bargaining games ([115] and [116]), however, we only consider three: egalitarian [56], Raiffa [85], and Nash [75]. All three begin with the disagreement payoff pair as the status quo and describe a method to arrive at an optimal feasible payoff pair, each with a different goal. The goal of the egalitarian method is to give each player the same incremental payoff from the disagreement payoff pair. The goal of the Raiffa method is to get each player as close as possible to their aspirational payoff. The Nash method seeks to maximize the combined value of the payoffs to the two players. We finish this section by providing a formal definition for each solution method and applying it to the County Fair bargaining game.

Definition 4.1.1. Suppose (N, O, o_d, u) is a bargaining game with the feasible payoff pairs $F = \{(u_1(o), u_2(o)) : o \in O\}$ and disagreement payoff pair $d = (d_1, d_2)$. The *egalitarian payoff pair* is the point $(d_1 + t, d_2 + t) \in F$ that maximizes $t \geq 0$. An *egalitarian outcome* is any outcome whose payoffs are the same as the egalitarian payoff pair.

In addition to the feasible and rational payoff pairs, Figure 4.4 shows (1) a dashed line with slope one emanating from the disagreement payoff pair, and (2) the intersection of this dashed line with the set of feasible payoff pairs that is as far from the disagreement payoff pair as possible. The idea here is that the players should receive the same incremental amount above their disagreement payoffs, and that incremental amount should be maximized.

We can compute the egalitarian payoff pair (a, b) by finding the intersection of the dashed line with parametric equation

$$(1 + t, 2 + t)$$

and the upper right feasible line segment L

$$(1 - s)(2.4, 5.6) + s(6.4, 1.6).$$

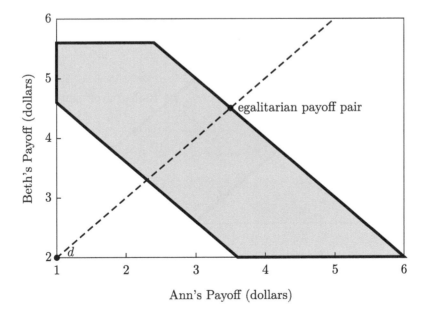

FIGURE 4.4 County Fair Egalitarian Payoff Pair.

This yields the equations

$$1 + t = 2.4(1 - s) + 6.4s$$
$$2 + t = 5.6(1 - s) + 1.6s.$$

Solving these equations, we obtain $(s, t) = (0.275, 2.5)$ which yields $(a, b) = (3.5, 4.5)$ as the egalitarian payoff pair. For Ann and Beth to obtain these payoffs, we solve the equation $u_B(x, m) = 4.5$ and $u_A(64 - x, 4 - m) = 3.5$. This results in $x = 16$ and $m = 2.9$ which corresponds to Beth obtaining 16 ounces of chocolate ice cream and \$2.90. Ann then receives 48 ounces of ice cream and \$1.10.

Definition 4.1.2. Suppose (N, O, o_d, u) is a bargaining game with the feasible payoff pairs $F = \{(u_1(o), u_2(o)) : o \in O\}$ and disagreement payoff pair $d = (d_1, d_2)$. The *aspiration payoff pair* (a_1, a_2) satisfies $a_i = max\{v_i : v \in F, v_1 \ge d_1, v_2 \ge d_2\}$. The *Raiffa payoff pair* is the point $(1 - t)d + ta \in F$ that maximizes $t \ge 0$. A *Raiffa outcome* is any outcome whose payoffs are the same as the Raiffa payoff pair.

In addition to the feasible and rational payoff pairs, Figure 4.5 shows (1) the aspiration payoff pair, (2) a dashed line determined by the disagreement and aspiration payoff pairs, and (3) the intersection of this dashed line with the set of feasible payoff pairs that is as far from the disagreement payoff pair as possible. A player's aspiration payoff is the largest payoff they could receive in an agreement with the other player. On the graph, the aspiration payoff pair is (a_1, a_2) where a_1 is the largest horizontal coordinate in the rational and feasible region and a_2 is the largest vertical coordinate in that region. In Figure 4.3 or Figure 4.5 we see that the aspiration payoffs for Ann and Beth are 6.0 and 5.6, respectively. Raiffa posits that the players should receive incremental amounts above their disagreement payoffs to move them the same proportion of the way towards their aspirations, and that proportion should be maximized.

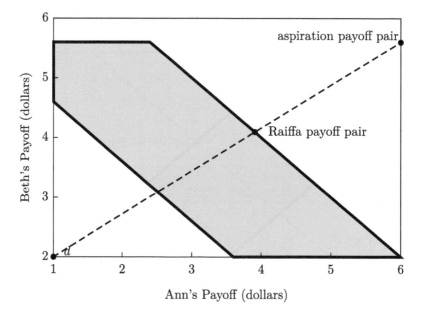

FIGURE 4.5 County Fair Raiffa Payoff Pair.

Again, the Raiffa payoff pair (a, b) lies on the intersection of two lines. Here, the dashed line is from $d = (1, 2)$ to the aspiration payoff pair $(6, 5.6)$ given parametrically as

$$(1 - t)(1, 2) + t(6, 5.6).$$

Finding the intersection of this line and L, the upper right feasible line segment, we obtain $(3.91, 4.09)$ as the Raiffa payoff pair. This corresponds to Beth obtaining 16 ounces of chocolate ice cream and \$2.49 and Ann obtaining 48 ounces of chocolate ice cream and \$1.51.

Definition 4.1.3. Suppose (N, O, o_d, u) is a bargaining game with the feasible payoff pairs $F = \{(u_1(o), u_2(o)) : o \in O\}$ and disagreement payoff pair $d = (d_1, d_2)$. The *Nash payoff pair* is the point $v \in F$ that maximizes the product $(v_1 - d_1)(v_2 - d_2)$. A *Nash outcome* is any outcome whose payoffs are the same as the Nash payoff pair.

In addition to the feasible and rational payoff pairs, Figure 4.6 shows (1) dashed level curves corresponding to equal products of incremental payoffs above the disagreement payoffs, and (2) the feasible payoff pair maximizing this product.

The Nash payoff pair (a, b) maximizes the product of the incremental payoffs above the disagreement payoffs on the upper right feasible line segment L, which can be expressed symbolically by

$$\max\{(a - 1)(b - 2) : a = 8 - b \text{ and } 1.6 \leq b \leq 5.6\}.$$

Substituting the constraint $a = 8 - b$ into $(a - 1)(b - 2)$, we obtain the function

$$f(b) = (8 - b - 1)(b - 2) = -b^2 + 9b - 14,$$

which we need to maximize. Solving

$$0 = f'(b) = -2b + 9$$

yields $b = 4.5$. Since $f''(b) = -2 < 0$, we verify that a maximum has been found.

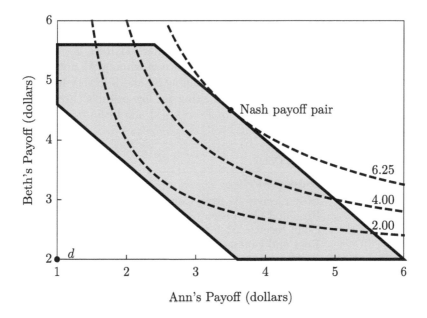

FIGURE 4.6 County Fair Nash Payoff Pair.

Hence, $(3.5, 4.5)$ is the Nash payoff pair, and this corresponds to Beth obtaining 16 ounces of chocolate ice cream and \$2.90. Ann obtains 48 ounces of ice cream and \$1.10.

Although the egalitarian and Nash solutions happen to be the same for the County Fair bargaining game, they lead to different solutions in general. In Section 4.3, we identify the pros and cons of choosing each of the methods.

4.2   BARGAINING IN STRATEGIC GAMES

In this section, we examine possibilities when players in a strategic game are permitted to make binding agreements. *Matching Coins* is a canonical two-player game in which one player is designated the matcher. Each player simultaneously reveals the heads or tails side of a coin. The matcher wins if the two coin sides revealed are the same and loses otherwise. If an amount of money is wagered and the two players are risk neutral with respect to money, the game is zero sum, and the Nash equilibrium has each player choosing each side with probability $1/2$.

Twisted Matching Coins. Larry suggests that Rose and Colin play Matching Coins with a small twist in the rules. As in normal Matching Coins, if Rose and Colin both show heads, Rose receives \$5 from Colin, and if Rose shows tails while Colin shows heads, Colin receives \$5 from Rose. The twist comes when Colin shows tails: Rose and Colin will not exchange money, but Larry will give \$2 each to Rose and Colin if Rose also shows tails, and Rose and Colin will each give Larry \$2 if Rose shows heads. With these small stakes, Rose and Colin are interested only in their own winnings, Colin is risk neutral, and Rose is risk loving.

Observe that once Larry has suggested the game, he has no strategic role. Hence, this scenario can be modeled by a two-player (Rose and Colin) strategic game. Let the utility of the worst outcome for each player be 0 and the utility of the best outcome be 10. Thus for Colin, $u_C(H, H) = 0$ and $u_C(T, H) = 10$

Colin's risk neutrality implies $u_C(H, T) = (7/10)u_C(H, H) + (3/10)u_C(T, H) = 3$ and $u_C(T, T) = (3/10)u_C(H, H) + (7/10)u_C(T, H) = 7$. Rose's risk-loving nature could be modeled in a number of ways. Here we will assume that (1) Rose is indifferent between losing \$2 and a lottery in which there is a 10% chance of winning \$5 and a 90% chance of losing \$5, and (2) Rose is indifferent between winning \$2 and a lottery in which there is a 50% chance of winning \$5 and a 50% chance of losing \$5. The possible (Rose, Colin) strategy profiles, associated outcomes and utilities for Rose $u_R(s)$ and Colin $u_C(s)$ are summarized in Table 4.1.

TABLE 4.1   Twisted Matching Coins

Strategy Profile s	Outcome o	$u_R(s)$	$u_C(s)$
(H,H)	Rose receives \$5 from Colin	10	0
(H,T)	Rose and Colin each pay \$2 to Larry	1	3
(T,H)	Colin receives \$5 from Rose	0	10
(T,T)	Rose and Colin each receive \$2 from Larry	5	7

The reader can verify that there is no pure-strategy Nash equilibrium, and

$$(\tfrac{1}{2}H + \tfrac{1}{2}T, \tfrac{2}{7}H + \tfrac{5}{7}T)$$

is the unique mixed-strategy Nash equilibrium. The Nash equilibrium payoff pair is $(25/7, 5)$, and the expected winnings are

$$\tfrac{1}{2}(\tfrac{2}{7}(\$5, -\$5) + \tfrac{5}{7}(-\$2, -\$2)) + \tfrac{1}{2}(\tfrac{2}{7}(-\$5, \$5) + \tfrac{5}{7}(\$2, \$2)) = (\$0, \$0).$$

Despite the twist, the game still has expected winnings of zero for Rose, Colin, and Larry and is therefore *fair* to all involved.

If either player is strategically risk adverse, they may adopt a prudential strategy. Rose's prudential strategy is

$$\tfrac{5}{14}H + \tfrac{9}{14}T$$

resulting in a security level of $25/7$, the same as her Nash equilibrium payoff. Rose may wish to ensure the $25/7$ expected payoff rather than risk Colin choosing a non-Nash equilibrium strategy that could result in Rose receiving a smaller expected payoff. Colin's prudential strategy is tails resulting in a security level of 3; however, if Rose thinks that Colin is strategically risk adverse, then it could be advantageous for her to reveal tails.

It is certainly mutually beneficial for Rose and Colin to both choose tails, resulting in each receiving \$2 from Larry and achieving payoffs $(5, 7)$. If Rose and Colin are permitted to communicate and make a binding agreement, they may agree to the strategy pair (T, T). Of course, Colin then has an incentive to change to heads, but he will not act upon this enticement if the agreement is binding.

If Rose and Colin have the opportunity to make a binding agreement, they will want to consider all possibilities and perhaps choose based on some fairness criterion, that is, Rose and Colin will play a bargaining game. Of course, the players could agree to some pure or mixed-strategy pair

$$((1 - p)H + pT, (1 - q)H + qT),$$

where $0 \leq p, q \leq 1$, resulting in the payoff pair

$$(1 - p)(1 - q)(10, 0) + (1 - p)q(1, 3) + p(1 - q)(0, 10) + pq(5, 7)$$
$$= (10 - 10p - 9q + 14pq, 10p + 3q - 6pq).$$

They may also agree to publicly roll a die and both will reveal heads if 1 is face up and both will reveal tails otherwise. This correlated strategy will be denoted

$$\tfrac{1}{6}(H, H) + \tfrac{5}{6}(T, T),$$

which would result in the probabilistic outcome

$$\tfrac{1}{6}(\$5, -\$5) + \tfrac{5}{6}(\$2, \$2)$$

and the expected payoff pair

$$\tfrac{1}{6}(10, 0) + \tfrac{5}{6}(5, 7) = (\tfrac{35}{6}, \tfrac{35}{6}).$$

This payoff pair could not be achieved through each player agreeing to a mixed strategy, would be better for each player than the Nash equilibrium payoff pair, and is better for Rose than both players agreeing to reveal tails.

In general, a *correlated strategy* is an assignment of probabilities to pure-strategy profiles. For example, $1/2(H, H) + 1/2(T, T)$ is a correlated strategy profile in which Rose and Colin agree to randomly select from the strategy profile (H, H) half the time and (T, T) half the time. This is different from the strategy profile $(1/2H + 1/2T, 1/2H + 1/2T)$ in which Rose and Colin individually randomly select between H and T. In the latter case, the strategy profile (H, T) actually occurs $1/4$ of the time and in the former it can never occur.

The expected payoff pair for a correlated strategy is the weighted sum of the payoff pairs corresponding to the pure-strategy profiles. Geometrically, the expected payoff pairs for correlated strategies involving two pure-strategy pairs lie on the line segment between the two pure-strategy payoff pairs, and the expected payoff pairs for correlated strategies involving three pure-strategy pairs lie in the filled triangle whose vertices are the three pure-strategy payoff pairs.

Figure 4.7 was created by first plotting the four Twisted Matching Coins pure-strategy payoff pairs given in Table 4.1. Assuming that the players can agree to any correlated

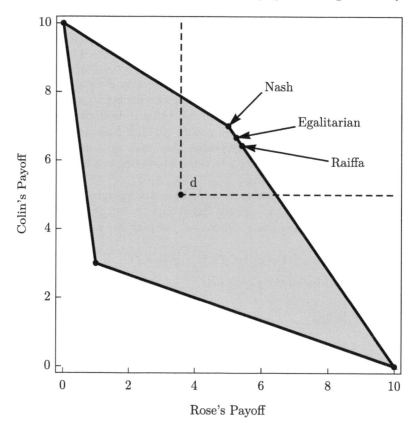

FIGURE 4.7 Twisted Matching Coins Bargaining Game.

strategy that they desire, they can obtain any payoff on the line segments between any two pure-strategy or correlated-strategy payoff pairs. Geometrically, the feasible payoff pairs then lie in the smallest convex region that includes all of the pure-strategy payoff pairs. Formally, we call this region the *convex hull* of the pure-strategy pairs. In the previous example, this is the shaded quadrilateral in Figure 4.7.

The disagreement payoff pair was chosen to be the unique Nash equilibrium payoff pair for the original strategic game, $(3.57, 5.0)$. Finally, the egalitarian, Raiffa, and Nash solutions to the resulting bargaining game were plotted. These solutions are also reported numerically in the first row of Table 4.2.

TABLE 4.2 Twisted Matching Coins Solutions

Disagreement Payoff Pair	Egalitarian	Raiffa	Nash
Nash Equilibrium	(5.24, 6.67)	(5.41, 6.43)	(5.00, 7.00)
Security Levels	(6.07, 5.50)	(5.98, 5.62)	(5.71, 6.00)

While it seems reasonable to model the set of feasible payoff pairs as those arising from correlated strategies, it is less clear how to choose the disagreement payoff pair. Since Twisted Matching Coins has a unique Nash equilibrium, it seems reasonable that the players may choose those strategies if there is a breakdown in the negotiations; however, it would also be reasonable to assume that the players might choose prudential strategies to ensure themselves a minimal payoff. Thus we might also choose the disagreement pair to be the security levels for each player, here $(3.57, 3.0)$. Table 4.2 compares the three solutions for the two disagreement payoff pair assumptions. We see that the solutions are sensitive to the choice of the disagreement payoff pair, and so this is a modeling assumption that requires close scrutiny.

The choice of the disagreement pair also adds technical nuances to the solution calculations when the set of efficient and rational payoff pairs is defined piecewise. Specifically, it may not always be apparent which segment to use when doing the calculations necessary to find a particular solution. For example, for the Twisted Matching Coins game, the efficient payoff pairs consist of (1) the line segment from $(0, 10)$ to $(5, 7)$, and (2) the line segment from $(5, 7)$ to $(10, 0)$. We can describe these two line segments parametrically by

$$(x, y) = (1 - s)(0, 10) + s(5, 7) \qquad 0 \leq s \leq 1 \qquad (4.1)$$
$$(x, y) = (1 - s)(5, 7) + s(10, 0) \qquad 0 \leq s \leq 1. \qquad (4.2)$$

Initially, it may not be clear on which of these two line segments a solution lies, and it will crucially depend upon the disagreement payoff pair. Suppose the disagreement payoff pair consists of the security levels $(3.57, 3.0)$. Then the egalitarian payoff pair lies on the line

$$(x, y) = (3.57, 3.0) + t(1, 1). \qquad (4.3)$$

Solving Equations 4.1 and 4.3, we obtain

$$x = 6.61, y = 6.04, s = 1.32, t = 3.04.$$

Since the obtained value for s violates the inequality $s \leq 1$, we know that the wrong line segment was chosen. Solving Equations 4.2 and 4.3, we obtain

$$x = 6.07, y = 5.50, s = 0.21, t = 2.50.$$

Since the obtained value for s satisfies the inequalities $0 \leq s \leq 1$, we know that the correct line segment was chosen.

Similarly, the Nash payoff pair maximizes $(x - 3.57)(y - 3.00)$ on one of the two line segments. For the first line segment, $y = -0.6x + 10$, we need to maximize

$$f(x) = (x - 3.57)(-0.6x + 10 - 3.00) = -0.6x^2 + 9.142x - 24.99.$$

Since $f'(x) = -1.2x + 9.142$ and $f''(x) = -1.2 < 0$, the maximum is at $x = 9.142/1.2 = 7.62$. Unfortunately, this value of x is not in the domain of the first line segment, we know that the wrong line segment was chosen. For the second line segment, $y = -1.4x + 14$, we need to maximize

$$f(x) = (x - 3.57)(-1.4x + 14 - 3.00) = -1.4x^2 + 16.00x - 39.27.$$

Since $f'(x) = -2.8x+16.00$ and $f''(x) = -2.8 < 0$, the maximum is at $x = 16.00/2.8 = 5.71$. Fortunately, because this value of x is in the domain of the second line segment, we know that the correct line segment was chosen. If this second solution had been to the left of the relevant domain, we would have chosen the intersection of the two line segments as our Nash solution, while if it had been to the right of the domain, we would have selected the endpoint $(10,0)$ as the Nash solution.

4.3 FAIRNESS PROPERTIES

In the previous sections, we defined and used three bargaining game solution methods: egalitarian, Raiffa, and Nash. In this section, we describe a variety of fairness properties for solution methods and identify which of these properties apply to our three methods. This provides a way to choose a solution method based upon general principles. From a modeling perspective, the formal properties we define give us a way of modeling the real-world concept of fairness, and the characterization theorems are the results of our mathematical analysis of our models in light of these properties.

We emphasize here the difference between a *solution* for a particular bargaining game and a *solution method* that can be used to determine solutions for any bargaining game in some large collection of bargaining games. The former could be obtained in an *ad hoc* manner based on biases or special circumstance, whereas the latter suggests uniformity of treatment. When we discuss fairness properties for a solution method, we mean that property applies to the method regardless of the bargaining game.

Our first fairness property suggests that players should not settle for an outcome if a different outcome is better for at least one player without being worse for any player. By better we mean that the outcome o_a *dominates* [resp., *strongly dominates*] the outcome o_b, which occurs if and only if the payoff pair $x = (u_1(o_a), u_2(o_a))$ dominates [resp., strongly dominates] the payoff pair $x = (u_1(o_b), u_2(o_b))$. Formally,

Definition 4.3.1. An outcome is *efficient* if there is no outcome that dominates it. A solution method is *efficient* on a set of bargaining games if it always chooses an efficient outcome.

Our second fairness property suggests that no player should agree to an outcome that is worse than having no agreement.

Definition 4.3.2. An outcome is *rational* if it is not dominated by the disagreement outcome. A solution method is *rational* if it always chooses a rational outcome.

Our third fairness property suggests that if all players have the same opportunities, then they should share equally.

Definition 4.3.3. A bargaining game with disagreement payoff pair d is *symmetric* if $d + (s_2, s_1)$ is the payoff pair for a rational outcome whenever $d + (s_1, s_2)$ is the payoff pair for a rational outcome. A solution method is *unbiased* if, whenever a game is symmetric with disagreement payoff pair d, the method yields an outcome with a payoff pair of the form $d + \lambda(1,1)$ for some $\lambda \geq 0$.

Our fourth fairness property suggests that having additional outcomes should harm no player.

Definition 4.3.4. A solution method is *strongly monotone* if, whenever a new bargaining game is obtained by adding more outcomes (and extending the definitions of the utility functions to these new outcomes without making any other change to the game), the solution method yields payoffs in the new game that are at least as large as the corresponding payoffs in the original game.

We now characterize the egalitarian method on the largest collection of bargaining games for which the egalitarian payoff pair exists and is undominated.

Theorem 4.3.1 (Egalitarian Characterization). Let $\mathcal{E}$ be the set of bargaining games for which the egalitarian solution exists and has no feasible outcome that dominates it. The egalitarian solution method on $\mathcal{E}$ is efficient, rational, unbiased, and strongly monotone. If a solution method on $\mathcal{E}$ is efficient and strongly monotone, then it chooses outcomes whose payoff pairs are the same as the egalitarian solution method [56].

Proof. We first prove that the egalitarian solution is efficient, rational, unbiased, and strongly monotone on $\mathcal{E}$. By the definition of $\mathcal{E}$, no outcome dominates the egalitarian solution; hence, the egalitarian solution is efficient. Since the egalitarian payoff pair is always of the form $d + \lambda(1,1)$ with $\lambda \geq 0$, the egalitarian payoff pair is not dominated by the disagreement payoff pair; hence, the egalitarian method is rational. Since the egalitarian payoff pair is always of the form $d + \lambda(1,1)$, and so is of that form for symmetric games, the egalitarian method is unbiased. Since the egalitarian payoff pair is of the form $d + \lambda(1,1)$ and maximizes λ, adding outcomes cannot reduce λ; hence, the egalitarian method is strongly monotone.

We now prove that if method X on $\mathcal{E}$ is efficient and strongly monotone, then X chooses outcomes whose payoff pairs are the same as the egalitarian solution method. Suppose $G_1 = (N, O, o_d, u)$ is a bargaining game in $\mathcal{E}$ with egalitarian outcome o_{eg}. By definition of the egalitarian method, there is a $\lambda \geq 0$ such that $u_1(o_{eg}) = u_1(o_d) + \lambda$ and $u_2(o_{eg}) = u_2(o_d) + \lambda$. Consider the bargaining game $G_0 = (N, \{o_d, o_{eg}\}, o_d, u)$. Since X is efficient, it must choose outcome o_{eg} for game G_0. Since X is strongly monotone, it must choose an outcome for game G_1 whose payoffs are at least as large as the payoffs for o_{eg}. Since by the definition of $\mathcal{E}$ no outcome dominates o_{eg}, the only outcome whose payoffs are at least as large as the payoffs for o_{eg} is o_{eg}. Hence, X chooses an outcome whose payoff pair is the same as the payoff pair for o_{eg}. □

If we are convinced that a bargaining game solution method should be efficient and strongly monotone, then the previous theorem tells us that we must use the egalitarian method. The Raiffa and Nash methods are characterized with different monotonicity properties and a scale invariance property. Before we present these properties and characterization theorems, we consider two examples that may lessen our enthusiasm for the egalitarian method and will motivate the consideration of different properties.

Figure 4.8 shows the disagreement payoff pair $d = (0,0)$ and feasible payoff pairs for two bargaining games.

The feasible payoff pairs for the first game are in the darker gray square

$$\{(x,y) : 0 \leq x \leq 5, 0 \leq y \leq 5\}.$$

The feasible payoff pairs for the second game add the lighter gray triangle

$$\{(x,y) : 0 \leq x, x + y \leq 10, 0 \leq y \leq 5\}.$$

The payoff pair for the egalitarian solution for both games is $(5,5)$. However, this seems somewhat unreasonable since by stretching the added triangle out to $(9999,0)$ instead of just $(10,0)$, the payoff pair $(5,5)$ would no longer seem fair. The additional payoff pairs in the stretched triangle provide higher payoff options to Rose but not to Colin. The egalitarian method is not sensitive to this change; however, we might believe that a solution method should be sensitive to such changes in the set of feasible payoff pairs. On the other hand, any payoff pair with a higher payoff for Rose causes Colin's payoff to decrease. This violates the

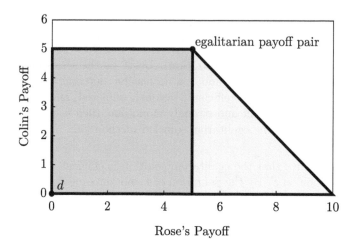

FIGURE 4.8 Strongly Monotone Example 1.

strongly monotone fairness property. We introduce a fifth fairness property that weakens the strongly monotone property and allows Rose's payoffs to increase when more outcomes are added that are beneficial only to her.

Definition 4.3.5. Suppose a bargaining game B' is obtained by adding more outcomes to the bargaining game B (and extending the definitions of the utility functions to these new outcomes without making any other change to the game) without changing player 1's aspirational payoff. A solution method is *individually monotone* if it yields a payoff to player 2 in B' that is at least as large as player 2's payoff in B.

Figure 4.9 shows the disagreement payoff pair $d = (0,0)$ and feasible payoff pairs for two bargaining games.

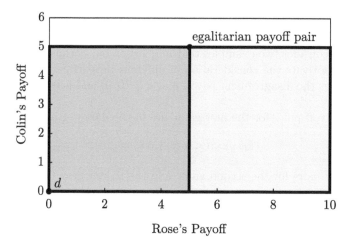

FIGURE 4.9 Strongly Monotone Example 2.

The feasible payoff pairs for the first game are in the darker gray square

$$\{(x,y) : 0 \le x \le 5, 0 \le y \le 5\}.$$

The feasible payoff pairs for the second game add the lighter gray square to form a rectangle

$$\{(x, y) : 0 \leq x \leq 10, 0 \leq y \leq 5\}.$$

The payoff pair for the egalitarian solution for both games is $(5, 5)$. Nonetheless, the second game can be considered the same as the first game with only the units of Rose's payoffs changing; they have been multiplied by 2. If payoffs are vNM utility functions, then such transformations should not change the outcomes of games. That is, Rose should still be able to reach her aspirational payoff. The egalitarian solution method is sensitive to a linear rescaling of vNM payoff functions. This is a violation of the following fairness property.

Definition 4.3.6. Suppose a bargaining game B' is obtained by applying a positive linear rescaling to a player's payoffs in game B. A solution method is *scale invariant* if it produces the same outcome for B' as for B.

Unlike the egalitarian method, the Raiffa method is scale invariant. It is in fact the only solution method that is efficient, unbiased, individually monotone, and scale invariant on a large set of bargaining games. Formally, we define a *free disposal* bargaining game as satisfying the following condition: if (p_1, p_2) is a rational payoff pair, then the payoff pairs (p_1, d_2) and (d_1, p_2) are in the set of feasible payoff pairs.

Theorem 4.3.2 (Raiffa Characterization). Let $\mathcal{R}$ be the set of free disposal bargaining games for which the Raiffa solution exists and has no feasible outcome that dominates it. The Raiffa solution method on $\mathcal{R}$ is efficient, rational, unbiased, individually monotone, and scale invariant. If a solution method on $\mathcal{R}$ is efficient, unbiased, individually monotone, and scale invariant, then it chooses outcomes whose payoff pairs are the same as the outcomes chosen by the Raiffa solution method [56].

The Nash solution method is characterized with a third monotonicity property. Whereas the strongly and individually monotone properties consider what should happen when payoff pairs are added to a game, the next property considers what should happen when certain payoff pairs are removed from a game.

Definition 4.3.7. A solution method is *independent of irrelevant payoff pairs* if it produces the same payoff pair p in a bargaining game B' obtained by removing payoff pairs other than p from the original game B.

Theorem 4.3.3 (Nash Characterization). Let $\mathcal{N}$ be the set of free disposal bargaining games for which the Nash solution exists. The Nash solution method on $\mathcal{N}$ is efficient, rational, unbiased, independent of irrelevant payoff pairs, and scale invariant. If a solution method on $\mathcal{N}$ is efficient, unbiased, independent of irrelevant payoff pairs, and scale invariant, then it chooses outcomes whose payoff pairs are the same as the outcomes chosen by the Nash solution method [75].

Proof. We first prove that the Nash solution is efficient, rational, unbiased, independent of irrelevant payoff pairs, and scale invariant on $\mathcal{N}$. If the payoff pair (w_1, w_2) dominates the Nash payoff pair (v_1, v_2) in the bargaining game (N, O, o_d, u), then

$$(w_1 - d_1)(w_2 - d_2) > (v_1 - d_1)(v_2 - d_2),$$

and since (v_1, v_2) maximizes $(z_1 - d_1)(z_2 - d_2)$ over feasible payoff pairs z, the payoff pair w must not be feasible; hence, the Nash solution is efficient. Similarly,

$$(v_1 - d_1)(v_2 - d_2) \geq (d_1 - d_1)(d_2 - d_2) = 0;$$

hence, the Nash method is rational. Since for a symmetric game $(z_1 - d_1)(z_2 - d_2)$ can have a unique maximum if and only if it is of the form $d + \lambda(1, 1)$, the Nash method is unbiased. If $(z_1 - d_1)(z_2 - d_2)$ is maximized over some set F of payoff pairs at v, then for any set $F' \subset F$ with $v \in F'$, the same function $(z_1 - d_1)(z_2 - d_2)$ is still maximized over the subset F' at v; hence, the Nash method is independent of irrelevant payoff pairs. If the player i's payoffs are changed by the positive linear function $f_i(z) = a_i z + b_i$, then their feasible payoffs will change by the same linear functions and the function to maximize may be rewritten as $a_1 a_2 (z_1 - d_1)(z_2 - d_2)$ in terms of the original payoffs, which has the same maximum; hence, the Nash method is scale invariant.

We now prove that if method X on $\mathcal{N}$ is efficient, unbiased, independent of irrelevant payoff pairs, and scale invariant, then X chooses outcomes whose payoff pairs are the same as the Nash solution method. Suppose $G_1 = (N, O, o_d, u)$ is a bargaining game in $\mathcal{N}$ with feasible payoff pairs

$$F = \{(u_1(o), u_2(o)) : o \in O\},$$

disagreement payoff pair $d = (d_1, d_2)$, Nash outcome o_N, and Nash payoff pair $p = (u_1(o_N), u_2(o_N))$. The disagreement and Nash payoff pairs appear in Figure 4.10. All feasible payoff pairs must be within the shaded region at or below the line tangent to

$$(x - d_1)(y - d_2) = (p_1 - d_1)(p_2 - d_2)$$

at p. For example, F may be the shaded region below the dotted line. Let G_2 have the

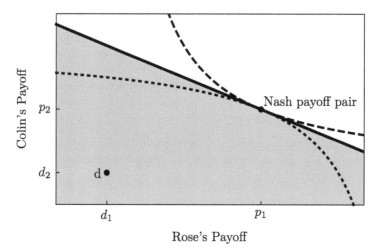

FIGURE 4.10 Bargaining Games G_1 and G_2.

feasible payoff pairs corresponding to the shaded region in Figure 4.10.

Let $G_3 = (N, O', (0, 0), v)$ where $O' = \{(x, y) : x + y \leq 2\}$ and $v(o) = o$ for all $o \in O$ as shown in Figure 4.11. Clearly, the Nash payoff pair for G_3 is $(1, 1)$. Also since G_3 is a symmetric game and X is unbiased and efficient, the X outcome for G_3 is also $(1, 1)$. Since $f_i(v) = d_i + (p_i - d_i)v$ transform the feasible payoff pairs of G_3 onto the feasible payoff pairs for G_2 and X is scale invariant, the X payoff pair for G_2 is the Nash payoff pair. Since the feasible payoffs for G_1 is a subset of those for G_2 and X is independent of irrelevant payoff pairs, the X payoff pair for G_1 is the Nash payoff pair. □

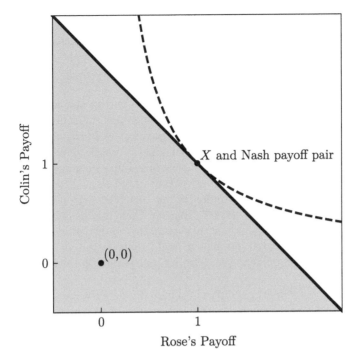

FIGURE 4.11 Bargaining Game G_3.

Before this section, we had introduced three solution methods. Our choice for which one to use could only be based upon our intuitions about which method description seemed most fair. This section has now given us a more rigorous way to compare the three solution methods. The egalitarian method is reasonable to use only if the players can agree on common scales for valuing for the outcomes. For example, money is often used as a common scale even if the players would price individual outcomes differently. If no such external standard for valuing the outcomes exist, but the payoffs are still vNM, we use the Raiffa or Nash method. The choice between these methods depends upon which monotonicity property seems to best capture the concept of fairness in the type of bargaining game under consideration. The importance we place on striving for aspirational payoffs vs. avoiding the influence of irrelevant outcomes determines, then, the method we use. In this last scenario, we detail the selection of a solution method.

Columbia River Watershed. The source of the Columbia River (located in North America) is in Canada, but it runs south into the United States before emptying into the Pacific Ocean. A 1961 treaty required Canada to build three dams and allowed an American reservoir to extend into Canada. In return, the United States agreed to pay for flood control in Canada and to provide over $120 million annually in electricity to the Canadians. This treaty expires in 2024. The United States claims that the dams have been paid for, and that giving away the electricity raises prices for Americans living in the Northwest. It would also like Canada to retrofit its dams to improve the ecosystem. Conversely, the Canadians claim that the dams continue to provide a navigable river downstream and that the adjustments to water levels to manage this harm the ecosystem upstream from the dams [30].

If Canada and the United States are both making decisions from an economic perspective, for example taking into account the monetary value of of electricity, navigation, or the ecosystem, then we have an external standard of value. Here, the egalitarian method can be used. The players then should try to make equally valuable incremental improvements above the default disagreement outcome.

If there is no external standard of value, but the parties can identify their individual aspiration payoffs, then they may be able to make incremental improvements to each country's payoff in a direction towards the aspiration payoff pair. In this case, the countries could negotiate towards the Raiffa solution. Introducing new possible agreements should only have a substantive effect upon the negotiations if new possibilities change the aspiration payoffs.

Finally, the US and Canada might agree that there is no external standard but any agreement between them should be independent of irrelevant possibilities. For example, changes in US trade negotiations with Europe or Canada's healthcare policies shouldn't influence the watershed agreement. In this case, the countries would be best served by the Nash solution. Any proposed agreement could then be tested by checking whether an expansion of possibilities (similar to going from G_1 to G_2 in Figure 4.10) results in a change of the fair outcome.

The modeling of this scenario is difficult. There are an extremely large number of financial distributions that could be agreed upon between the two countries and many unknown consequences to the ecosystem. The disagreement outcome involves returning to a default agreed to under the expiring treaty. It is likely to be difficult for each country to determine a utility function over all possible outcomes, and there are likely to be incentives to publicly misrepresent preferences. Nonetheless, our models of fairness allow each country to decide what they value in an agreement and work towards what they deem a fair solution.

EXERCISES

4.1 Determine which of the three payoff pairs described in Section 4.1 (egalitarian, Raiffa, and Nash) aligns with each of the following fairness beliefs:

   a. The collective good should be maximized.

   b. Everyone should be given equal amounts.

   c. Everyone should achieve the same proportion of their goals.

4.2 Is there any situation in which the egalitarian and Raiffa payoff pairs are equal? Are all three ever equal?

4.3 Suppose that in the County Fair Prize scenario, Ann and Beth value the chocolate ice cream equally, but that Beth still has no use for more than 16 ounces of ice cream. Create new utility functions $u_A(x, m)$ and $u_B(x, m)$ and graph them.

4.4 **Acme Industrial.** The employees and owners at ACME Industrial are in the midst of negotiating a new labor contract. At this time, there are three issues central to the negotiations: (a) a wage increase, with the owners offering the previous contract's 2% and the workers wanting 5%; (b) how to share medical insurance costs, with the owners wanting workers to continue to pay 30% of the total premium and the workers wanting to pay only 20%; and (c) job security, with the owners wanting the ability to lay off workers "at will" and the workers insisting on the current buy-out plan. An arbiter asked the two sides to weight these issues by allocating 100 points among them. The results are shown in Table 4.3. To the best of our knowledge, Allen [2] was to first to apply the Nash bargaining solution in a methodical manner to collective bargaining. This exercise is similar to the development in [41].

TABLE 4.3 Worth of Each Issue to Workers and Owners

Issue	Worker Worth	Owner Worth
Wages	40	10
Insurance	20	40
Security	40	50
Total	100	100

a. Using the points as payoffs, explain why $(60, 50)$ is the payoff pair for the outcome in which the workers get the raise that they want and keep the lower premium while the owners obtain the right to let workers go "at will." We denote this payoff pair (WI, S). Use this argument to determine payoff pairs for each of the eight possible outcomes in which for each issue one side wins completely.

b. Graph the region of feasible payoff pairs by plotting the payoff pairs found in (a) and determining their convex hull. Mark (40,50) as the disagreement payoff pair. Comment on whether $(40, 50)$ is appropriate as the disagreement payoff pair.

c. For each of the following payoff pairs (Workers, Owners), determine the resolution of the issues that result in the payoff pair.

   i. $(40, 50)$
   ii. $(20, 95)$
   iii. $(84, 32)$

d. Find the egalitarian payoff pair and a corresponding outcome.

e. Find the Raiffa payoff pair and a corresponding outcome.

f. Find the Nash payoff pair and a corresponding outcome.

4.5 **Divorce Settlement**. Brams and Taylor [15] present a fictional divorce settlement based on [57] and give valuations to each partner on three critical issues. Mary and Tom (we have switched the names) must agree on custody of their son John, alimony payments from Mary to Tom between 0 and 50% of Mary's current salary, and ownership of the house. The judge who will make the decision asked each person to assign a percentage indicating the importance of each issue to them, with the results shown in Table 4.4.

TABLE 4.4 Worth of Each Issue to Mary and Tom

Issue	Mary Worth	Tom Worth
Custody	25	65
Alimony	60	25
House	15	10
Total	100	100

    a. State explicitly the assumptions being made by the judge in asking Tom and Mary to assign percentages as done in Table 4.4.

    b. Using the percentages as payoffs, explain why $(40, 25)$ is the payoff pair for the outcome in which Mary wins custody of the child, is awarded the house, but must pay alimony. We denote this payoff pair as (CH,A). Use this argument to determine payoff pairs for each of the eight possible outcomes.

    c. Graph the region of feasible payoff pairs by plotting the payoff pairs found in (a) and determining their convex hull. Mark (50,50) as the disagreement payoff pair. Discuss whether $(50, 50)$ is appropriate as the disagreement payoff pair.

    d. For each of the following payoff pairs (Mary, Tom), determine the resolution of the issues that result in the payoff pair.

        i. $(85, 10)$

        ii. $(87.5, 32.5)$

        iii. $(12, 95)$

    e. Find the egalitarian payoff pair and a corresponding outcome.

    f. Find the Raiffa payoff pair and a corresponding outcome.

    g. Find the Nash payoff pair and a corresponding outcome.

    h. Compare the three outcomes. Which one yields the fairest resolution? Justify your answer.

4.6 Suppose the disagreement payoff pair is $(0, 0)$ and the set of feasible payoff pairs is $F = \{(x, y) : (x/a)^2 + (y/b)^2 \leq 1, 0 \leq x, 0 \leq y\}$ for some positive real numbers a and b.

    a. Find the egalitarian payoff pair.

    b. Find the Raiffa payoff pair.

    c. Find the Nash payoff pair.

    d. Describe the relationship among the solutions as a and b are varied.

4.7 Suppose the disagreement payoff pair is $(0, 0)$ and the set of feasible payoff pairs is the convex hull of $(0, 0)$, $(1, 0)$, $(0, 1)$ and (x_1, y_1) where $0 < x_1 \leq 1, 0 < y_1 \leq 1$, and $2 < x_1 + y_1$.

    a. Find the egalitarian payoff pair.

    b. Find the Raiffa payoff pair.

    c. Find the Nash payoff pair.

    d. Describe the relationship among the solutions as x_1 and y_1 are varied.

4.8 Model the Prisoner's Dilemma strategic game whose payoff matrix is given in Table 4.5 as a bilateral bargaining game. Include a graph of the feasible payoff pairs and justify the disagreement payoff pair used. Find the egalitarian, Raiffa, and Nash payoff pairs and corresponding correlated strategies. Compare the three outcomes. Which one yields the fairest resolution? Justify your answer.

TABLE 4.5   Prisoner's Dilemma Payoff
Matrix

Rose/Colin	Cooperate	Defect
Cooperate	(3, 3)	(0, 5)
Defect	(5, 0)	(1, 1)

TABLE 4.6   Twisted Matching Coins Solutions

Disagreement Payoff Pair	Egalitarian	Raiffa	Nash
Nash Equilibrium	(5.24, 6.67)	(5.41, 6.43)	(5.00, 7.00)
Security Levels	(6.07, 5.50)	(5.98, 5.62)	(5.71, 6.00)

4.9   This exercise references Table 4.6, which summarizes solutions to the Twisted Matching Coins games.

    a.   Assume players choose the Nash equilibrium strategies if a binding agreement cannot be reached. Find the egalitarian payoff pair and an associated correlated strategy.

    b.   Find the Raiffa payoff pair and an implementing correlated strategy assuming players choose the Nash equilibrium strategies if a binding agreement cannot be reached.

    c.   Find the Nash payoff pair and an implementing correlated strategy assuming players choose the Nash equilibrium strategies if a binding agreement cannot be reached.

    d.   Find the egalitarian payoff pair and an implementing correlated strategy assuming the disagreement payoff pair are the security levels.

    e.   Find the Raiffa payoff pair and an implementing correlated strategy assuming the disagreement payoff pair are the security levels.

    f.   Find the Nash payoff pair and an implementing correlated strategy assuming the disagreement payoff pair are the security levels.

4.10   Consider the game Twisted Matching Coins. We obtained the outcome table in Table 4.1 by assuming each player was only interested in the money she or he won or lost, Rose was risk loving, and Colin was risk neutral. We also arbitrarily set each player's utility to 0 and 10 for the worst and best outcomes. For each of the following assumptions, obtain a similar outcome table.

    a.   Both players were risk neutral.

    b.   Both players were risk averse.

    c.   Rose was risk averse while Colin was risk neutral.

4.11   In the Twisted Matching Coins scenario, we assumed that Rose is risk-loving. Model the scenario if Rose is risk-neutral and the disagreement payoff pair is the Nash equilibrium.

*4.12 For the Twisted Matching Coins strategic game, show that no pair of mixed strategies yields the same expected payoff pair as obtained with the correlated strategy (1/6)(Heads, Heads) + (5/6)(Tails, Tails).

4.13 Consider the fairness property efficient in the County Fair prize scenario. Describe a payoff pair that is not efficient and explain why Ann and Beth are unlikely to settle for that payoff pair.

4.14 Figure 4.2 shows a feasible payoff pair of $(7.2, 0)$. Explain why Ann and Beth are unlikely to agree on that payoff pair. Which fairness property would that agreement violate?

4.15 In the County Fair Prize scenario, given that Ann and Beth have never met each other and only have three minutes to make a decision, it is probable that they will simply split the ice cream and $4 equally. Calculate Ann and Beth's utilities for this allocation. For each fairness property described in Section 4.3, determine whether this allocation shows or does not show a violation.

*4.16 Prove the Raiffa Characterization Theorem. Hint: You will need to use some ideas from the proofs of the Egalitarian and Nash Characterization Theorems.

*4.17 **Adams-Onis Treaty**. The Tejas border between the United States and Spain was settled by the Adams-Onis Treaty of 1819. Identify the issues of concern to each country, and model the scenario as a bilateral negotiation. Which of the methods of this section give a solution closest to what was actually agreed upon? [106]

*4.18 **Tennessee-Georgia Border**. The states of Georgia and Tennessee are currently locked in a border dispute regarding a very small strip of land near the Tennessee River. Model this dispute as a bilateral negotiation and use one of the methods of this section to propose a solution [111], [20], and [23].

CHAPTER 5

Sequential Play

Sequential games model scenarios in which the players act sequentially, usually taking turns according to some rule. As with strategic games, each player has complete information about each other's strategies and preferences. We introduce the basic solution concepts, techniques, and key theorems. Further, we build models in a variety of contexts demonstrating a range of options in the number of players and the turn-taking process.

A play of a sequential game consists of a sequence of actions taken by the players. Thus, we can describe a general game in terms of these sequences. The game begins with an empty sequence, known as the *empty history* and ends with a sequence called a *terminal history*. Terminal histories determine the outcomes of the game, just like strategy profiles determine outcomes in strategic games. As we did in that chapter, we will often use the terms interchangeably. Subsequences of play that are obtained along the way are known as *non-terminal histories*. We use the phrasing "history" rather than sequence to place emphasis on the fact that when a player takes an action, the history of previous actions taken is known. Thus, formally, we have the following definition.

Definition 5.0.1. A *sequential game* consists of the following:

1. A set $N = \{1, 2, \ldots, n\}$ of at least two players.

2. A set O of terminal histories. We define a *history* to be an ordered list (possibly empty) of player actions; a *terminal history* is a history with no actions following it.

3. A *player function*, P, which assigns a player to every non-terminal history.

4. A rule stating that starting with the empty history, if a non-terminal history is reached, the player assigned to it selects an action to append to the current history.

5. Utility functions $u_i : O \to \mathbb{R}$ that specify the preferences among terminal histories for each player $i \in N$.

At every non-terminal history, the assigned player must choose an action. In order to describe a strategy for a player in a sequential game, then, we must identify an action for every possible non-terminal history to which the player was assigned. This differs from strategies in strategic games which simply identified the single action each player would take.

5.1   SEQUENTIAL GAMES

In the early 2000s, media-streaming products were introduced by a few companies to compete with cable TV. Within a few years, several other companies joined the market with similar products. Bierman and Fernandez [12] created a model of competing software companies developing new products. Here we adapt their model for a fictitious scenario in which one company is attempting to clone another's media streaming device in order to capture market share.

Media Streaming. PlumProducts is about to launch a media-streaming device for modern TVs and is deciding how aggressive to be with its marketing options. Seriously aggressive marketing costs more but is more likely to saturate the market quickly. A competitor, SeeTV, is working on a clone of PlumProducts' device which will require a year to bring to market. SeeTV must decide whether to continue this research in order to enter the market. However, SeeTV can clone this product better and more cheaply if it can recruit some of PlumProducts' engineers to work for SeeTV. Knowing this, PlumProducts can choose to include restraining clauses in its engineers' contracts barring them from working for competitors if they leave PlumProducts. However, there is a cost to these clauses because they require PlumProducts to pay their engineers higher salaries in exchange. (This scenario is based on [12].)

In this scenario, we consider PlumProducts and SeeTV to be the players. We next need to determine the chronology of player actions. To start, PlumProducts can choose to include or not include restraining clauses in its engineers' contracts (Restrain or Open). In either case, SeeTV will later choose to enter or not enter the market with its own product (In or Out). Finally, if SeeTV chooses to enter the market, PlumProducts will choose to be Aggressive or Passive with its marketing. This chronology determines the possible terminal histories, outlined in Table 5.1. The sequential nature of the game is nicely captured in the

TABLE 5.1   Media Streaming
Terminal Histories

Terminal history h
Restrain, In, Aggressive
Restrain, In, Passive
Restrain, Out
Open, In, Aggressive
Open, In, Passive
Open, Out

game tree illustrated in Figure 5.1. Table 5.2 lists non-terminal histories and the player who acts at each subhistory.

TABLE 5.2   Media Streaming Subhistories

Non-terminal history h	**Acting Player** $P(h)$
$\emptyset$	PlumProducts
Restrain	SeeTV
Open	SeeTV
Restrain, In	PlumProducts
Open, In	PlumProducts

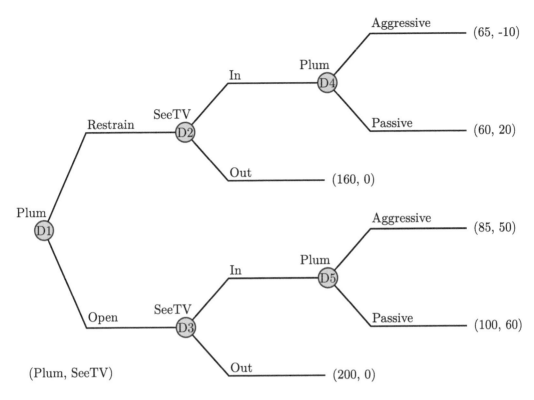

FIGURE 5.1 Media Streaming Game Tree.

In the game tree, the shaded circles, called *nodes*, correspond to non-terminal histories. For example, the node labeled "D5" corresponds to the non-terminal history "Open, In." The label above the D5 node indicates that it is PlumProducts' turn to act. PlumProducts can choose to either be Aggressive or Passive, as indicated by the labels on the line segments, called *edges*, emanating from the D5 node to the right. In the former case, the terminal history is "Open, In, Aggressive." In the latter case, the terminal history is "Open, In, Passive."

To continue modeling this scenario, we need to quantify the various financial components. How we do this will affect the fidelity and flexibility of our model. Here we will choose values that are not supported by data, but seem reasonable for the scenario. This makes our model flexible, but reduces the fidelity.

We assume that the market will support sales sufficient to generate $200 million in net revenue. The restraining clauses will cost PlumProducts an additional $40 million if it chooses to use them, and the Aggressive marketing option costs an additional $25 million. With the Aggressive option, PlumProducts obtains a 65% market share with the restraining clause and a 55% share without it. When choosing the Passive option, PlumProducts only obtains a 50% market share. It costs SeeTV $40 million to develop the product if it can hire PlumProducts' engineers and $80 million if it cannot.

Since we may assume that each company's preference is to maximize its expected profit, Table 5.3 shows the payoffs to each company associated with each terminal history, and these payoffs are also included as the terminal nodes in the Figure 5.1 game tree. For PlumProducts, its payoff for the first terminal history is determined by finding 65% of $200 million and then subtracting the additional $40 million for the restraining clauses and $25 million for the aggressive marketing. Similarly, SeeTV's payoff for the fifth terminal history is found by subtracting its $40 million development costs from 50% of $200 million. The remainder of the entries are determined in a similar manner.

TABLE 5.3   Media Streaming Payoffs (in thousands of dollars)

Terminal History h	$u_{\mathbf{PlumProducts}}(h)$	$u_{\mathbf{SeeTV}}(h)$
Restrain, In, Aggressive	65	−10
Restrain, In, Passive	60	20
Restrain, Out	160	0
Open, In, Aggressive	85	50
Open, In Passive	100	60
Open, Out	200	0

To construct a strategy for one of the companies, we select an action available to the company for each of the non-terminal histories assigned to the company by the player function, as outlined in Table 5.2. Thus we can describe a strategy for SeeTV by selecting an action at each of its two non-terminal histories from top to bottom. One such strategy is (Out, In), which indicates that SeeTV will stay out of the market if PlumProducts writes a restraining clause into its employee contracts but will enter into the market if PlumProducts does not write a restraining clause into its employee contracts. SeeTV has three other strategies: (In, In), (In, Out), and (Out, Out).

Similarly, PlumProducts is assigned three non-terminal histories, so each one of its strategies must describe what action to take at each of them left to right and then top to bottom. Thus the strategy (Restrain, Aggressive, Passive) indicates that PlumProducts should include the restraining clause and if SeeTV enters the market, employ an aggressive marketing plan. The Passive part of this strategy gives an action for decision point D5 even though choosing Restrain in the game tree indicates PlumProducts will never arrive at D5. However, the definition of strategy requires us to identify an action for each decision point. This can be helpful, for example, if PlumProducts fails to include the restraining clause in one engineer's contract and SeeTV recruits that particular engineer. Our strategy then covers this scenario. By assigning one of two actions to each of the 3 decision points, we see that PlumProducts has a total of eight strategies.

Using the strategies that we have identified for each player, we can construct a strategic game model for this scenario, as displayed in Table 5.4. From this table, we can identify the four Nash equilibria in pure strategies, whose payoffs are both boxed as best response payoffs. (There are other Nash equilibria in mixed strategies, but they are not essential for our discussion.) This model is unsatisfying, however, since it does not capture the sequential nature of the scenario.

TABLE 5.4   Media Streaming Strategic Game

(PlumProducts, SeeTV)	(In, In)	(In, Out)	(Out, In)	(Out, Out)
(R, A, A)	(65, −10)	(65, −10)	([160], [0])	(160, [0])
(R, A, P)	(65, −10)	(65, −10)	([160], [0])	(160, [0])
(R, P, A)	(60, [20])	(60, [20])	([160], 0)	(160, 0)
(R, P, P)	(60, [20])	(60, [20])	([160], 0)	(160, 0)
(O, A, A)	(85, [50])	([200], 0)	(85, [50])	([200], 0)
(O, A, P)	([100], [60])	([200], 0)	(100, [60])	([200], 0)
(O, P, A)	(85, [50])	([200], 0)	(85, [50])	([200], 0)
(O, P, P)	([100], [60])	([200], 0)	(100, [60])	([200], 0)

An analysis based on the game tree is often more illustrative than the strategic game matrix. Beginning at the end of the game tree in Figure 5.2, we see that PlumProducts will select Aggressive at node D4 because the payoff of 65 with Aggressive is larger than the payoff of 60 with Passive, and PlumProducts will select Passive at node D5 because the payoff of 100 with Passive is larger than the payoff of 85 with Aggressive. These choices are indicated by making the edges thicker and labeling the nodes with the payoff vectors.

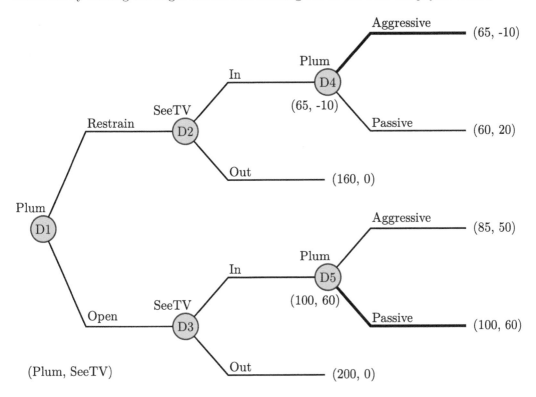

FIGURE 5.2 Media Streaming Backward Induction: Step 1.

Given this information, SeeTV should select Out at node D2 because the payoff of 0 with Out is larger than the payoff of −10 with In, and SeeTV should select In at node D3 because the payoff of 60 with In is larger than the payoff of 0 with Out. These choices are indicated in Figure 5.3 with additional thickened edges and labeled nodes.

And now, anticipating SeeTV's strategy, PlumProducts should select Restrain at node D1 because the payoff of 160 with Restrain is larger than the payoff 100 with Open. This choice is indicated in Figure 5.4 with an additional thickened edge and labeled node. Thus, the strategy pair ((Restrain, Aggressive, Passive), (Out, In)) is the strategy profile that should result from players attempting to maximize their payoffs. Note that this is one of the Nash equilibria from the strategic game version, and we argue that this is precisely the single Nash equilibrium that should result from this game.

Although PlumProducts' choice of Restrain at the beginning of the game precludes the need to make choices at nodes D3 and D5, our analysis of the best responses by the players required us to consider potential actions at all nodes. This is the reason we insist that a strategy specify actions at all nodes, even unreachable ones.

Before considering our next scenario, we generalize what we have just done.

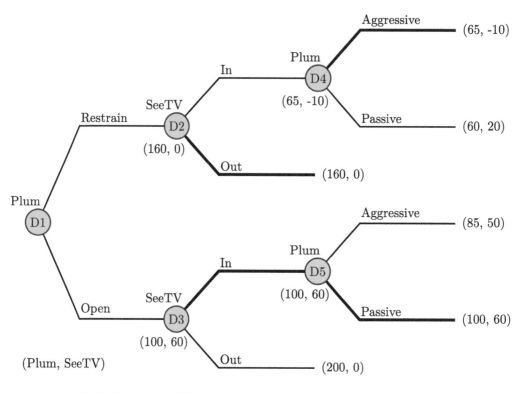

FIGURE 5.3 Media Streaming Backward Induction: Step 2.

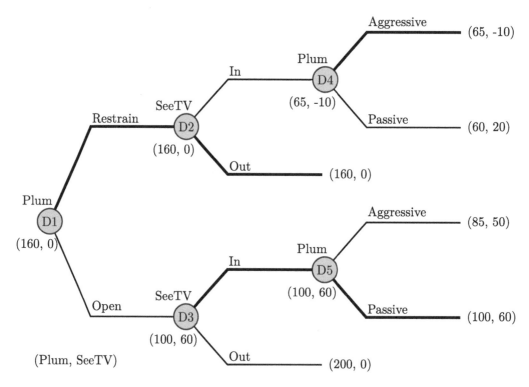

FIGURE 5.4 Media Streaming Backward Induction: Step 3.

Definition 5.1.1. A *game tree* is a visual representation of a sequential game which displays all ways that a game could be played. Each internal node in the tree represents a non-terminal history and is labeled by the player function to indicate which player is uniquely able to move from that position. Each of the player's legal moves at a given node is indicated by an edge leading from that node to another node. There is an identified root node which is the initial game position, and the terminal nodes are labeled with the outcomes and payoffs for that play of the game.

We can also generalize the process used in the previous discussion to find a meaningful Nash equilibrium.

Definition 5.1.2. *Backward induction* is the following algorithm: Label all of the terminal nodes of the game tree with the payoff vectors. In an inductive process, select any unlabeled node which has all of its directed edges extending to nodes labeled with payoff vectors. We call the unlabeled node here the parent and the labeled nodes children. For each *parent*, select a *child* with the largest payoff for the player assigned to the parent node. Mark the edge from the parent to this child and label the parent with the payoff vector from the selected child. Repeat this process until all nodes are labeled. The marked edges for each player describe their *backward induction strategy*, and the path of marked edges starting at the root lead to a *backward induction outcome*.

The backward induction strategy profile is a Nash equilibrium. Zermelo in 1913 [128] was the first to prove that such a strategy profile always exists.

Theorem 5.1.1 (Zermelo's). Every sequential game with a finite number of nodes has a backward induction solution, that is, a strategy profile and outcome.

Proof. We proceed by induction. Let n denote the maximum number of steps needed to reach a terminal node from the root node. If $n = 1$, then we know the backward induction solution: pick the action that yields the highest payoff.

Now suppose that any game with $n - 1$ as the maximum number of steps has a backward induction solution, and consider a game in which the maximum number of steps is n. At the parent of each terminal node in this game, select a child whose payoff vector maximizes the payoff for the player who is assigned to the parent and assign this action to the player. Now consider the game in which this set of parent nodes includes the terminal nodes of the game tree. This game tree has a maximum depth of $n - 1$ and hence has a backward induction solution by hypothesis. To each of the strategies that are part of this backward induction solution, we append the action taken at the appropriate parent nodes to obtain a backward induction solution to the original game. □

U.S. Appellate Court. Since 1973, three-judge panels of the U.S. Appellate Court system have been able to render decisions via unpublished opinions. These opinions provide a ruling in the case being reviewed but, unlike published opinions, do not enter the body of common law and hence provide no precedent for future cases. This means the unpublished opinion need not be followed in any future cases the court might hear. While there are serious questions as to the legal, social, and political consequences of allowing unpublished opinions, implementation of the system is straightforward when there is a consensus among the members of the three-judge panel. However, when there is a dissenting opinion resulting in a two-to-one split on the resolution of the case, it is more complicated. Most federal circuit courts either explicitly require publication when there is a dissenting opinion or permit the dissenter to force publication because non-publication requires unanimous agreement by the three judges on the panel [22].

Here we consider a scenario in which three judges have reviewed a case and there is a two-to-one split on the decision of the case. We have two players: the dissenting judge, and the majority (the other two judges, viewed as one player here).

The dissenting judge then has two options: he may write a dissenting opinion which forces the publication of the ruling with his opinion or he may offer to join the majority opinion on the condition that the ruling, also called an opinion here, goes unpublished so it cannot be used as precedent. This partial control over publication of the court's opinion by a dissenting judge results in a strategic decision-making environment amendable to game theoretic modeling. If the dissenter does not offer to join the majority, then both the majority and dissenting opinions are published and the case enters the record as precedent. If the dissenter offers to join the majority with the intent of having the opinion unpublished, the majority must choose between achieving consensus, but without publication, or opting to publish its opinion anyway, with the dissenting opinion attached.

This results in the three following terminal histories:

A Dissent. This forces publication with both the majority and dissenting opinions and the case becomes precedent.

B Offer to Join, Publish. Here the dissenter offers to join the consensus with the condition of non-publication, but the majority decides to publish anyway, and the case becomes precedent with the dissenting opinion on the record. This also demonstrates an underlying disrespect by the majority for the dissenter since they ignored his request to join them and leave the opinion unpublished.

C Offer to Join, Don't Publish. This results in consensus among the judges, but creates an opinion which is not precedent.

The preferences of the dissenter and the majority among these three outcomes can be influenced by their own ideology or how much value they place on keeping the ruling from becoming precedent. For example, an ideological dissenter whose top priority is to get his opinion in the record would rank the outcomes A, B, C. A dissenter with a weak argument might rank the outcomes C, A, B. His first goal is to bury his weak opinion, but if his opinion is published he'd at least like to avoid disrespect.

Similarly, the majority may also be driven by either ideology or by the quality of their opinion. An ideological majority might have the preference order of B, A, C since they would like their opinion published but also want the internal satisfaction of disrespecting the dissenter. On the other hand, a majority with a weak opinion might have the preference order C, B, A since it would like to bury its opinion, but is willing to show disrespect to the dissenter if it is forced to publish.

We analyze the case when an ideological dissenter is faced with an ideological majority. The dissenter ranks the outcomes A, B, C while the majority ranks the outcomes B, A, C. By backward induction, the majority will choose to publish if given a choice. Given this, in the next stage of backward induction, the dissenting judge will choose not to give the majority this choice, leading to the strategy profile (Dissent, Publish) resulting in outcome A. In contrast, if a dissenter with a weak argument is faced with a majority with a weak opinion, backward induction leads to the strategy profile (Offer to Join, Don't Publish) resulting in outcome C.

Frequently, the game tree representation of a sequential game can be rather cumbersome because of the number of edges and nodes. We will use the next scenario to explore an alternate approach to completing a backward induction analysis of a game without a full game tree.

Patent Race. Dutta [29] describes a scenario in which companies R and S are competing to develop a new product for the market. The distance each is from completing the development can be measured in discrete steps, so at any point we can quantify the number of steps n left in the project. The maximum number of steps either company can take at once is 3, and it costs $2 million, $7 million, and $15 million to take one, two, or three steps, respectively, at one time. The first company to develop the product gets the patent, which is worth $20 million, and a company quits the competition when it is no longer profitable to play. The two companies alternate moves so that they can observe each other and manage their investment dollars.

The empty history of this game consists of some ordered pair (r, s) that indicates how many steps companies R and S, respectively, are away from applying for the patent. If company R has the first move in this game, then its first action is to select from moving to (r, s), $(r - 1, s)$, $(r - 2, s)$, and $(r - 3, s)$. Company S then has four actions at each of these new positions. Figure 5.5 shows a partial game tree when R and S are 3 and 4 steps away from completion (Ellipses indicate where more of the tree would need to be added to complete the tree. All images used with permission.)

It is important to note that since an ordered pair may be reached in several different ways, several nodes may be labeled with the same ordered pair. While nodes, and not ordered pairs, correspond to non-terminal histories, it suffices to perform backward induction based on the ordered pairs.

Let the number of steps that R and S still need to take to reach completion be represented as the coordinates of a point on the grid as illustrated in Figure 5.6. Completion for company R means reaching the vertical axis and completion for company S means reaching the horizontal axis. The goal therefore of company R is to move left in the location space and that of company S is to move down.

To begin our Backward Induction analysis, consider a position (r, s) in which $r \leq 3$ and $s \leq 3$, as illustrated by the gray square in Figure 5.6. In this case, whichever company has the first move should take sufficient steps to end, and win, the game. We call this region a trigger zone, as the first mover has a winning strategy.

Now consider a position (r, s) in which $r > 3$ and $s \leq 3$. Company R cannot complete the research in one turn, and observes that company S can complete the research on its next turn. Thus, its best option is to drop out of the patent race, leaving company S to complete the research in s moves to minimize its cost (two single-step moves cost less than one two-step move, and three single step moves cost less than one three-step move). Therefore the region described by $r > 3$ and $s \leq 3$ is a safety zone for company S. Similarly, the region described by $r \leq 3$ and $s > 3$ is a safety zone for company R.

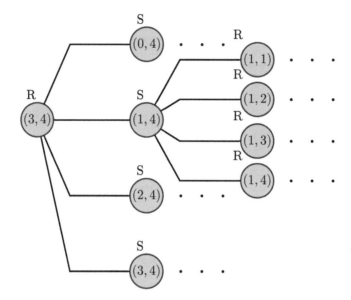

FIGURE 5.5 Patent Race Partial Game Tree. Reproduced from Dutta, Prajit K., *Strategies and Games: Theory and Practice*, Figure 12.1, 1999 Massachusetts Institute of Technology, reprinted by permission of The MIT Press.

Now suppose that we are in a position (r, s) such that $3 < r \le 5$ and $3 < s \le 5$ and that it is company R's turn to move. It can move into its safety zone for a cost of at most \$7 million and then complete the research in three steps for an additional \$6 million, yielding a profit of \$7 million. We note that company R cannot move into its safety zone in one step from position $(6, 5)$ and still make a profit. (The cost is \$21 million.) Since the same analysis applies to company S, we have identified this region as a second trigger zone.

This also creates a second pair of safety zones. Positions (r, s) such that $3 < r \le 5$ and $5 < s$ is a safety zone for company R, and there is a symmetric one for company S.

We continue in this fashion to identify the six trigger zones and their corresponding safety zones, as illustrated in Figure 5.7. The first mover from a position in the nth trigger zone spends what is needed to move, profitably, into its $(n-1)$th safety zone.

5.2 SUBGAME PERFECT EQUILIBRIA

In the media streaming example of the previous section, we observed that the Nash equilibrium found through backward induction in a sequential game might only be one of several found in the strategic form of the game. In this section, we will use the following simple scenario to understand this phenomena and to define a more refined equilibrium concept which accounts for the difference.

> **Political Challenger**. After the 2017 National Women's March, the "First we march, then we run" campaign inspired many women who had never been involved in politics to run for local, state or even national offices. In some cases, these women ran against long time incumbents. Thus, before the election the challenger had to decide whether or not to enter the race. The long-term incumbent then had to decide whether to withdraw from the race or fight the challenger.

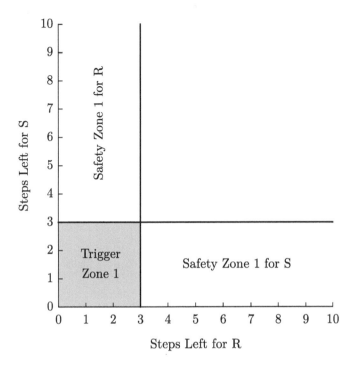

FIGURE 5.6 Patent Race Analysis - Step 1. Reproduced from Dutta, Prajit K., *Strategies and Games: Theory and Practice*, Figure 12.4, 1999 Massachusetts Institute of Technology, reprinted by permission of The MIT Press.

Figure 5.8 models this scenario as a sequential game. The utilities, to the challenger and the incumbent, are simply their ordinal preferences among the three terminal histories. It is certainly the case that the challenger would most prefer the terminal history (In, Withdraw), but it is less clear how she would rank the other two terminal histories. Similarly, the incumbent would most prefer the terminal history (Out), but again, his preferences for the other two histories are less clear. We have assigned utilities in the form (Challenger, Incumbent) in Figure 5.8, but other utility functions are possible. Figure 5.9 reveals the backward induction strategy profile is (In, Withdraw).

Table 5.5 displays the strategic form of this game and best response payoffs from which we can see that there are two Nash equilibria: (In, Withdraw) and (Out, Fight). The former is also the backward induction strategy profile. The latter involves the willingness of the incumbent to threaten to choose a less preferred outcome in order to convince the challenger to stay out of the race.

TABLE 5.5 Primary Strategic Game

(Challenger, Incumbent)	Withdraw	Fight
In	($\boxed{2}$, $\boxed{1}$)	(0, 0)
Out	(1, $\boxed{2}$)	($\boxed{1}$, $\boxed{2}$)

Whenever we model this scenario as a strategic game, we assume the challenger and the incumbent will always see the full strategy chosen by the other player. But when modeled as a sequential game, the challenger never sees the action of the incumbent when the challenger

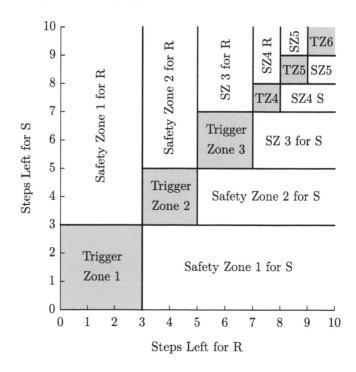

FIGURE 5.7 Patent Race Analysis - Steps 2-6. Reproduced from Dutta, Prajit K., *Strategies and Games: Theory and Practice*, Figure 12.6, 1999 Massachusetts Institute of Technology, reprinted by permission of The MIT Press.

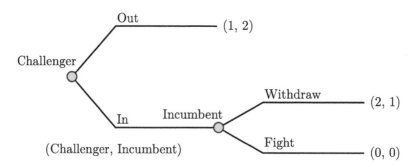

FIGURE 5.8 Political Challenger Game Tree.

chooses Out. Thus choosing Out is not necessarily logical for the challenger, as they do not know what the incumbent would do if they had selected In. One way to give the players of a sequential game the same information as they have in the strategic game version is to imagine many repetitions of the game with different strategy profiles, allowing all players to observe the others' behavior. Based on past behavior, then, the challenger would know that the incumbent would choose Fight if the challenger chooses In. However, in a one-time sequential game, the incumbent would not actually choose Fight when the challenger selects In because it leads to a sub-optimal outcome. We are led to the conclusion that (Out, Fight), while a Nash equilibrium in the strategic game, does not meet some more robust understanding of equilibrium in a sequential game.

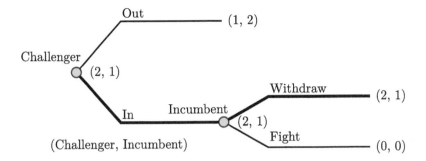

FIGURE 5.9 Political Challenger Game Tree Backward Induction.

We begin developing this more robust understanding with the preliminary definition of a *subgame*. A subgame $G(h)$ of the sequential game G, beginning at the non-terminal history h, consists of the players in the game, any terminal histories for which h is its initial part, and the player function and preferences inherited from the full game. The game G is a subgame of itself, and there is a proper subgame for each non-terminal history of the game.

Informally, a subgame perfect equilibrium is a strategy profile $s^* = (s_1^*, s_2^*, ..., s_n^*)$ such that in no subgame can any player i do better by choosing a strategy different from s_i^* given that all other players j are playing s_j^*. Formally, we have the following definition which uses the notation introduced in Chapter 3 for the utility obtained by replacing individual player's strategy's s_i within a strategy profile s. Specifically, in the definition, $O_h(r_i, s_{-i}^*)$ replaces player i's strategy with r_i and leaves all other player's following the strategies given in s^*.

Definition 5.2.1. The strategy profile s^* is a *subgame perfect equilibrium* if

$$u_i(O_h(s^*)) \geq u_i(O_h(r_i, s_{-i}^*))$$

for every strategy r_i of player i where u_i is player i's utility function and $O_h(s)$ is the terminal history consisting of non-terminal history h followed by the actions as directed by the strategy profile s.

The following theorem proves the existence of subgame perfect equilibria and provides a method for finding them.

Theorem 5.2.1 (Finding Subgame Perfect Equilibria). In a sequential game, subgame perfect equilibria are precisely the backward induction strategy profiles.

Proof. Consider any final decision node, that is, a parent whose children are all terminal nodes and a non-terminal history that ends at this node. The player assigned to this node will pick the action which yields the highest payoff; a best response by definition. Thus step one of the backward induction process results in Nash equilibria in all of the subgames generated by such non-terminal histories.

At a penultimate decision node, there are two decision makers, the penultimate one and the final one. We already know the best response of the final decision maker. The penultimate decision maker can select a best response from the inferred final decision. This is precisely what happens in the backward induction process. Thus backward induction yields a Nash equilibrium on each subgame generated by non-terminal histories ending in a penultimate decision node.

Continuing in this manner, we see that backward induction yields a Nash equilibrium on each subgame of the original game. □

We examine one more scenario to further illustrate the relationship between subgame perfect equilibria found via backward induction and the more general Nash equilibria found through an analysis of a strategic game. Here we have a specific question in mind when we build and analyze the model.

> **Voting Order.** Three local council members must vote on whether to give themselves a raise. A simple majority $(2 > 1)$ is required to pass the raise, which will apply to all three council members. The raise itself has value b to each of them, and there is a cost c of extra expenses in the next election campaign for each member who votes yes. We have $b > c$, otherwise all three members would vote no. It is reasonable to wonder if there is any value in being able to vote first [72].

The game tree representation of this scenario can be quickly constructed, with the assumption that council member 1 votes first, council member 2 votes second, and council member 3 votes last. It is illustrated in Figure 5.10.

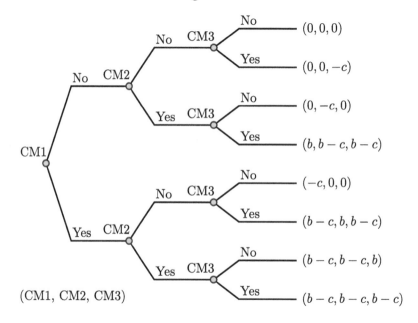

FIGURE 5.10 Voting Order Game Tree.

Performing backward induction in Figure 5.11, we obtain the subgame perfect equilibrium, that is, the strategy profile

$$(\text{No}, (\text{Yes}, \text{No}), (\text{No}, \text{Yes}, \text{Yes}, \text{No})).$$

Here, council member 1 votes No, council member 2 votes Yes at the upper node and votes No at the lower node, and council member 3 votes No at the top and bottom nodes and votes Yes at the middle two nodes. This results in the outcome where council member 1 votes No and the other two council members vote Yes, with the corresponding payoff vector $(b, b - c, b - c)$. Thus, it is best to be the first voter.

The same result can be discerned from the strategic game version of this scenario, but it takes a bit more work. First, constructing the strategic form of the game is nontrivial as council member 1 has two strategies available (No and Yes), council member 2 has four

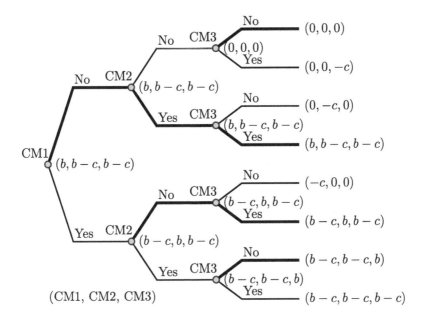

FIGURE 5.11   Voting Order Game Backward Induction.

strategies available (e.g., NY indicates No at the upper node and Yes at the lower node), and council member 3 has sixteen strategies available (e.g., YNNN indicates Yes at the top node and No at the three lower nodes). This results in a game with 128 outcome cells, as displayed in Table 5.6.

With a bit of work, we can establish that there are 36 Nash equilibria in pure strategies, as shown in boxes in Table 5.6.

The strategy profile (N, YN, NYYN) corresponds to the backward induction strategies and so is subgame perfect. Since there were no ties for best responses during the backward induction process, it must be that in 35 of these Nash equilibria, there are subgames in which the players' strategies are not in a Nash equilibrium. For example, consider the Nash equilibrium (Y, YY, NNNN). Council member 1 is not likely to vote Yes since her payoffs in the equilibria in which she votes Yes are smaller than those in which she votes No. Hence if council members 2 and 3 pre-commit to the strategies given, they are led to the payoffs $(0, -c, 0)$ which are suboptimal for all of the council members. None of the equilibria in which council member 1 votes Yes are subgame perfect. Most of those in which she votes No are not either. To demonstrate the latter point, consider the equilibrium (N,YY,YYYY). If the first council member were to vote No, the second council member would prefer a strategy in which he were voting No, selecting either NY or NN.

Sequential games can also be used to model scenarios in which the players have a continuum of strategy options available to them. To explore this situation, we return again to our Cournot Duopoly scenario, with a variation given in 1934 by Stackelberg [120]. Many game theory texts refer to this variation as a Stackelberg Duopoly. We name it Sequential Duopoly because it leads to a sequential game and use the method for depicting continuous choices in game trees found in [112].

TABLE 5.6 Voting Order Strategic Game

Boxed (best-response) payoffs are shown in **bold**.

1: N

(1, 2, 3)	2: NN	2: NY	2: YN	2: YY
3: NNNN	(**0**, **0**, **0**)	(0, **0**, **0**)	(**0**, −c, 0)	(0, −c, 0)
3: NNNY	(**0**, **0**, **0**)	(0, **0**, **0**)	(**0**, −c, 0)	(0, −c, 0)
3: NNYN	(0, **0**, **0**)	(0, **0**, **0**)	(0, −c, 0)	(0, −c, 0)
3: NYNN	(**0**, 0, **0**)	(0, 0, **0**)	(**b**, **b-c**, **b-c**)	(**b**, **b-c**, **b-c**)
3: YNNN	(**0**, **0**, −c)	(0, **0**, −c)	(**0**, −c, 0)	(0, −c, 0)
3: NNYY	(0, **0**, **0**)	(0, **0**, **0**)	(0, −c, 0)	(0, −c, 0)
3: NYNY	(**0**, 0, **0**)	(0, 0, **0**)	(**b**, **b-c**, **b-c**)	(**b**, **b-c**, **b-c**)
3: YNNY	(**0**, **0**, −c)	(0, **0**, −c)	(**0**, −c, 0)	(0, −c, 0)
3: NYYN	(0, 0, **0**)	(0, 0, **0**)	(**b**, **b-c**, **b-c**)	(**b**, **b-c**, **b-c**)
3: YNYN	(0, **0**, −c)	(0, **0**, −c)	(0, −c, 0)	(0, −c, 0)
3: YYNN	(**0**, 0, −c)	(0, 0, −c)	(**b**, **b-c**, **b-c**)	(**b**, **b-c**, **b-c**)
3: NYYY	(0, 0, **0**)	(0, 0, **0**)	(**b**, **b-c**, **b-c**)	(**b**, **b-c**, **b-c**)
3: YNYY	(0, **0**, −c)	(0, **0**, −c)	(0, −c, 0)	(0, −c, 0)
3: YYNY	(**0**, 0, −c)	(0, 0, −c)	(**b**, **b-c**, **b-c**)	(**b**, **b-c**, **b-c**)
3: YYYN	(0, 0, −c)	(0, 0, −c)	(**b**, **b-c**, **b-c**)	(**b**, **b-c**, **b-c**)
3: YYYY	(0, 0, −c)	(0, 0, −c)	(**b**, **b-c**, **b-c**)	(**b**, **b-c**, **b-c**)

1: Y

(1, 2, 3)	2: NN	2: NY	2: YN	2: YY
3: NNNN	(−c, 0, 0)	(**b-c**, **b-c**, **b**)	(−c, 0, 0)	(**b-c**, **b-c**, **b**)
3: NNNY	(−c, 0, 0)	(**b-c**, **b-c**, b − c)	(−c, 0, 0)	(**b-c**, **b-c**, b − c)
3: NNYN	(**b-c**, **b**, **b-c**)	(**b-c**, b − c, **b**)	(**b-c**, **b-c**, **b-c**)	(**b-c**, b − c, **b**)
3: NYNN	(−c, 0, 0)	(**b-c**, **b-c**, **b**)	(−c, 0, 0)	(b − c, **b-c**, **b**)
3: YNNN	(−c, 0, 0)	(**b-c**, **b-c**, **b**)	(−c, 0, 0)	(**b-c**, **b-c**, **b**)
3: NNYY	(**b-c**, **b**, **b-c**)	(**b-c**, b − c, b − c)	(**b-c**, **b-c**, **b-c**)	(**b-c**, b − c, b − c)
3: NYNY	(−c, 0, 0)	(**b-c**, **b-c**, b − c)	(−c, 0, 0)	(b − c, **b-c**, b − c)
3: YNNY	(−c, 0, 0)	(**b-c**, **b-c**, b − c)	(−c, 0, 0)	(**b-c**, **b-c**, b − c)
3: NYYN	(**b-c**, **b**, **b-c**)	(**b-c**, b − c, **b**)	(b − c, **b-c**, **b-c**)	(b − c, b − c, **b**)
3: YNYN	(**b-c**, **b**, **b-c**)	(**b-c**, b − c, **b**)	(**b-c**, **b-c**, **b-c**)	(**b-c**, b − c, **b**)
3: YYNN	(−c, 0, 0)	(**b-c**, **b-c**, **b**)	(b − c, **b-c**, **b-c**)	(b − c, **b-c**, **b**)
3: NYYY	(**b-c**, **b**, **b-c**)	(**b-c**, b − c, b − c)	(b − c, **b-c**, **b-c**)	(b − c, b − c, b − c)
3: YNYY	(**b-c**, **b**, **b-c**)	(**b-c**, b − c, b − c)	(**b-c**, **b-c**, **b-c**)	(**b-c**, b − c, b − c)
3: YYNY	(−c, 0, 0)	(**b-c**, **b-c**, b − c)	(−c, 0, 0)	(b − c, **b-c**, b − c)
3: YYYN	(**b-c**, **b**, **b-c**)	(**b-c**, b − c, **b**)	(b − c, **b-c**, **b-c**)	(b − c, b − c, **b**)
3: YYYY	(**b-c**, **b**, **b-c**)	(**b-c**, b − c, b − c)	(b − c, **b-c**, **b-c**)	(b − c, b − c, b − c)

> **Sequential Duopoly.** We consider the duopoly scenario in which Firm 1 moves first by selecting a production level and then Firm 2 moves second, with knowledge of what Firm 1's decision was.

Figure 5.12 illustrates the situation. The shading between the "0" and "∞" branches for each firm indicates that there are an infinite number of actions available to each player at each decision point: any nonnegative number indicating the production level of the corresponding firm. A generic choice for Firm 1 is labeled Q_1 and the resulting choice for Firm

2 is labeled Q_2. Notice that Firm 2 has a choice for each choice of Firm 1 and the resulting total production level is $Q = Q_1 + Q_2$.

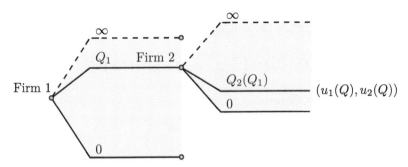

FIGURE 5.12 Sequential Duopoly Game Tree.

We do not need to reconstruct all of our analysis from Chapter 3, so we begin with Firm 2's best response function,

$$Q_2 = \frac{1}{2}(a - bc - Q_1),$$

where a and b are the positive parameters representing the amount that could be given away for free and the rate at which demand falls as price increases, respectively, and c is the cost per unit.

This function now represents Firm 2's strategy: take the production level that Firm 1 selects and use the best response function to compute Firm 2's production level. Because Firm 2 is the last firm to move, this strategy is precisely the one indicated by the backward induction method. To each of the infinitely many parents of terminal nodes, Firm 2 is assigning an optimal move.

Turning to Firm 1, its profit function is now

$$u_1(Q_1, Q_2) = \left(\frac{a}{b} - \frac{Q_1 + Q_2}{b} \right) Q_1 - cQ_1$$

$$= \left(\frac{a}{b} - \frac{Q_1 + \frac{1}{2}(a - bc - Q_1)}{b} \right) Q_1 - cQ_1$$

$$= \frac{1}{2b}(a - bc - Q_1)Q_1.$$

Since Firm 1 knows what action Firm 2 will be taking, its utility function can depend on just Q_1. Thus, Firm 1's best response is to maximize its profit by choosing the optimal value of Q_1. Taking the derivative of u_1 with respect to Q_1, we obtain

$$u_1'(Q_1) = \frac{1}{2b}(a - bc - 2Q_1).$$

Setting this equal to zero and solving for Q_1, we obtain

$$Q_1 = \frac{a - bc}{2}.$$

Substituting this result into Firm 2's best response function, we obtain its optimal action in response,

$$Q_2 = \frac{a - bc}{4}.$$

Note that the order pair

$$\left(\frac{a-bc}{2}, \frac{a-bc}{4}\right)$$

is an outcome of the game, while the subgame perfect equilibria that yields this outcome is

$$(Q_1, Q_2) = \left(\frac{a-bc}{2}, \frac{1}{2}(a-bc-Q_1)\right).$$

The resulting expected payoffs are

$$\left(\frac{(a-bc)^2}{8b}, \frac{(a-bc)^2}{16b}\right).$$

In comparison with their profit $(a-bc)^2/(9b)$ when quantities were chosen simultaneously, the leader (Firm 1) is obtaining a larger payoff, the follower (Firm 2) is obtaining a smaller payoff, and the joint profit is smaller.

5.3   COMBINATORIAL GAMES

We have seen by now that game theory is much more than the study of how to play everyday games like video games, Tic-Tac-Toe, or chess. However, the study of those games also falls within the realm of game theory. Further, since such games almost always require sequential moves by the players, we can use the methods of this chapter in their analysis. Unfortunately, many games, like Monopoly, poker, and soccer, involve chance and/or hidden information, so we exclude them from our analysis for now and focus on combinatorial games.

A *combinatorial game* is a two-player sequential game that has well-defined positions and alternating moves between players. Further, either (1) one player wins obtaining a payoff of 1 and the other player loses obtaining a payoff of -1, or (2) there is a tie with each player obtaining a payoff of 0. As usual, we assume that players act to obtain the highest payoff. (Notice that we exclude the possibility that a player may prefer to lose, such as when a parent is playing a game with their young child.)

Perhaps the most well-known combinatorial game is *Tic-Tac-Toe* which is played on a 3-by-3 grid of squares as shown in the leftmost drawing in Figure 5.13, where the outermost edges are not drawn. The players alternately mark an unmarked square with the first player

FIGURE 5.13   One Play of Tic-Tac-Toe.

using "X" and the second player using "O." The winner is the first player to obtain three of their marks in a row, column, or diagonal. Figure 5.13 shows one possible play of the game with the player using "X" winning with their fourth mark.

Versions of Tic-Tac-Toe have been played since antiquity although the name we use was not adopted until the twentieth century in the United States [122]. Tic-Tac-Toe is considered a children's game since it does not take an adult long to realize how to play the game to ensure at least a tie. We examine this game in order to illustrate the strengths and weaknesses of the techniques we have developed. By Zermelo's Theorem, there is a backward induction equilibrium. Since the only possible outcomes are a win for X (the player using "X"), a win for O (the player using "O"), or a tie, exactly one of the following must hold: (1) there is a winning strategy for X, (2) there is a winning strategy for O, or (3) there are strategies that ensure at least a tie for both X and O.

Included within the game theorist's notion of rationality is a player's ability to handle any size calculation. For sequential games a rational player must be able to construct the game tree and find its backward induction strategies. A brute force approach to this with Tic-Tac-Toe involves the nine initial branches for X, the eight subsequent branches for O, the seven subsequent branches for X, and so forth. This suggests that there could be a many as $9! = 362,880$ terminal histories. Of course, some plays of Tic-Tac-Toe will end before the ninth move, and we could use the symmetries of the board to exclude equivalent moves such as shown in Figure 5.14. Even here we have only fully considered the first moves of X

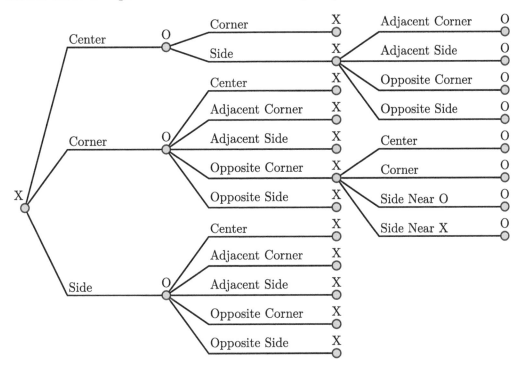

FIGURE 5.14 Start of a Tic-Tac-Toe Game Tree.

and O and a couple of the next possible moves for X. It looks to be nearly impossible to construct the game tree by hand.

Despite the difficulty of constructing and analyzing the game tree for Tic-Tac-Toe, most adults can fairly quickly come to conjecture that there are strategies that ensure at least a tie for both X and O. We can describe these conjectured strategies using the language and pictures of game trees. Figure 5.15 provides a way to visualize the following description of a strategy that ensures X will obtain at least a tie. Mark the center square initially (board 1A). If O's first mark is a corner square, reorient the board so that it is the top-left corner square and mark the top-right corner square (board 1B). If O's second mark is any square other than the bottom-left square, mark the bottom-left square to win (board 1C). If O's second mark is the bottom-left square, mark the middle-left square to block (board 2C). If O's third mark is any square other than the middle-right square, mark the middle-right square to win (board 2D). If O's third mark is the middle-right square, mark the bottom-center

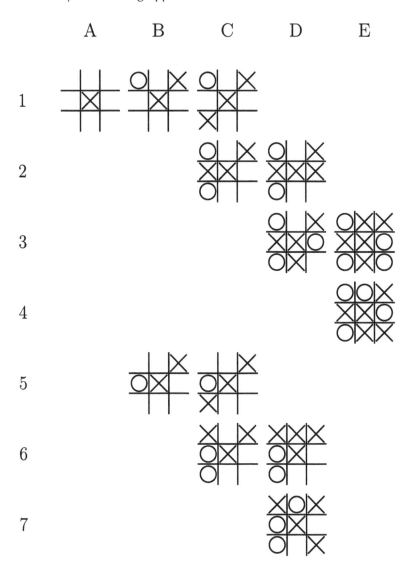

FIGURE 5.15 X Strategy Schematic.

(board 3D). If O's fourth mark is the bottom-right square, mark the top-center square to win (board 3E). If O's fourth mark is the top-center square, mark the bottom-right square to tie (board 4E).

If O's first mark is a side square, reorient the board so that it is the middle-left side square and mark the top-right corner square (board 5B). If O's second mark is any square other than the bottom-left square, mark the bottom-left square to win (board 5C). If O's second mark is the bottom-left square, mark the top-left square to block (board 6C). If O's third mark is any square other than the top-center square, mark the top-center square to win (board 6D). If O's third mark is the top-center square, mark the bottom-right square to win (board 7D).

A similar approach can be used to show that O has a strategy that ensures at least a tie, although it takes a lot more work.

Our application of Zermelo's Theorem to Tic-Tac-Toe applies to all combinatorial games. The definition of combinatorial games specifies that the only outcomes are win, lose, and tie. Therefore, in each combinatorial game, exactly one of the following exists: (1) a winning strategy for the first player, (2) a winning strategy for the second player, or (3) strategies for each player that ensures at least a tie. Despite the complexity of a combinatorial game, Zermelo's Theorem tells us that there is a well-defined and predetermined outcome when the game is played between two rational players. Finding the strategies that lead to this outcome, however, can be quite difficult.

Versions of our next game have also been played since antiquity. *Nim* is played on a board consisting of columns of circles. Players alternate marking one or more unmarked circles in a single column. The player to make the last mark is the winner. Figure 5.16

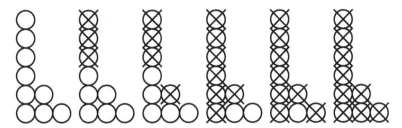

FIGURE 5.16   One Play of $[6, 2, 1]$-Nim.

shows an example of a Nim game consisting of columns of six, two, and one circles, denoted $[6, 2, 1]$-Nim in which Rose marks three of the circles in the first column, Colin then marks one of the circles in the second column, Rose then marks the remaining circles in the first column, Colin then marks the one circle in the third column, and Rose finally marks the remaining circle in the second column to win.

For $[6, 2, 1]$-Nim, the first player has nine essentially different initial moves: mark 1–6 of the circles in the first column, mark 1–2 circles in the second column, or mark the one circle in the third column. In general, the number of possible moves for a player is the number of unmarked circles remaining. Once again, the size of the game tree can be enormous, but Zermelo's Theorem once again tells us that for any particular game of Nim, either the first player has a winning strategy or the second player has a winning strategy.

While the construction and backward induction of game trees for Nim is prohibitive, it is possible to describe winning strategies. It is helpful to first consider some special cases. If there is only one column, then the first player has the winning strategy of marking all of the circles on their first turn.

If there are two columns of unequal size, the first player again has a winning strategy: mark enough circles in one column so that the number of unmarked circles in each column is the same. This works because when the other player takes their turn, they must make the columns have unequal numbers of unmarked circles, where upon the first player can again make the number of unmarked circles in each column the same. Thus, no matter what the other player does, the first player will always have a valid move and will therefore be the player to make the last mark. Of course, if the two columns are of equal size, then this strategy of evening up the number of unmarked circles can be adopted by the second player to ensure a win.

This "evening up" idea can be extended to Nim games with any number of columns if we first write each number of unmarked circles as a sum of powers of two. For example, in $[6, 2, 1]$-Nim, we can write $6 = 4 + 2$ and lay out the sums as in Table 5.7. Looking at the

TABLE 5.7   Nim Sums for
$[6, 2, 1]$-Nim

4	0	0
2	2	0
0	0	1

rows of the table, we can see that there is an odd number of 4s, an even number of 2s, and an odd number of 1s. Our goal will be be make a move that results in even numbers for each power of two. We can even up the 4s by removing 4 from the first column and even up the 1s by adding 1 into the first column, a net change of a removal of 3 from the first column, that is, Rose should mark three circles in the first column. This results in Table 5.8. Now if

TABLE 5.8   Nim Sums for
$[3, 2, 1]$-Nim

2	2	0
1	0	1

Colin marks one circle in the first column, the rows of 1s will then have only one positive number in it. If Colin marks two circles in the first column, the rows of 2s will then have only one positive number in it. If Colin marks three circles in the first column, the rows of 2s and 1s will then each have only one positive number in them. Similar statements can be said about any other legal move for Colin. Thus, when it is Rose's turn, she will again be faced with a nim-sum table with some rows with an odd number of positive numbers, and she can again find a legal move to even up the rows again.

In general, we will call a row in a nim-sum table odd [resp., even] if it has an odd [resp., even] number of positive numbers in it. If there is at least one odd row in the nim-sum table, then the player whose turn it is can make all of the rows even by starting with the highest odd row, choosing a column in that row having a positive number, removing that number from that column, and for each remaining odd row either removing or adding back the corresponding number depending upon whether that number is already present or not. For example, in $[26, 19, 11, 9, 6]$-Nim, the nim-sum table is shown in Table 5.9.

TABLE 5.9   Nim Sums for
$[26, 19, 11, 9, 6]$-Nim

16	16	0	0	0
8	0	8	8	0
0	0	0	0	4
2	2	2	0	2
0	1	1	1	0

The rows containing 16s and 2s are even while the rows containing 8s, 4s, and 1s are odd. There are three columns (first, third, and fourth) containing a positive number in the 8s row. If we choose the third or fourth column, then 8 should be removed, 4 should added back, and 1 should be removed, resulting in a net removal of 5. If we choose the first column, then 8 should be removed, 4 should added back, and 1 should be added back, resulting in a net removal of 3.

If every row of the nim-sum table is even, then any legal move will change at least one of the positive numbers in the corresponding column to zero, and so the corresponding row will now be odd. Thus, a player faced with a nim-sum table containing at least one odd row can ensure a win by making a move that makes all of the new nim-sum table rows even, forcing the other player to create a new nim-sum table with at least one odd row.

We now turn our attention to a board game designed by David Gale in 1961 and first marketed by Hasbro. *Bridg-it* is played on a game board consisting of an $n \times (n + 1)$ rectangular grid of white dots interlaced with an $(n + 1) \times n$ rectangular grid of black dots, as seen in Figure 5.17. Rose's goal is to connect white dots to form a bridge from the left

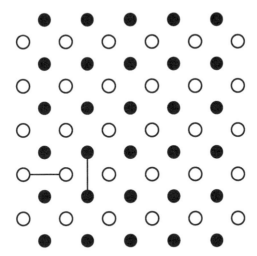

FIGURE 5.17 Bridg-It Game Board with $n = 5$ after Two Turns.

to the right of the grid. Colin's goal is to connect black dots to form a bridge from the top to the bottom of the grid. On a given player's move, they place a "bridge" between two adjacent dots of their color. Bridges are not allowed to intersect.

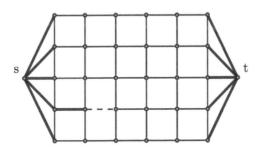

FIGURE 5.18 Bridg-It with $n = 5$ Shannon Graph after Two Turns.

Figure 5.18, called a Shannon graph, provides a different but equivalent way of describing Bridg-It. The white circles in the Bridg-It board correspond to the unlabeled circles in the Shannon graph. White placing a bridge between two adjacent white dots in the Bridg-It board corresponds to White claiming one of the edges (by bolding) in the Shannon graph. Black placing a bridge between two adjacent black dots in the Bridg-It board blocking White from making a connection between two white dots is equivalent to Black removing the corresponding edge in the Shannon graph. White wins once there is a path of claimed edges between the left and right sides of the Shannon graph. Black wins once it is impossible for White to win. By adding the labeled circles s and t along with the already claimed edges connecting s to the left side of the Shannon graph and connecting t to the right side of the Shannon graph, Bridg-It is equivalent to White trying to form a path from s to t with claimed edges and Black trying to make that impossible by removing edges.

This equivalent representation suggests that it may be interesting to generalize the game by maintaining the same rules but varying the graph on which the game is played. A *Shannon switching game* [96] is a two-player combinatorial game played on a graph with two identified non-adjacent vertices, s and t and potentially some already claimed edges. For convenience, we name the players Rose and Colin. Rose's objective is to construct a path of claimed edges connecting s and t. Colin's objective is to remove unclaimed edges so as to prohibit Rose from achieving her objective. One of the players is designated to move first, and then they alternate turns claiming and removing edges until either Rose reaches her objective or is prohibited from doing so. In the former case, Rose wins and in the latter, Colin wins.

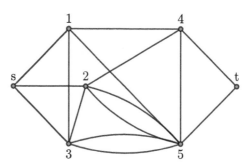

FIGURE 5.19 Example of a Graph $G = (V, E)$.

To further analyze Shannon switching games, we will first formalize some concepts already introduced informally. A *graph* $G = (V, E)$ consists of a set V of vertices, together with a multiset E containing pairs of vertices. Pictorially, vertices are represented by (labeled) circles and edges are represented by lines between two circles. Figure 5.19 is a pictorial representation of the graph with

$$V = \{s, 1, 2, 3, 4, 5, t\} \text{ and}$$
$$E = \{\{s, 1\}, \{s, 2\}, \{s, 3\}, \{1, 3\}, \{1, 4\}, \{1, 5\}, \{2, 3\}, \{2, 4\},$$
$$\{2, 5\}, \{2, 5\}, \{3, 5\}, \{3, 5\}, \{4, 5\}, \{4, t\}, \{5, t\}\}.$$

Corresponding to Figure 5.19, the edge $\{2, 5\}$ and the edge $\{3, 5\}$ each appear twice in the multiset E.

Game trees are also examples of graphs having additional labeling and an implied directionality. Two vertices are *adjacent* if they share an edge. An edge and a vertex are *incident* to each other if the vertex is part of the edge. A *path* connecting vertices x and y in a graph is a alternating sequence of incident vertices and edges beginning at x and ending at y. One path in the Figure 5.19 graph is

$$1, \{1, 3\}, 3, \{3, 2\}, 2, \{2, s\}, s, \{s, 1\}, 1, \{1, 4\}, 4.$$

Note that $\{2, 3\} = \{3, 2\}$, and we have used the later representation in the illustration of a path to make the correspondence to the intuitive movement more easily discerned. A *cycle* is a path with at least one edge but not traversing an edge more than once that connects a vertex to itself. One cycle in the Figure 5.19 graph is

$$4, \{4, t\}, t, \{t, 5\}, 5, \{5, 4\}, 4.$$

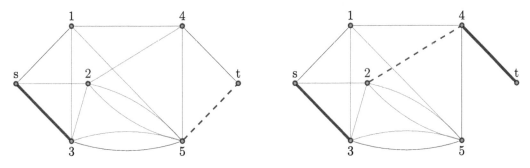

FIGURE 5.20 After Two and Four Turns in a Shannon Switching Game.

Suppose Rose starts the Shannon switching game of the graph in Figure 5.19 by claiming $\{s, 3\}$ followed by Colin removing $\{5, t\}$. In the left graph of Figure 5.20, Rose's claim is indicated by thickening the edge, and Colin's removal is indicated by dashing the edge. Perhaps Rose next chooses to claim $\{4, t\}$ and Colin then removes $\{2, 4\}$. The right graph in Figure 5.20 shows the state of the game after these turns. Notice in this graph the edge $\{5, t\}$ is not included in the graph since it was removed during the prior turn.

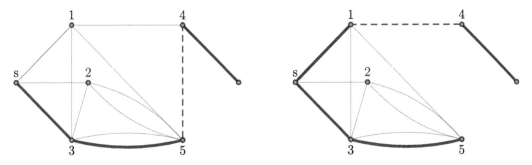

FIGURE 5.21 After Six and Eight Turns in a Shannon Switching Game.

The left graph of Figure 5.21 shows the state of the game after Rose claims one of the $\{3, 5\}$ edges and Colin removes $\{4, 5\}$. The right graph of Figure 5.21 shows the state of the game after Rose claims $\{s, 1\}$ and Colin removes $\{1, 4\}$. At this point, Colin has won because it would be impossible for Rose to build a path from s to t with the remaining edges. Notice that had Rose claimed $\{1, 4\}$ instead of $\{s, 1\}$, Rose could have then won on her next move by claiming either $\{s, 1\}$ or $\{1, 3\}$. We claim that Rose can ensure a win even if Colin starts the game.

We begin our analysis of Shannon switching games by introducing several additional graph theoretic notions. A *subgraph* $G' = (V', E')$ of a graph $G = (V, E)$ satisfies $V' \subseteq V$, $E' \subseteq E$, and $v \in V'$ whenever $e \in E'$ and v is incident to e. A *spanning tree* of a graph is a subgraph containing all of the vertices of the graph and contains a unique path connecting any two of those vertices. Figure 5.22 illustrates two spanning graphs of the graph G given in Figure 5.19. If a graph has a sufficient number of edges, it is possible that it has two edge-disjoint spanning trees, in which case we say that the graph is *positive*. Two trees with the same set of vertices are said to be *coextensive*. For example, $T_1^{(0)}$ and $T_2^{(0)}$ are edge-disjoint coextensive spanning trees, showing that G is positive.

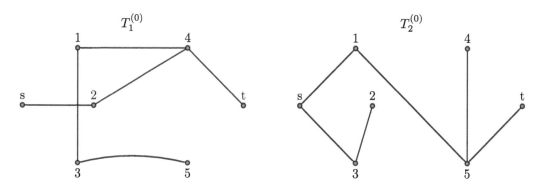

$T_1^{(0)}$ $T_2^{(0)}$

FIGURE 5.22 Edge-Disjoint Spanning Trees.

Theorem 5.3.1. For Shannon switching games with no edges already claimed at the beginning of the game, Rose has a winning strategy whether playing first or second if and only if the graph is positive.

There are many papers providing proofs of this theorem, including Lehman [63], who first published a very general solution, and Brualdi [19] who proved only the "if" direction in a more specific solution. All of them involve a significant amount of technical details that will distract us from our goal of understanding strategies, so we opt to present the strategy solution via an example on the graph G of Figure 5.19 that uses the methods and notation presented in [19].

We should note that if Rose has a winning strategy when playing second, then she has a winning strategy when playing first by playing an initial random move and then following the strategy as a second mover (ignoring the edge she has initially captured). If the strategy would require her to select that initial move at some point, she should make a new random move. Therefore, in what follows, Colin moves first; Rose has identified the two edge-disjoint spanning trees $T_1^{(0)}$ and $T_2^{(0)}$.

Rose's plan is to claim edges in one or the other tree in sequence to construct new, increasingly overlapping spanning trees that connect s and t.

We denote the trees with the notation $T^{(i)}$ because they guide Rose's actions at her turn i. We will adjust the tree diagrams to show the choices she's making in accordance with her strategy. Part of her strategy is to only choose edges from the identified trees, and we'll mark those claimed edges indicating she has claimed it in the game. As needed, we'll add edges claimed in one tree to the other. When Rose no longer considers an edge as a choice, she will remove it from the tree even though that edge may remain in the game graph G for Colin to remove.

Suppose Colin begins by removing the edge $\{3,5\}$, which is in neither of the trees identified by Rose. Rose now claims any edge in either of the trees, say $\{s,3\}$, and then revises the pair of trees as shown in Figure 5.23. Since $\{s,3\} \in T_2^{(0)}$, there is no revision to this tree other than to mark the claimed edge, that is, $T_2^{(1)} = T_2^{(0)}$. The claimed edge is added to $T_1^{(0)}$ creating the cycle

$$s, \{s,2\}, 2, \{2,4\}, 4, \{4,1\}, 1, \{1,3\}, 3, \{3,s\}, s,$$

and an unclaimed edge from this cycle, say $\{s,2\}$ is removed from $T_1^{(0)}$ to obtain $T_1^{(1)}$. In revising this tree, we have used theorems that state that adding an edge to a spanning tree will always create a cycle and that removing any edge of that cycle will yield another spanning tree. The two spanning trees $T_1^{(1)}$ and $T_2^{(1)}$ have exactly one edge in common, the claimed edge.

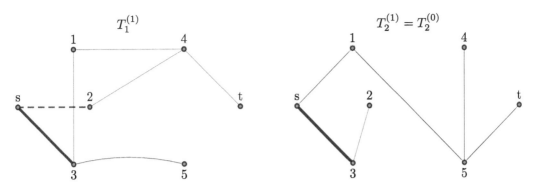

FIGURE 5.23 Trees after Colin Removes $\{3,5\}$ and Rose Claims $\{s,3\}$.

Colin now might remove edge $\{1,5\}$. This edge is in the spanning tree $T_2^{(1)}$ and its removal partitions the vertices into those with a path to vertex 1, $A = \{s,1,2,3\}$, and those with a path to vertex 5, $B = \{4,5,t\}$, and there is no edge in $T_2^{(1)}$ that is incident to one vertex in A and one vertex in B other than $\{1,5\}$ (this is a general property of spanning trees that we do not prove here). Since Rose had not previously claimed edge $\{1,5\}$, it must not be in the other spanning tree $T_1^{(1)}$, but since $T_1^{(1)}$ is a spanning tree, there must be a unique path in $T_1^{(1)}$ from vertex 1 to vertex 5,

$$1, \{1,3\}, 3, \{3,5\}.$$

Since this path goes from 1 to 5, there must be an edge in it that is incident to a vertex in A and a vertex in B: here $\{3,5\}$. As noted previously, such an edge cannot be in $T_2^{(1)}$, and adding it to $T_2^{(1)}$ after removing $\{1,5\}$ will result in a new spanning tree (another general property of spanning trees that we do not prove here). Rose must claim one of these edges, which in this case means she must claim $\{3,5\}$. Now $T_2^{(1)}$ is revised by removing the edge $\{1,5\}$ removed by Colin and adding the edge $\{3,5\}$ claimed by Rose resulting in $T_2^{(2)}$ while the other tree is left unchanged, except to mark the claimed edge, that is, $T_1^{(2)} = T_1^{(1)}$ illustrated in Figure 5.24. The two spanning trees have exactly two edges in common, the claimed edges.

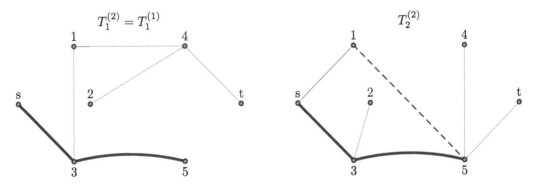

FIGURE 5.24 Trees after Colin Removes $\{1,5\}$ and Rose Claims $\{3,5\}$.

Next suppose Colin removes edge $\{1,3\}$. Since $\{1,3\} \in T_1^{(2)}$, Rose notes that the edge removal partitions the vertices into the sets $A = \{1,2,4,t\}$ and $B = \{s,3,5\}$, finds the unique path $1, \{1,s\}, s, \{s,3\}, 3$ in $T_2^{(2)}$ from vertex 1 to vertex 3, and claims an edge $\{s,1\}$ in that path with a vertex in A and a vertex in B. Rose updates the trees to obtain $T_1^{(3)}$ and $T_2^{(3)}$ illustrated in Figure 5.25. The two spanning trees have exactly three edges in common, the claimed edges.

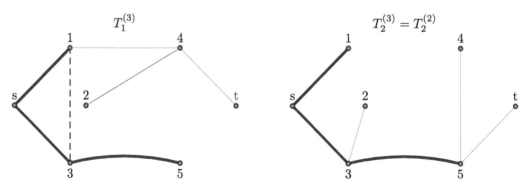

FIGURE 5.25 Trees after Colin Removes $\{1,3\}$ and Rose Claims $\{s,1\}$.

Now Colin might remove edge $\{5,t\}$, to which Rose responds by claiming edge $\{4,t\}$ to obtain $T_1^{(4)}$ and $T_2^{(4)}$ illustrated in Figure 5.26. The two spanning trees have exactly four edges in common, the claimed edges.

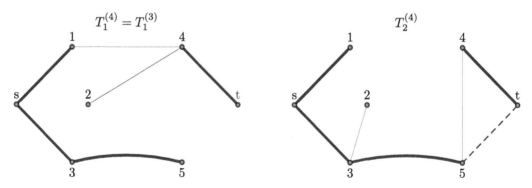

FIGURE 5.26 Trees after Colin Removes $\{5,t\}$ and Rose Claims $\{4,t\}$.

Finally, Colin attempts a block by removing edge $\{1,4\}$, to which Rose responds by claiming edge $\{4,5\}$ to obtain $T_1^{(5)}$ and $T_2^{(5)}$ illustrated in Figure 5.27. Rose has created a path from s to t and won the game.

This process must end in at most $n-1$ steps as that is the number of edges in a spanning tree on n vertices, and therefore the method describes a winning strategy for Rose playing second.

For some insight into how the converse is proved, consider the following. Given a sequence of winning moves by Rose, let Colin block the last move. Because a winning strategy exists for Rose, there is an alternate move available to Rose; insert that move. Working backwards in this fashion, one can construct a winning sequence of moves disjoint from the original sequence. Thus there are two disjoint spanning trees within the graph. For a full development of a more specific solution than Lehman, see [66].

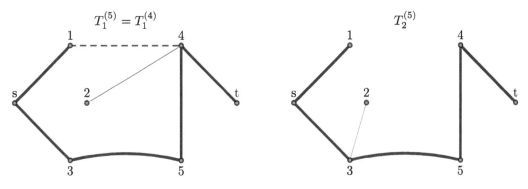

FIGURE 5.27 Trees after Colin Removes $\{1,4\}$ and Rose Claims $\{4,5\}$.

Since Rose wins games that are positive, it is reasonable to describe the games in which Colin wins when playing first or second as *negative*. Other games, described as *neutral*, are neither positive nor negative, and the first player to move will have a winning strategy. The following theorem categorizes these latter two types of games.

Theorem 5.3.2. Let G be a graph with distinct, non-adjacent vertices s and t. Let $\overline{G}$ be the graph obtained by adding the edge $\overline{e} = \{s, t\}$. Then

1. The Shannon switching game played on G is a neutral game if and only it is not a positive game but the game played on $\overline{G}$ is.

2. The Shannon switching game played on G is a negative game if and only if neither of the games played on G and $\overline{G}$ are positive.

We can apply the above theorems to Bridg-It. We first show that the Shannon graph for Bridg-It, illustrated in Figure 5.18, is not positive. The graph has $n(n+1)+2 = n^2+n+2$ vertices. Since the $2n$ edges incident to s or t are already claimed, the leftmost and rightmost vertical edges cannot be used to construct a spanning tree containing the already claimed edges. So, there are n^2 horizontal $+ (n-1)^2$ vertical $+ 2n$ incident to s or $t = 2n^2+1$ usable edges. To create two trees with nonintersecting edges except for the already claimed ones, we would need to use $2(n^2+n+2-1)-2n = 2n^2+2$ edges. Since there are only $2n^2+1$ usable edges, the Shannon graph for any Bridg-It game is not positive.

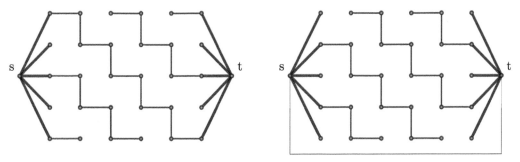

FIGURE 5.28 Bridg-It with $n = 5$ Shannon Graph is Positive.

The two disjoint spanning trees shown in Figure 5.28 show that adding the edge $\{s, t\}$ to the Shannon graph creates a positive graph. By the above theorem, the Shannon graph is neutral and the first player in Bridg-It has a winning strategy. The first player's first move is to claim one of the unclaimed edges in the left spanning tree. This will add that edge to the claimed edges in the right spanning tree, allowing edge $\{s, t\}$ to be removed from the right spanning tree. The first player can now follow the winning strategy outlined earlier.

Hex, invented independently by Piet Hein and John Nash in the 1940s, was marketed by the Parker Brothers company as a board game in the 1950s [123]. It is played on an $n \times n$ rhombus-shaped board consisting of hexagons, as seen in Figure 5.29. Each of the players

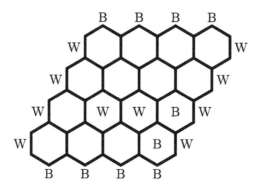

FIGURE 5.29   4-by-4 Hex Game Board after Two Turns.

is attempting to construct a path of adjacent hexagons from one side of the board to the opposite side. Rose plays white pieces and is attempting to connect the left and right edges, while Colin is attempting to construct a path from the top to the bottom sides consisting of black pieces.

We have found strategies in Tic-Tac-Toe that ensure each player at least a tie, in Nim that ensures a win depending on whether the initial nim-sum table contains odd rows, and in Shannon switching games depending on whether certain graphic theoretic properties are satisfied. Since there is always a winner in a game of Hex, Zermelo's Theorem tells us that either Rose or Colin has a winning strategy for each size of the game board. Nash was able to prove which player had the winning strategy without providing the strategy.

Theorem 5.3.3 (Nash's Hex). In Hex, Rose has a strategy to ensure a win when she moves first.

> *Proof.* First observe that one of the players always wins in Hex. To see this, imagine the playing board to be made out of paper. Whenever Rose moves, she claims the space of her choice. Whenever Colin moves, he cuts out the space of his choice. Suppose we repeat this until no one can move any more. We pick up the playing board in our hands, holding the left and right edges, and then pull our hands apart. Either the paper stops us, in which case there must be a path from left to right and so Rose wins; or nothing stops us, in which case there are cut-out spaces that form a path between the top and the bottom of the board, and so Colin wins. Clearly, one of the two must occur.
>
> By Zermelo's Theorem, either Rose or Colin has a strategy to ensure a win. Assume, by way of contradiction, that it is Colin that has a winning strategy. Rose can employ the following strategy: make a random initial move and then follow Colin's winning strategy on left/right spaces in subsequent moves. If this requires selecting the space chosen in the initial random move, make another random move. With this strategy, Rose has adopted Colin's second move winning strategy for herself and therefore has

a winning strategy. We now have a contradiction since both players cannot have a winning strategy, and therefore Rose, and not Colin, must have the winning strategy.

□

In summary, combinatorial games can be modeled as sequential games, and so by Zermelo's Theorem, one of the two players has a winning strategy or both players have strategies ensuring at least a tie. Additional mathematics, including graph theory and number theory, are often required to determine which player has the winning strategy and to find what that strategy is. Winning strategies have been determined for many, but not all, such games. For Tic-Tac-Toe, we found strategies to ensure at least a tie for both players. For Nim and Bridg-It, we determined the winning strategies. Finally, for Hex, we proved a winning strategy exists for the first player without finding it. Checkers is perhaps the most complicated combinatorial game for which optimal strategies have been obtained, but there are games, such as Chess, for which it is still unknown which player has the winning strategy. Combinatorial game theory is a very broad and deeper studies can be found in [11], and [99], [18].

5.4 MORAL HAZARD

When we think of risk-aversion and risk-neutrality, we often think about insurance. People who are monetarily risk-adverse often purchase insurance policies to protect themselves or their properties against large losses. Even though the expected value of paying $100 is the same as having a 1% chance of paying $10,000, for many, paying a guaranteed monthly premium is preferable to having even a small chance of paying a much larger sum. We then wonder if having security against having to pay a larger sum changes people's behavior in anyway. Insured drivers may not drive as cautiously or may leave their cars unlocked knowing they won't be responsible for large damages or losses. Insurance companies must take this into account when calculating premiums or must find ways to reduce riskier behavior amongst their insured clients.

The change in behavior produced by passing some of the risk or responsibility from an individual to another player is known in game theory as *moral hazard*. The individual is called an agent and the other party assuming some of the risk is known as the principal. Insurance is one example with the insurance company acting as the principal for clients who are agents. Another common example studied is that of managers attempting to influence their employees behavior through incentive programs related to sales or profits. In this section, we explore the circumstances under which scholarships, particularly those given out by federal governments, can result in moral hazard.

Free College. In the primaries leading up to the 2016 United States presidential election, the topic of "free college" became a widely discussed issue. The topic has many political, economic, and social implications, but we are interested in answering one question: does free college encourage or discourage students in putting forth their best efforts? [60], [54], [16], and [31]

A key characteristic of a situation that involves moral hazard is that the principal cannot influence the agent's action in any binding way. In many situations the principal is able to observe only the final results. We can see, then, how a large federal government handing out scholarships to individual students spread out across a country could be in this situation.

We are exploring then, whether the government, in the role of the principal, may enable the student, in the role of the agent, to engage in activities that do not support the intentions of the government. Specifically, rather than working hard to succeed, a student may choose to put minimal effort into their studies and have a higher risk of failure since there is no or little cost to them for the classes.

Rather than consider the entire college career of a student, we build a simpler model that considers a single course and winnow the levels of accomplishment to two: success (S) or failure (F). The cost of the course (tuition, books, and supplies) is a fixed amount, K. The government (G) reimburses the student (A) for part of the cost of the course (an amount less than K), for the entire cost of the course (the amount K), or even more (an amount greater than K as in a fellowship), and may also make this amount depend on whether the student succeeds or fails. Let r_S and r_F be the amount that the government will reimburse in the two cases, respectively. Completely free college occurs when $r_S = r_F = K$.

The student can choose how much effort they are willing to put into succeeding in this course and, again for simplicity's sake, we limit their options to high effort (H), low effort (L), or no effort (N) by not taking the course. If the student has chosen to take the course, chance (C) then determines success or failure with the probability of success beings p_H or p_L depending on whether the student chose to put in high or low effort. While these probabilities will vary by the individual skills and talents of a student, we can assume that students can pass and that hard work increases their chances of getting a passing grade ($0 < p_L < p_H$). Figure 5.30 summarizes these rules of play.

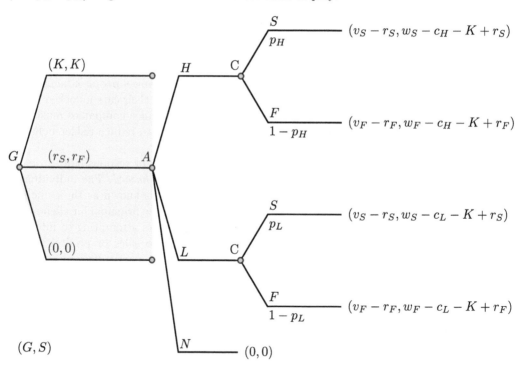

FIGURE 5.30 Free College Game Tree.

Finally, we need to establish the payoffs to the government and student for each of the five possible outcomes. In the insurance company-driver and business owner-manager illustrations, the benefits and costs are clearly given in monetary terms. Here the government has a social value associated with student success and students have a personal, not necessarily strictly monetary, value associated with course success. However, we must quantify these values in some manner in order to create a game theoretic model. We assume they can find monetary equivalents for these benefits.

We will arbitrarily set their payoffs to zero when the student chooses to not take the course and scale the remaining payoffs in units of money. We will assume that both players are monetarily risk neutral and different aspects of their payoffs are separable and additive. For the government, the monetary worth for student success and failure is v_S and v_F, and as previously described, their cost for student success and failure is r_S and r_F. For the student, the monetary worth for success and failure is w_S and w_F, the monetary cost for high and low effort is c_H and c_L, the cost for taking the course is K, and the reimbursement received from the government for success and failure is r_S and r_F. Thus, if the government chooses reimbursement amounts r_S and r_F for success and failure, the student chooses a high level of effort, and the student is successful, then the government's payoff is $v_S - r_S$ and the student's payoff is $w_S - c_H - K + r_S$. Similar calculations yield the payoffs for each outcome as shown in Figure 5.30.

We make the reasonable assumptions that the government does not view the student taking the course as detrimental but prefers success ($0 \leq v_F < v_S$), there may be intrinsic worth to the student to take and pass the course ($0 \leq w_F \leq w_S$), and effort is costly for the student ($0 < c_L < c_H$). Recall that we have also assumed that students can pass and that hard work increases their chances of getting a passing grade ($0 < p_L < p_H$).

Given the government's action (r_S, r_F) and the student's action H, L, or N, the government's expected payoff is

$$u_G((r_S, r_F), H) = p_H(v_S - r_S) + (1 - p_H)(v_F - r_F)$$
$$u_G((r_S, r_F), L) = p_L(v_S - r_S) + (1 - p_L)(v_F - r_F)$$
$$u_G((r_S, r_F), N) = 0.$$

Without its assistance ($r_H = r_L = 0$), government most prefers the student to take the course and put in a high level of effort, because our previous parameter assumptions imply $p_H(v_S - v_F) + v_F > p_L(v_S - v_F) + v_F > 0$, which with the additional $r_H = r_L = 0$ assumption, implies $u_G((0,0), H) > u_G((0,0), L) > u_G((0,0), N)$. This means that the government can contemplate giving some amount of assistance to encourage the student to take the course and put in a high level of effort.

Given the government's action (r_S, r_F) and the student's action H, L, or N, the student's expected payoff is

$$u_A((r_S, r_F), H) = p_H(w_S - c_H - K + r_S) + (1 - p_H)(w_F - c_H - K + r_F)$$
$$= (p_H w_S + (1 - p_H)w_F - c_H - K) + p_H r_S + (1 - p_H)r_F$$
$$= h + p_H r_S + (1 - p_H)r_F$$
$$u_A((r_S, r_F), L) = p_L(w_S - c_L - K + r_S) + (1 - p_L)(w_F - c_L - K + r_F)$$
$$= (p_L w_S + (1 - p_L)w_F - c_L - K) + p_L r_S + (1 - p_L)r_F$$
$$= l + p_L r_S + (1 - p_L)r_F$$
$$u_A((r_S, r_F), N) = 0$$

where $h = p_H w_S + (1 - p_H)w_F - c_H - K$ and $l = p_L w_S + (1 - p_L)w_F - c_L - K$ are the student's intrinsic (without government reimbursement) net benefit for high and low effort. Observe that

$$h \geq l \text{ if and only if } (p_H - p_L)(w_S - w_F) \geq c_H - c_L.$$

While our parameter assumptions tell us that $p_H - p_L > 0$, $w_S - w_F \geq 0$, and $c_H - c_L > 0$, they do not tell us which of h and l is larger. If there is no intrinsic worth to the student passing instead of failing the course ($w_S = w_F$) and no government assistance ($r_H = r_L = 0$), then $h < l$ and the student has no incentive to put in a high amount of effort (i.e., $u_A((0,0), H) < u_A((0,0), L)$). On the other hand, if passing the course is sufficiently valuable and effort is sufficiently effective in comparison with the cost of effort (i.e., $h > 0$ and $h > l$), then the student will take the course and put in a high level of effort even without government assistance.

We now want to determine the government's best course of action. In a subgame perfect equilibrium, the student's strategy a chooses for each government action (r_S, r_F) the action $a(r_S, r_F)$ among H, L, and N that maximizes $u_A((r_S, r_F), a(r_S, r_F))$, and the government's strategy chooses the action (r_S, r_F) that maximizes $u_G((r_S, r_F), a(r_S, r_F))$. Since $u_G((r_S, r_F), H)$ and $u_G((r_S, r_F), L)$ are strictly decreasing functions of r_S and r_F, and $u_G((r_S, r_F), N)$ is a constant, the only thing that prevents the government from reducing r_S and r_F to 0 is that the student's action choice depends on r_S and r_F. Our approach then to finding the subgame perfect equilibrium is to solve (using computer software such as Mathematica or WolframAlpha.com) the linear programs

$$\begin{aligned} \max \quad & u_G((r_S, r_F), H) \\ \text{s.t.} \quad & u_A((r_S, r_F), H) \geq u_A((r_S, r_F), L) \\ & u_A((r_S, r_F), H) \geq 0 \\ & r_H \geq 0, r_L \geq 0, \end{aligned}$$

and

$$\begin{aligned} \max \quad & u_G((r_S, r_F), L) \\ \text{s.t.} \quad & u_A((r_S, r_F), L) \geq u_A((r_S, r_F), H) \\ & u_A((r_S, r_F), L) \geq 0 \\ & r_H \geq 0, r_L \geq 0. \end{aligned}$$

If the value of the first linear program is larger than the value of the second linear program and the value of the student not taking the course (0), then the subgame perfect equilibrium will have the government announce the reimbursement policy found to be optimal in the first linear program and the student will take the course and put in a high level of effort. Similarly, if the value of the second linear program is larger than the value of the first linear program and the value of the student not taking the course (0), then the subgame perfect equilibrium will have the government announce the reimbursement policy found to be optimal in the second linear program and the student will take the course and put in a low level of effort. Finally, if the value of both linear programs are non-positive, then the government will announce $(0,0)$ as its reimbursement policy and the student will choose to not take the course.

TABLE 5.10   Free College Equilibria

ID	v_S	v_F	w_S	w_F	p_H	p_L	c_H	c_L	K	h	l	r_S^*	r_F^*	e^*	u_G^*	u_A^*
1	any	0	8	0	0.75	0.50	1	0	2	3	2	0	0	H	$0.75v_S$	3
2	16	0	4	0	0.75	0.50	2	1	2	-1	-1	1	1	H	11	0
3	20	0	4	0	0.75	0.50	8	4	2	-7	-4	12	0	H	6	2
4	16	0	4	0	0.75	0.50	8	4	2	-7	-4	8	0	L	4	0
5	6	0	4	0	0.75	0.50	8	4	2	-7	-4	0	0	N	0	0

Table 5.10 provides the results for five different sets of parameter values. Only the values of success (v_S and w_S) and the effort costs (c_H and c_L) differ. When the value of success for the student is high relative to the effort costs, as in row one, the student finds that it is valuable to put in a high amount of effort (e^*) without government assistance (r_S^* and r_F^*).

When the value of success for the student (w_S) is low relative to the effort costs, as in the remaining remaining rows, the student is unwilling to take the course unless there is government assistance. If the government's value for success (v_S) is sufficiently high, it is optimal for the government to provide assistance, as for rows two through four. Sometimes the assistance can be unconditional (as in the second row); at other times assistance conditioned on whether the student's success is enough to motivate him or her to put in a high level of effort (as in the third row); and sometimes conditional assistance is only enough to motivate the student to take the course but put in only a low amount of effort (as in the fourth row). Finally, if the government's value for success is sufficiently small (as in the fifth row), government does not have an incentive to provide assistance.

The above information does not tell us whether moral hazard exists in the free college scenario. For the third set of parameters, if the government were to reimburse the direct monetary cost of the course regardless of whether the student succeeds or fails (i.e., $r_S = r_F = K = 2$), then the student would choose to not take the course. If the government were to reimburse regardless of whether the student succeeds or fails (i.e., $r_S = r_F$) the resulting equilibrium results in a reimbursement of 4 and the student taking the course with a low level of effort. Thus, there is moral hazard in this circumstance if the government chooses to ignore whether the student has succeeded or failed. Of course, this level of moral hazard is analogous to stating that there is moral hazard with vehicle insurance if premiums were the same for every driver, regardless of their driving history or other factors related to their likely level of vigilance. The government does not need to ignore the easily observable success or failure of the student. We now ask whether a reimbursement based on the observable outcome of the course could remove all moral hazard.

To answer this question, we must consider what would happen if the government could directly observe the student's effort level and choose reimbursements based on this

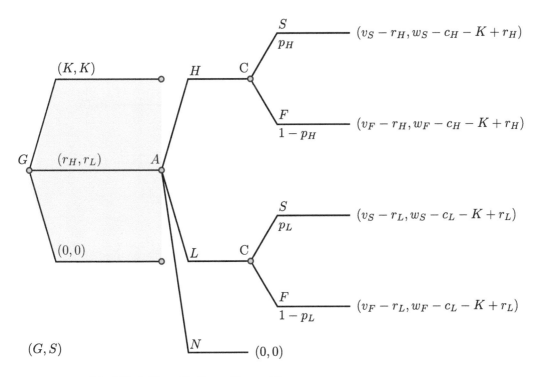

FIGURE 5.31 Modified Free College Game Tree.

information. In fact, the U.S. federal government requires reporting on the last date of attendance by any student receiving federal financial aid; this is its proxy for student effort.

Figure 5.31 summarizes the modified rules of play (reimbursements are based on the student's level of effort rather than the student's success or failure) and the resulting expected payoffs. For the fourth set of parameters in Table 5.10, the equilibrium results in the government choosing $(r_H, r_L) = (7, 0)$, the student choosing the action H, and the expected payoff to the government being 5. We can see that in this circumstance, the government can obtain a larger payoff by direct observation of the student's behavior, and so moral hazard remained.

In summary, there are circumstances for which government financial assistance is irrelevant, circumstances for which government financial assistance leads to better results from the government's perspective without moral hazard, and circumstances under which moral hazard exists, even if the government makes its reimbursements based on the easily observed success or failure of the student. In making government policy with respect to college financial assistance, it would be beneficial to determine the circumstances likely to hold.

EXERCISES

5.1 Show the computations for the utilities for each player given in Table 5.3 for the Media Streaming game. The first and fifth are done in the text.

5.2 Consider the sequential game tree illustrated in Figure 5.32.

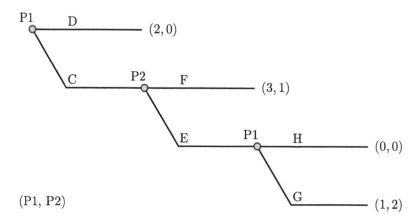

FIGURE 5.32 Two-Two-Two Game Tree.

a. List all of the terminal histories and the corresponding player payoffs.

b. List the non-terminal and corresponding acting players.

c. Find the backward induction solution.

d. Identify the strategies available to each player and construct a strategic game model.

e. Find the Nash equilibria of the strategic game.

f. In what way do the Nash equilibria that are not the backward induction solution involve players choosing actions that lower their own payoffs? Why might they do so?

5.3 Consider the game illustrated in the game tree in Figure 5.33.

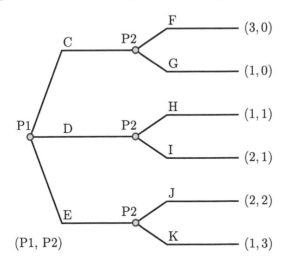

FIGURE 5.33 Three-Two Game Tree.

a. List the three strategies available to P1 and the eight strategies available to P2.

b. Find the six backward induction solutions for the game, written as ordered pairs of strategies.

c. How do you interpret the fact that there are multiple backward induction solutions?

5.4 Create the game tree for the U.S. Appellate Court scenario.

5.5 In the U.S. Appellate Court model, what preferences might a dissenter with a pragmatic approach who does not want to have to write and publish his opinion have? What preferences might a majority with a pragmatic approach have?

5.6 Do the analysis of the U.S. Appellate Court model when an ideologue dissenter is faced with a majority with a weak opinion. Consider how each would rank the outcomes and find the backward induction solution.

5.7 Model the U.S. Appellate Court scenario as a sequential game in which the majority moves first by asking the dissenter to join them in an unpublished opinion. Give the game tree and preferences and provide your analysis.

5.8 How one might interpret the use of von Neumann-Morgenstern utilities in the U.S. Appellate Court model?

5.9 Consider the Patent Race in the case that the patent is only valued at $12 million for company R.

a. Show that the first trigger zone is the region in which $r \leq 2$ and $s \leq 3$.

b. Show that the first safety zone for company R is the region in which $r \leq 2$ and $s > 3$.

c. Determine the second trigger zone.

*5.10 Build a model and draw the location space for the Patent Race for the case in which the two companies have different costs associated with taking steps.

5.11 Consider the Political Challenger scenario. While it was clear that the challenger first preferred the terminal history (In,Withdraw), and the incumbent first preferred the terminal history (Out), the order of the remaining terminal histories for each player was not as clear.

a. Using the utilities in Table 5.11 where the Challenger's preferences have changed from the original game, create a game tree and find the backward induction solution.

TABLE 5.11   Political Challenger Alternate
Utilities

Terminal History h	$u_{Ch}(h)$	$u_{Inc}(h)$
Out	0	2
In, Withdraw	2	1
In, Fight	1	0

b. Create a utility table like Table 5.11 where only the Incumbent's preferences have changed from the original game, create a game tree, and find the backward induction solution.

c. Finally, create a utility table where both players' preferences have changed from the original game, create a game tree, and find the backward induction solution.

d. Describe the similarities and differences among the four possible games, including the original.

5.12 For each of the three possible utility tables given in the previous problem, for the Senate Primaries scenario, show the strategic game, and find the Nash equilibria. Describe the similarities and differences among the four possible games, including the original.

5.13 Argue that in any sequential game, the backward induction strategy profile is in fact a Nash equilibrium.

5.14 Set the parameters of the Sequential Duopoly scenario to $a = 100$, $b = 1$, and $c = 10$ and analyze the resulting model. Compare the solution with the solution to the corresponding Cournot duopoly model from Chapter 3 with the same parameters. If different, is the solution to the duopoly model also a subgame perfect equilibrium?

5.15 How does the Sequential Duopoly model change when considering a Sequential Oligopoly with n players instead of 2?

5.16 Suppose in Tic-Tac-Toe that X's first mark is the top-left square.

a. Show that if O's first mark is the center square, X can ensure a win.

b. Show that if O's first mark is the top-right square, X can ensure a win.

c. Show that if O's first mark is the bottom-right square, X can ensure a win.

*5.17 Show that in Tic-Tac-Toe, O has a strategy that will ensure that O at least ties.

5.18 For each of the following nim games, determine whether the first or second player has a winning strategy. If the first player has a winning strategy, state a move that is consistent with that strategy. If the second player has a winning strategy, state a move that is consistent with that strategy if the first player marks one circle in the first column on their turn.

a. $[7, 5, 4, 3]$-nim.

b. $[25, 16, 11, 3, 1]$-nim.

c. $[16, 16, 10, 8, 2]$-nim.

d. $[16, 16, 10, 4, 3]$-nim.

5.19 *Poison* is played on a board with $n \geq 1$ non-overlapping circles. The two players take turns marking one or two unmarked circles. The player who marks the last circle loses. Determine for each value of n which player has a winning strategy and describe the winning strategy. Hint: Start by examining small values of n and then generalize.

5.20 Consider the Shannon switching game played on the graph in Figure 5.34.

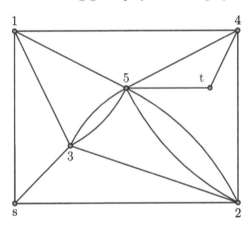

FIGURE 5.34 Shannon Switching Game: Graph 1.

a. Show that the graph is positive by finding two disjoint spanning trees.

b. Suppose Colin moves first by removing edge $s3$. Update the spanning trees and identify which edge Rose should claim as part of her winning strategy.

c. Suppose Colin moves next by removing edge $t4$. Update the spanning trees and identify which edge Rose should claim next as part of her winning strategy.

d. Finish the game by choosing moves for Colin and having Rose respond each time with the winning strategy. Give the list of edges that creates Rose's winning path.

5.21 Consider the Shannon switching game played on the graph in Figure 5.35. Who has a winning strategy? Why?

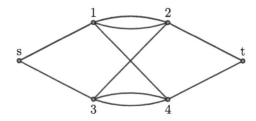

FIGURE 5.35 Shannon Switching Game: Graph 2.

5.22 Consider the Shannon switching game played on the graph in Figure 5.36. Who has a winning strategy? Why?

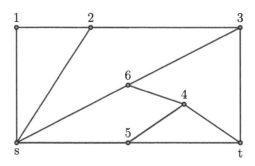

FIGURE 5.36 Shannon Switching Game: Graph 3.

5.23 Nash's Hex Theorem guarantees a winning strategy for the first player. In this exercise, we find that winning strategy for simple games of Hex.

    a. Give the winning strategy for 1×1 Hex.

    b. Can Rose, as the first player playing left to right, win a 2×2 game of Hex? If so, give her winning strategy as a sequence of hexagons to capture. If not, explain why not.

    c. Can Colin, as the second player playing top to bottom, ever win a 2×2 game of Hex? Is there a strategy that guarantees his win or does it depend on Rose's moves?

    d. For the 3×3 game of Hex, describe a winning strategy for Rose without constructing a tree diagram.

    e. For the 4×4 Hex game, describe a winning strategy for Rose without constructing a tree diagram.

5.24 *Chomp* begins with a rectangular array of cookies (represented by circles). The players take turns choosing a cookie, each time removing that cookie and all of the cookies in the rectangle of which it is the lower left corner (representing a chomp of the cookies by marking all of the circles corresponding to the removed cookies). The bottom-left cookie contains poison, and so the player to eat it loses.

a. Who has the winning strategy in 3-by-3 chomp? Describe the strategy.

b. Describe the winning strategy for the first player in 2-by-n (2 rows by n columns) chomp. Explain why it is a winning strategy.

c. Prove that if there is more than one cookie at the beginning of the game, the first player has a winning strategy in chomp. Hint: Use an argument similar to the one used to prove that the first player has a winning strategy in Hex.

5.25 The game Y is played on a three-sided board of hexagonal spaces as shown in Figure 5.37. Each player attempts to construct Y-shaped paths that connect the three sides.

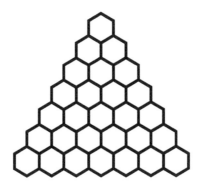

FIGURE 5.37 A Y Board.

a. Copy the board above and show a winning path.

b. By reflecting the Y board with the winning path along one side, create a four-sided board. This should resemble a Hex Board. Does the reflected winning Y path give a winning Hex path?

c. Starting with a Hex board as in part (b), create a winning Hex path. By reversing the process above, project half of the Hex board onto the other half creating a Y board. Does the projection onto the Y board of the winning Hex path create a winning Y path?

d. Explain how to use a winning strategy in Hex to obtain a winning strategy in Y.

5.26 How might a policy of paying management bonuses based on profit margins of a company create an instance of moral hazard?

5.27 How might a parent requiring a child to wear a helmet when riding a bicycle create an instance of moral hazard?

5.28 Moral hazards are considered when underwriting insurance, particularly fire insurance, and are addressed by certain policy exclusions. For example, underwriters are hesitant to insure vacant and unoccupied buildings because of the possibility that the owner will be tempted to intentionally start a fire to obtain an insurance recovery. What exclusions might reduce this moral hazard?

5.29 Colleges are pressured to have students succeed (i.e., graduate), but employers in the U.S. complain that the students are under-prepared. This is possibly an alternate description of grade inflation. Is this a case of moral hazard created by the government applying pressure on the colleges to ensure students grades allow them to graduate?

5.30 If a government is paying for college courses, does a university have any incentives to restrain the growth in tuition costs of its courses?

*5.31 Many states provide lottery-sponsored college scholarship programs (e.g., Georgia Hope Scholarships). Investigate one of these programs. In particular, explain how the program mitigates the potential for moral hazard.

*5.32 Describe a new scenario in which two or more persons, companies, or nations must choose actions in a sequential manner. Model the scenario as a sequential game. Find the subgame perfect equilibria and perhaps the Nash equilibria of the corresponding strategic game. Interpret the solution(s) in the context of the original scenario.

*5.33 Describe a combinatorial game not already described in the text. For at least small instances of the game, determine which one of the players has a winning strategy or whether both have strategies that can ensure at least a tie. If possible, fully describe these strategies.

CHAPTER 6

Missing Information

Sometimes players find themselves in a game in which some aspects are hazy. They may not know precisely the preferences of the other players, what actions the other players may have already taken, or even the actions available to them. We examine how players can choose appropriate actions with this lack of information and how a modeler can predict the likely outcome of a game when the players have different information about the game being played.

Until now, we have assumed that players in a game know the rules, the possible outcomes, and each other's preferences over outcomes. Such games are said to have *complete information*. Sometimes we have also assumed that players choose their actions sequentially and know all actions taken previously. Such games are said to have *perfect information*. Sequential games have both complete and perfect information and have no actions left to chance. Strategic games are imperfect information games with complete information. An extensive game is a way to model scenarios that have imperfect or incomplete information in some way.

Definition 6.0.1. An *extensive game* consists of the following:

1. A set $N = \{1, 2, \ldots, n\}$ of at least two players.

2. A set O of terminal histories.

3. A partition of non-terminal histories into information sets such that each non-terminal history within an information set has the same set of possible actions after it.

4. A player function, P, which assigns a player, or chance, to each information set.

5. For each information set assigned to chance, there is a probability distribution that chance uses to select its action.

6. Starting with the empty history, if a non-terminal history is reached, chance or the player assigned to the information set containing it selects an action to append to the current history.

7. Utility functions $u_i : O \to \mathbb{R}$ that specify the preferences among terminal histories for each player $i \in N$.

As can be seen from the definition, an extensive game includes all aspects of a sequential game, but adds chance and information sets which may contain more than one non-terminal history. Because of their sequential nature, we can use game trees as a visual representation for extensive games. Each node in the tree corresponds to a history: terminal nodes correspond to terminal histories and are labeled with the player utilities, and the other nodes correspond to non-terminal histories and are grouped together within dashed boxes to indicate information sets, which are labeled with chance or the player who chooses an action there. As with sequential games, actions correspond to the edges emanating from each node, and the same edges emanate from each node within a dashed box information set. We assume that a player assigned to an information set only knows that they are somewhere within the dashed box, but does not know specifically at which node, when choosing an action.

When all of the information sets contain a single history and no information set is assigned to chance, the extensive game is identical to a sequential game. Having an information set assigned to chance incorporates uncertainty. An information set containing two or more histories incorporates imperfect information: the players do not know some of the history of actions before making a choice and do not necessarily know the direct effects of an action. Combining an initial chance choice with non-singleton information sets models scenarios with incomplete information.

A player in an extensive game has *perfect recall* if for every history in an information set for that player, the player has made the same sequence of action choices leading up to that information set. Intuitively, perfect recall means the player never forgets their previous actions. An extensive game has perfect recall if every player in that game does. If the partners in a card game such Bridge or the members of a soccer team are modeled as a single player, then the game will not have perfect recall because each Bridge partner only knows their own cards and each team member can only see certain parts of the soccer field. Even individual persons often forget certain choices they made in the past. Except for a couple of exercises, we will only examine extensive games with perfect recall.

6.1   IMPERFECT INFORMATION

In this section, we will focus on games with imperfect information. Recall, these occur when players do not know one or more actions taken by the other players and so find themselves in an information set with more than one history.

FoodPro. A local food processing plant, FoodPro, needs a small repair done. While they would normally call Ben's contracting company, the repair seems small enough that Suzette, the FoodPro manager, decides to explore another contractor (possibly as an alternative for future work) by first asking Mark's contracting company for a bid. Mark, who would like to have the work, can decide not to place a bid or to place a bid of some amount. After receiving a response from Mark, Suzette tells Ben whether Mark has submitted a bid, but not the amount of the bid if one was submitted, and asks Ben for a bid. Since the project is small, Ben does not really want the work; however, he wants to keep FoodPro as a customer over the long term and so is concerned about how his current actions will affect his relationship with FoodPro. Suzette plans to accept the lower of the two bids, but if she receives bids of similar amounts, she will choose Ben, her regular repair contractor, over Mark.

Assuming that Suzette has already made her decisions about how to carry out the bidding process, Mark and Ben are the players in this game. The game is sequential in nature, and Ben has incomplete information about Mark's action. To create a simple model, we assume Mark can choose one of three actions: not bid (No), bid low (Lo), or bid high

(Hi). Ben only knows whether Mark bids or does not bid; if Mark does bid, Ben does not know whether he bid high or low. Since Ben is not interested in doing the work for a low price, we assume he chooses between two actions: not bid (No) or bid high (Hi). Assigning vNM utilities consistent with this discussion, we obtain the game tree in Figure 6.1 and Table 6.1.

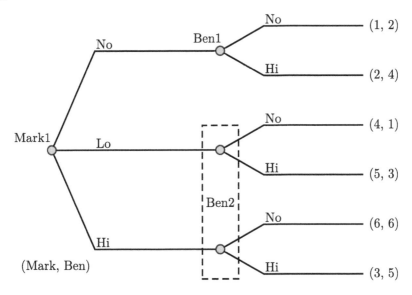

FIGURE 6.1 FoodPro Game Tree.

TABLE 6.1 FoodPro Preferences

Terminal History h	Utility Pair $(u_{\mathbf{Mark}}(h), u_{\mathbf{Ben}}(h))$
(No, No)	$(1, 3)$
(No, Hi)	$(2, 1)$
(Lo, No)	$(4, 2)$
(Lo, Hi)	$(5, 4)$
(Hi, No)	$(6, 6)$
(Hi, Hi)	$(3, 5)$

We can see from the game tree the non-terminal histories are partitioned into three information sets, listed in Table 6.2. Only Ben faces an information set, Ben2, with more than one node (non-terminal history). Here Ben does not know which sequence of actions has occurred and is making his decision whether to choose No or Hi without knowing whether Mark has bid Hi or Lo.

TABLE 6.2 FoodPro Information Partition and Player Function

Label	Information Set I	Player $P(I)$
Mark1	$\{()\}$	Mark
Ben1	$\{(\text{No})\}$	Ben
Ben2	$\{(\text{Lo}), (\text{Hi})\}$	Ben

It is important to consider players' overall strategies, rather than just individual actions. In FoodPro, one pure strategy for Ben is to always choose No. Another is to choose No when Mark Bids (at information set Ben2) and Hi if Mark does not bid (at information set Ben1). Mark has three pure strategies, corresponding to the actions at his only non-terminal history: No, Lo, or Hi. For strategic games, we allowed players to adopt mixed strategies. While we could allow players in extensive games to adopt such mixed strategies, it is typically sufficient and more natural to allow players to randomize action choices at each information set based on probability distributions. We call such strategies behavior strategies and formally define them below.

Definition 6.1.1. A *pure strategy* for player i is a function s_i which assigns to each of the player's information sets a possible action. A *behavior strategy* for player i is a function s_i which assigns to each of the player's information sets a probability distribution over possible actions. If s is used to denote a pure or behavior strategy, we will use $s_i(I)$ or simply $s(I)$ to denote the action or probability distribution over actions chosen by player i at the assigned information set I, and $s_i(a|I)$ or simply $s(a|I)$ will denote the probability that player i will choose action a at information set I.

In the FoodPro example, one pair of behavior strategies is for Mark to choose between No, Lo, and Hi with equal probability and for Ben to choose No or Hi with equal probability at Ben1 but choose Hi over No twice as often at Ben 2. This gives us the strategy profile $s = (s_{Mark}, s_{Ben})$ with

$$s_{\text{Mark}}(\text{Mark1}) = (1/3)\text{No} + (1/3)\text{Lo} + (1/3)\text{Hi}$$

and

$$s_{\text{Ben}}(\text{Ben1}) = (1/2)\text{No} + (1/2)\text{Hi}$$
$$s_{\text{Ben}}(\text{Ben2}) = (1/3)\text{No} + (2/3)\text{Hi}.$$

For the strategy profile s, Mark's (expected) payoff is

$$(1/3)((1/2)(1) + (1/2)(2)) + (1/3)((1/3)(4) + (2/3)(5)) + (1/3)((1/3)(6) + (2/3)(3))$$
$$= 61/18 \approx 3.39$$

and Ben's payoff is

$$(1/3)((1/2)(2) + (1/2)(4)) + (1/3)((1/3)(1) + (2/3)(3)) + (1/3)((1/3)(6) + (2/3)(5))$$
$$= 32/9 \approx 3.56.$$

Viewing behavior strategies as mixed strategies, where each action is chosen with the probability given in the behavior strategy, any extensive game can be modeled as a strategic game, called the extensive game's *corresponding strategic game*. Since sequential games are a type of extensive game (those in which information sets each consist of a single node and no information set is assigned to chance), we know that the corresponding strategic game can have Nash equilibria that are inconsistent with our interpretation of utilities as revelatory of player actions. For sequential games, we defined and used a stronger solution concept, subgame perfect equilibrium, that required player strategies to be best responses from each non-terminal history of the game. In an extensive game, players may not know at which node they find themselves within an information set, and so it would be impossible to determine the best response in the same way. However, if each player has a belief about how likely each history in the information set is to have occurred, they can determine a best response based on these beliefs.

Definition 6.1.2. A *belief system* for an extensive game is a function, β, that assigns a probability distribution over histories in each information set not assigned to chance. Thus, $\beta(I)$ denotes the probability distribution on the nodes in information set I. Further, $\beta(h|I)$ denotes the probability of history h in the information set I with respect to the belief system β.

A belief system models what we assume about the players' understanding of what has happened in the game up to I. When the player assigned to I makes a choice of actions, that player uses the probabilities $\beta(h|I)$ for all $h \in I$, so these probabilities should reflect that player's beliefs about how likely it is that each h has occurred. (Occasionally, we will say "player i's belief system at I" for the probability distribution $\beta(I)$.)

For the FoodPro game, we have assumed Ben knows when Mark has bid and we now assume in that case that Ben believes Mark is equally likely to have bid Hi as Lo. Thus we construct a belief system to assign the uniform probability distribution to Ben2, denoted by $(1/2)(\text{Lo}) + (1/2)(\text{Hi})$. In information sets with only one node, for example Ben1, Ben knows with certainty that No has occurred. Hence, one belief system for the FoodPro game is

$$\beta(\text{Mark1}) = (1)()$$
$$\beta(\text{Ben1}) = (1)(\text{No})$$
$$\beta(\text{Ben2}) = (1/2)(\text{Lo}) + (1/2)(\text{Hi}).$$

From the belief system β we can see

$$\beta(\text{Hi}|\text{Ben2}) = 1/2.$$

The belief system is represented on the game tree in Figure 6.2 with the probability associated with each history in the Ben2 information set shown below the corresponding node. We did not label the nodes in information sets Mark1 and Ben1 because those probabilities are 1.

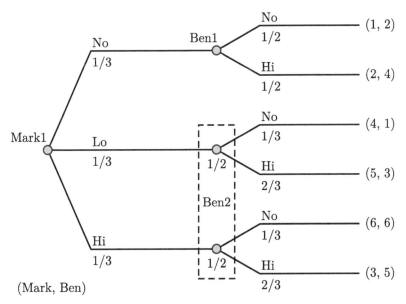

FIGURE 6.2 FoodPro Game Tree with Assessment (s, β).

When describing the final solution of an extensive game, we will include both strategies and beliefs.

Definition 6.1.3. An *assessment* is a pair (s, β) consisting of a profile of behavior strategies, $s = (s_1, \ldots, s_n)$, together with a belief system, β.

One assessment for the FoodPro game is therefore

$$(s, \beta) = ((s_{\text{Mark}}, s_{\text{Ben}}), \beta)$$

where
$$
\begin{aligned}
s(\text{Mark1}) &= (1/3)\text{No} + (1/3)\text{Lo} + (1/3)\text{Hi} \\
s(\text{Ben1}) &= (1/2)\text{No} + (1/2)\text{Hi} \\
s(\text{Ben2}) &= (1/3)\text{No} + (2/3)\text{Hi} \\
\beta(\text{Ben2}) &= (1/2)(\text{Lo}) + (1/2)(\text{Hi}).
\end{aligned}
$$

This assessment is represented in Figure 6.2 with the probabilities displayed below the appropriate edges and nodes. We are now ready to formally extend the notion of best response to an extensive game.

Definition 6.1.4. Suppose (s, β) is an assessment for an extensive game. Player i's strategy is a *best response* at the information set I if it maximizes the player's utility at I given the beliefs $\beta(I)$ and the other players' strategies. The assessment (s, β) is *sequentially rational* if each player's strategy is a best response at each information set to which either the player or chance is assigned.

The assessment (s, β) given above is not sequentially rational because the payoff to Ben beginning at the singleton information set Ben1 is not a best response: the expected payoff of $(1/2)(2) + (1/2)(4) = 3$ is less than the expected payoff from strategy $s'(\text{Ben1}) = \text{Hi}$ of 4. Additionally, Ben's payoff from his behavior strategy at Ben2 is not a best response either:

$$(1/2)((1/3)(1) + (2/3)(3)) + (1/2)((1/3)(6) + (2/3)(5)) \approx 3.83$$

is less than his expected payoff of

$$(1/2)(3) + (1/2)(5) = 4$$

had he selected $s'(\text{Ben2}) = \text{Hi}$ instead. Notice that Ben calculates his expected payoff at Ben2 by using his beliefs about at which history he is, not what is actually happening before the information set.

But let's assume for the moment that Ben revises his behavior strategy to

$$
\begin{aligned}
s'(\text{Ben1}) &= \text{Hi} \\
s'(\text{Ben2}) &= \text{Hi}
\end{aligned}
$$

so that it is a best response at each information set for Ben given his beliefs (note that Ben's best response does not depend on Mark's strategy choice). Mark's strategy is not a best response to Ben's new behavior strategy since

$$s'(\text{Mark1}) = \text{Lo}$$

would yield a better payoff: 5 rather than

$$(1/3)(2) + (1/3)(5) + (1/3)(3) \approx 3.33.$$

In fact, since against Ben's strategy s'_{Ben}, Mark obtains payoffs of 1, 5, and 3 by choosing strategies No, Lo, and Hi, respectively, $s'_{\text{Mark}}(\text{Mark1})$ is Mark's unique best response. As already observed, $s'_{\text{Ben}}(\text{Ben1})$ and $s'_{\text{Ben}}(\text{Ben2})$ are best responses for Ben given his beliefs and regardless of Mark's strategy choice. Since both players have best responses at each information set, the new assessment

$$(s', \beta) = ((s'_{\text{Mark}}, s'_{\text{Ben}}), \beta)$$

where

$$
\begin{aligned}
s'(\text{Mark1}) &= \text{Lo} \\
s'(\text{Ben1}) &= \text{Hi} \\
s'(\text{Ben2}) &= \text{Hi} \\
\beta(\text{Ben2}) &= (1/2)(\text{Lo}) + (1/2)(\text{Hi})
\end{aligned}
$$

is sequentially rational.

We would also like each player's beliefs to be consistent with the given strategy profile. That is, the probability that a player assigns to a node within an information set should be the same as the probability with which that history is obtained given that the information set is reached if each player follows their strategy. Continuing to generalize the subgame perfect concept from sequential games, we would like for this consistency condition to hold on all subgames, but because of the information sets, it is no longer meaningful to say that a subgame exists rooted at any non-terminal history.

Definition 6.1.5. A *subgame* G' of an extensive game G consists of a non-terminal history h, called the *root*, all histories H from G that start with h, and all other aspects of G that relate to H (i.e., players, actions, terminal histories, information sets, player function, chance probability distributions, and utilities) for which each information set I of G is either completely inherited by G' (i.e., $I \cap H = I$) or mutually exclusive with G' (i.e., $I \cap H = \emptyset$).

Observe that for each information set to be either completely inherited by or mutually exclusive of the subgame, its root must be in an information set alone. In the FoodPro game, there are two subgames: the whole game rooted at the node labeled Mark1, and the subgame rooted at the node labeled Ben1.

Definition 6.1.6. Let $\Pr_{G'}(h|s)$ denote the probability that h is reached in the subgame G' given that the strategy profile s is used. An assessment (s, β) of a game G achieves *consistency of beliefs* if

$$\beta(h|I) = \frac{\Pr_{G'}(h|s)}{\sum_{k \in I} \Pr_{G'}(k|s)} \tag{6.1}$$

for each history h in each information set I in each subgame G' of G for which $\Pr_{G'}(k|s) > 0$ for some history $k \in I$.

Observe that the consistency Equation 6.1 always holds at a singleton information set $I = \{h\}$ because in such a situation $\beta(h|I) = 1$ and $\Pr_{G'}(h|s) = \sum_{k \in I} \Pr_{G'}(k|s)$. Also observe that if the consistency Equation 6.1 holds for all but one history in an information set, then the consistency Equation 6.1 holds for the remaining history. Thus, to verify that an assessment achieves consistency of beliefs, it is sufficient to check the consistency Equation 6.1 at all but one history in each non-singleton information set in which some history is reached with positive probability.

In the FoodPro game, since the subgame rooted at the node labeled Ben1 has no non-singleton information sets, the consistency of beliefs condition holds vacuously on it. For the (sub)game G rooted at Mark1, there is one non-singleton information set to consider: Ben2. Since

$$\frac{\Pr_G((\text{Lo})|s)}{\Pr_G((\text{Lo})|s) + \Pr_G((\text{Hi})|s)} = \frac{1/3}{1/3 + 1/3} = \frac{1}{2} = \beta((\text{Lo})|\text{Ben2}),$$

the assessment (s, β) achieves consistency of beliefs. In contrast, the assessment (s', β) does not achieve consistency of beliefs because

$$\frac{\Pr_G((\text{Lo})|s')}{\Pr_G((\text{Lo})|s') + \Pr_G((\text{Hi})|s')} = \frac{1}{1+0} = 1 \neq \frac{1}{2} = \beta((\text{Lo})|\text{Ben2}).$$

For any information set that is reached via player strategies with positive probability in some subgame, consistent beliefs can be defined by directly calculating the probability ratios from the behavior strategies. If an information set is not reached via player strategies with a positive probability in any subgame, then any probability distribution on that information set will be consistent.

We can now extend our notion of subgame perfect equilibria to extensive games.

Definition 6.1.7. An assessment (s, β), is a *weak sequential equilibrium* if it is both sequentially rational and achieves consistency of beliefs.

Thus, the assessment (s, β) in FoodPro is not a weak sequential equilibrium since it is not sequentially rational, even though it achieves consistency of beliefs. The assessment (s', β) is not a weak sequential equilibrium because, even though it is sequentially rational, it does not achieve consistency of beliefs. Finally, the assessment

$$(s', \beta') = ((s'_{\text{Mark}}, s'_{\text{Ben}}), \beta')$$

where

$$\begin{aligned}
s'(\text{Mark1}) &= \text{Lo} \\
s'(\text{Ben1}) &= \text{Hi} \\
s'(\text{Ben2}) &= \text{Hi} \\
\beta'(\text{Ben2}) &= (\text{Lo})
\end{aligned}$$

displayed in Figure 6.3 is a weak sequential equilibrium. The assessment (s', β') is sequentially rational because (1) given Ben's strategy $s'(\text{Ben1}) = s'(\text{Ben2}) = \text{Hi}$, Mark's best

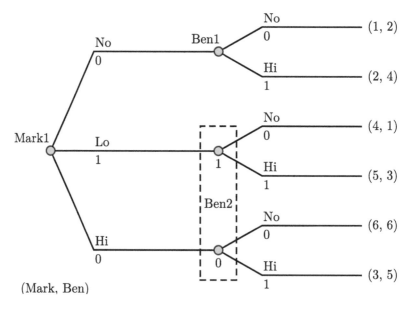

FIGURE 6.3 FoodPro Game Tree with Assessment (s', β').

response at Mark1 is Lo since it yields a payoff of 5 versus payoffs of 2 or 3 obtained by choosing No or Hi, (2) Ben's best response at Ben1 is Hi because it yields a payoff of 4 versus a payoff of 2 by choosing No, and (3) given the belief system $\beta'(\text{Ben2}) = (\text{Lo})$, Ben's best response at Ben2 is Hi since it yields a payoff of $(1)(3) + (0)(5) = 3$ versus a payoff of $(1)(1) + (0)(6) = 1$ by choosing No. It also achieves consistency of beliefs because for the subgame G rooted at Mark1

$$\frac{\Pr_G((\text{Lo})|s')}{\Pr_G((\text{Lo})|s') + \Pr_G((\text{Hi})|s')} = \frac{1}{1+0} = 1 = \beta'((\text{Lo})|\text{Ben2}).$$

In the FoodPro scenario, imperfect information arose because one player's actions were unknown to the other player. Other information besides previous actions could also be privately held by some players; since this information could affect outcomes and payoffs this also leads to an imperfect information game. We assume in this case that the other players have a probabilistic understanding of any privately held information. A canonical scenario is a card game in which each player knows the cards they hold but cannot see the cards held by the other players. Poker is such a card game, and many believe that to play well requires an understanding of the psychological aspects of risky play. We will consider a highly simplified version of Poker for which we will show that bluffing can be a completely rational, and not at all risky, strategy in which to engage.

One Card Poker. Rose and Colin each place one token in the empty kitty. Rose is dealt either a hi or lo card, each being equally likely. Rose knows what card has been dealt to her but Colin does not. Rose can either fold or raise. If Rose folds, the game is over and Colin receives the two tokens from the kitty (a net winning of one token). If Rose raises, she adds another token to the kitty, and Colin can either fold or see. If Colin folds, the game is over and Rose receives the three tokens from the kitty (a net winning of one token). If Colin sees, he adds another token to the kitty, and Rose can either fold or raise. If Rose folds, the game is over and Colin receives the four tokens from the kitty (a net winning of two tokens). If Rose raises, she adds another token to the kitty, and Colin can either fold or see. If Colin folds, the game is over and Rose receives the five tokens from the kitty (a net winning of two tokens). If Colin sees, he adds another token to the kitty, and the card is revealed. The six tokens in the kitty (a net winnings of three tokens) go to Rose if the card is hi and to Colin if the card is lo.

The extensive game tree, along with an arbitrary assessment, is shown in Figure 6.4. The player names are abbreviated by their first letters. Observe the addition of an information set, consisting of the root history (), that is assigned to chance and the required probabilities shown below the edges emanating from the root node. Rose's strategy is specified by the probabilities $a = \Pr(\text{Fold}|\text{R1})$, $b = \Pr(\text{Fold}|\text{R2})$, $c = \Pr(\text{Fold}|\text{R3})$, and $d = \Pr(\text{Fold}|\text{R4})$. Colin's strategy is specified by the probabilities $e = \Pr(\text{Fold}|\text{C1})$ and $f = \Pr(\text{Fold}|\text{C2})$. The belief system is specified by the probabilities $p = \beta((\text{Lo, Raise})|\text{C1})$ and $q = \beta((\text{Lo, Raise, See, Raise})|\text{C2})$. We have assumed that both players are self-interested and risk-neutral in their winnings, hence, player payoffs are given by the number of tokens won.

We now investigate under what conditions the arbitrary assessment will be a weak sequential equilibrium. To be sequentially rational, Rose must choose her strategy to maximize her utility at the information sets Chance, R1, R2, R3, and R4 given Colin's strategy choice following these. Colin must choose his strategy to maximize his utility at the information sets Chance, C1, and C2 given his beliefs at, and Rose's strategy choices after, these.

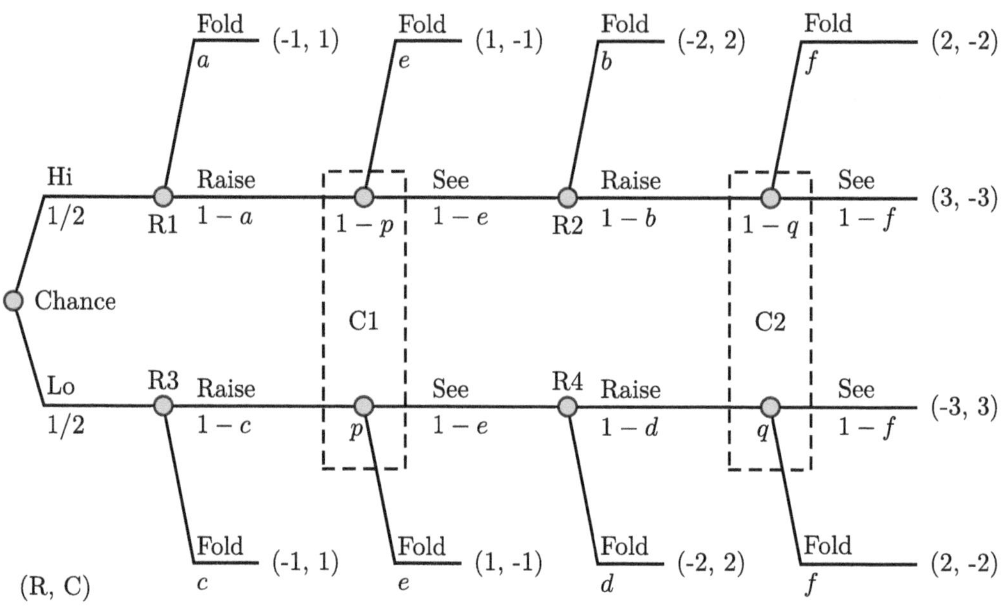

FIGURE 6.4 One Card Poker Game Tree.

The only subgame is the entire game because the only singleton information sets after the root of the game are R1, R2, R3, and R4, and each of these would inherit part but not the entire information set C2. Hence, an assessment that achieves the consistency of beliefs criterion only places a restriction on p (resp., q) if information set C1 (resp., C2) is reached with positive probability.

To find a weak sequential equilibrium, we first search for any actions that are clearly part of any best response. There are two information sets at which the utility maximizing actions are easy to determine. At information set R2, Rose's choice of Fold results in a payoff of -2, whereas a choice of Raise leads to a payoff of $f(2) + (1-f)(3) > -2$ regardless of the value of probability f. Thus, Rose's unique best response at R2 is Raise, and so $b = 0$ in any weak sequential equilibrium. At information set R1, Rose's choice of Fold leads to a payoff of -1 and, given that $b = 0$, Raise leads to a payoff of $e(1) + (1-e)(f(2) + (1-f)(3)) > -1$ regardless of the values of the probabilities e and f. Therefore, at R1, Rose's best response is Raise, and so we also have $a = 0$ in any weak sequential equilibrium. These restrictions are displayed in Figure 6.5, where the two edges corresponding to zero probability actions have been removed.

Rose's best responses at R3 and R4 are not so straightforward, and there are a variety of ways to continue the analysis. Since it's easiest to analyze pure actions, we examine at which nodes, if any, Rose or Colin could employ pure actions and still achieve consistency of beliefs.

Since $a = 0$, the top node of information set C1 is reached with positive probability, and achieving consistency of beliefs implies

$$\frac{\text{Pr(top node in C1 is reached)}}{\text{Pr(C1 is reached)}} = \beta(\text{Hi}, \text{Raise})$$

and

$$\frac{\text{Pr(bottom node in C1 is reached)}}{\text{Pr(C1 is reached)}} = \beta(\text{Lo}, \text{Raise})$$

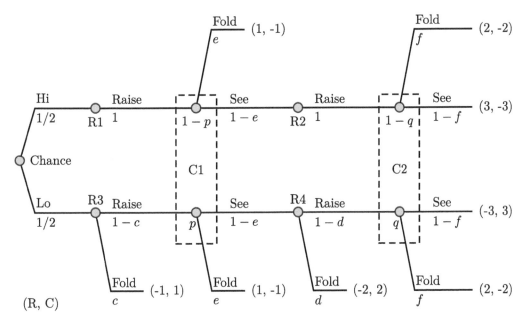

FIGURE 6.5 One Card Poker Game Tree after the First Step of the Analysis.

which together imply

$$\frac{\text{Pr(bottom node in C1 is reached)}}{\text{Pr(top node in C1 is reached)}} = \frac{p}{1-p}.$$

Then

$$\frac{(1/2)(1-c)}{1/2} = \frac{p}{1-p}$$

which implies

$$p = \frac{1-c}{2-c} \le \frac{1}{2}. \tag{6.2}$$

since $0 < c < 1$.

Suppose $e = 1$, meaning Colin chooses the pure action Fold at C1, then Rose's best response at R3 is Raise ($c = 0$) to obtain a payoff of 1 rather than Fold with a payoff of -1. By Equation 6.2, then $p = (1-0)/(2-0) = 1/2$ to achieve consistency of beliefs. But Colin's choice of Fold at C1 yields a payoff of -1. If he instead chooses See at both C1 and C2 (with $p = 1/2$), he receives a payoff of $(1 - 1/2)(-3) + (1/2)((1 - d)(3) + d(2)) = -d/2 > -1$; hence, Fold at C1 is not a best response for Colin, contradicting our assumption that $e = 1$. Thus, $e < 1$ in any weak sequential equilibrium.

Suppose $c = 1$, meaning that Rose chooses the pure action Fold at R3. This forces $p = (1 - 1)/(2 - 1) = 0$. In this case Colin's best response at information set C1 is Fold to obtain a payoff of -1 rather than See which yields a payoff of $(1-f)(-2)+f(-3) < -1$, and so we must have $e = 1$, which we have already shown is not possible in a weak sequential equilibrium. Thus, we must also have $c < 1$.

Since $c < 1$, the top node in information set C2 is reached with positive probability, and achieving consistency of beliefs here implies through similar calculations to Equation 6.2 that

$$\frac{(1/2)(1-c)(1-e)(1-d)}{(1/2)(1-e)} = \frac{q}{1-q}$$
$$\Rightarrow$$
$$(1-c)(1-d) = \frac{q}{1-q}. \tag{6.3}$$

At information set C2, Colin's choice of Fold yields a payoff of -2 regardless of the value of q and his choice of See yields a payoff of $(1-q)(-3) + q(3) = -3 + 6q$. These payoffs are equal if and only if $q = 1/6$, suggesting that we should consider the three cases $q < 1/6$, $q > 1/6$, and $q = 1/6$.

Case 1. Suppose $q < 1/6$. Then the payoff for See is less than the payoff for Fold, making Colin's unique best response at C2 to choose Fold and $f = 1$. At information set R4, Rose's choice of Raise yields a payoff of 2 (since $f = 1$) and her choice of Fold yields a payoff of -2; hence, Rose's unique best response at R4 is Raise, and so $d = 0$. At information set R3, Rose's choice of Fold yields a payoff of -1 and Raise yields a payoff that is some weighted average of 2 (if Colin chooses See at C1) and 1 (if Colin chooses Fold at C1); hence, Rose's unique best response at R3 is Raise, making $c = 0$. By Equation 6.2, it follows that $p = 1/2$. Examining information set C1, we see that Colin's choice of Fold yields a payoff of -1, his choice of See and then Fold at C2 yields a payoff of -2, and his choice of See at both C1 and C2 yields a payoff of $(1 - 1/2)(-3) + (1/2)(3) = 0$; hence, Colin's unique best response is to choose See at both C1 and C2, leading to $e = f = 0$. But we have already shown that when $q < 1/6$, $f = 1$. This contradiction shows that $q < 1/6$ cannot result in a weak sequential equilibrium.

Case 2. Suppose $q > 1/6$. Colin's unique best response at C2 is See, and so $f = 0$. At information set R4, Rose's choice of Raise yields a payoff of -3 (since $f = 0$) and her choice of Fold yields a payoff of -2; hence, Rose's unique best response at R4 is Fold, and so $d = 1$. By Equation 6.3, it follows $q = 0 < 1/6$, a contradiction. Thus, we cannot have $q > 1/6$ in any weak sequential equilibrium.

Case 3. The first two cases allow us to conclude that any weak sequential equilibrium must have $q = 1/6$, so we continue our analysis with this assumption. Then any choice of f is a best response by Colin at information set C2. By Equation 6.3, $(1-c)(1-d) = 1/5$ which implies $d \neq 1$, and Rose's best response at R4 is not the pure strategy Fold. Since at R4 Rose's choice of Fold yields a payoff of -2 and of Raise yields a payoff $(1-f)(-3) + f(2) = -3 + 5f$, it follows that $f \geq 1/5$.

We suppose first that $f > 1/5$. The unique best response for Rose at R4 is to choose Raise, making $d = 0$. By Equations 6.3 and 6.2, $c = 4/5$ and $p = 1/6$. Then with the belief $p = 1/6$, at C1, Colin's choice of Fold yields a payoff of -1 and his choice of See yields a payoff of $(5/6)(f(-2) + (1-f)(-3)) + (1/6)((1-f)(3) + f(-2)) = -2$. Thus Fold ($e = 1$) is Colin's unique best response. But we have already shown $e < 1$ in any weak sequential equilibrium and so we cannot have $f > 1/5$. Therefore, we must have $f = 1/5$ when $q = 1/6$. In this case, any value of d is a best response by Rose at information set R4 since both Fold and Raise yield the same payoff of -2.

Recalling from earlier in the proof that $e < 1$ and $c < 1$, we suppose $e = 0$, and examine Rose's pure choices at R3. Fold yields a payoff of -1 while Raise yields a payoff of $(1-d)((1-1/5)(-3) + (1/5)(2)) + d(-2) = -2$; hence, Rose's unique best response would be Fold, making $c = 1$, a contradiction.

Thus, $0 < e < 1$ which implies that Colin's payoffs choosing Fold and See at C1 must be equal, that is,

$$-1 = (1-p)((1/5)(-2) + (1-1/5)(-3)) + p((1-d)((1-1/5)(3) + (1/5)(-2)) + d(2))$$

which implies $p = 3/8$. By Equation 6.2, $c = 2/5$ and by Equation 6.3, $d = 2/3$. Since $0 < c < 1$, Rose's payoffs for choosing Raise and Fold at R3 must be the same, giving

$$(1 - e)((1 - 2/3)((1 - 1/5)(-3) + (1/5)(2)) + (2/3)(-2)) + e(1) = -1$$

and so $e = 1/3$.

At this point we have proved that the only possible weak sequential equilibrium must have

$$a = 0, \ b = 0, \ c = 2/5, \ d = 2/3, \ e = 1/3, \ f = 1/5, \ p = 3/8, \ \text{and} \ q = 1/6,$$

giving the final game tree in Figure 6.6. It is straightforward to check that the assessment given by these values is, in fact, a weak sequential equilibrium.

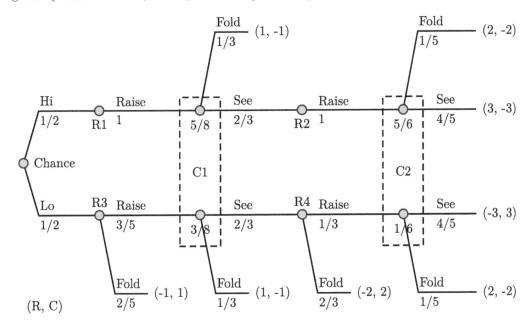

FIGURE 6.6 One Card Poker Weak Sequential Equilibrium.

When this assessment is used, we can examine Rose's payoffs at each information set, that is, her payoff if the game were completed from that information set. Rose's payoffs are

$$(1/5)(2) + (4/5)(3) = 14/5 \text{ at R2}$$
$$(1/3)((4/5)(-3) + (1/5)(2)) + (2/3)(-2) = -2 \text{ at R4}$$
$$(1/3)(1) + (2/3)(14/5) = 33/15 \text{ at R1}$$
$$(3/5)((2/3)(-2) + (1/3)(1)) + (2/5)(-1) = -1 \text{ at R3}$$
$$(1/2)(33/15) + (1/2)(-1) = 3/5 \text{ at the beginning of the game.}$$

Because the game is zero-sum, Colin's payoff at the beginning of the game is $-3/5$. Thus, at the start of the game, Rose has an expected payoff of $3/5$ and Colin of $-3/5$.

In conclusion, we have found a (unique) weak sequential equilibrium for the One Card Poker game, and in that equilibrium, Rose with a lo card will bluff by initially raising with probability $3/5$ and subsequently raising with probability $1/3$. The use of this bluff by Rose leaves Colin uncertain about which card Rose is holding, and so he folds with positive probability even when Rose holds a lo card. The asymmetry brought about by the private information Rose possesses results in her obtaining a positive payoff in what would otherwise be a fair zero-sum game.

The necessity for Rose to adopt a strategy in which she bluffs occurs even with the assumption of risk-neutrality relaxed. Therefore, in a standard game of Poker with self-interested players, we would expect bluffing to be part of any weak sequential equilibrium. Since both players possess private information, it is less clear who, if anyone, possesses an advantage. Of course, the bluff probability and intensity (in standard poker, the amount by which one raises can vary) depends sensitively on the rules and payoffs, but it should now be clear that bluffing can be part of a rational strategy, rather than something motivated solely by psychology.

Political Challenger Take 2. In the Political Challenger scenario in the previous chapter, a challenger was deciding whether or not to enter a race against a long time incumbent. In that scenario we assumed the incumbent made his decision after the challenger entered the race. Here we add complexity to the situation. We assume the challenger will spend a certain amount of time, money, and effort before announcing whether she will enter or stay out of the campaign. Afterwards, the incumbent can choose to retire from office or fight for reelection with some level of effort. The incumbent does not know how much preparation went into the challenger's announcement; however, it is reasonable to speculate that the more preparation that went into an announcement to enter the campaign, the more likely it is that the challenger can defeat the incumbent in the election.

We will simplify the challenger's actions to either announcing she will stay out of the campaign (Out), entering the campaign with little preparation (Unready), or entering the campaign with a large amount of preparation (Ready). The incumbent's actions will be simplified to either retire from office (Withdraw) or fight for reelection (Fight) if the challenger has announced that she will enter the campaign. If the challenger decides to not enter the campaign, the incumbent will be assumed to easily gain reelection. Figure 6.7 presents

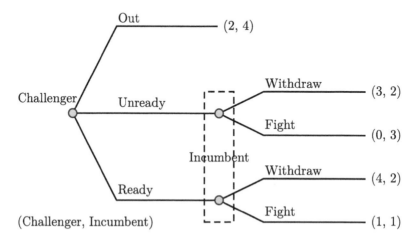

FIGURE 6.7 Political Challenger Take 2 Game Tree.

the resulting extensive game model for this scenario with one possible specification of utilities. We are assuming that the challenger is most interested in winning the election (most facilitated by the incumbent withdrawing from the campaign) and secondarily interested in preparing for any decision she makes. For the incumbent, we are assuming that he wants to win the election while minimizing campaigning effort; hence, the easy reelection for the history (Out) is most preferred, the likely reelection with some effort against an unprepared challenger with history (Unready, Fight) is the second most preferred, and the likely loss in a tough campaign against a prepared challenger with history (Ready, Fight) is the least preferred. These descriptions make the challenger's and incumbent's payoffs ordinal but not vNM.

One weak sequential equilibrium is the assessment in which the challenger selects the strategy Out and the incumbent selects the strategy Fight given the belief that the challenger has selected Unready. Note that in this assessment, the incumbent's information set is not reached with a positive probability and, hence, he can assign any belief he wants to the histories that are contained in it. However, this assessment cannot happen in play because the incumbent's belief in the challenger choosing Unready conflicts with the fact that Ready strongly dominates Unready for the challenger. A more reasonable weak sequential equilibrium has the challenger select Ready and the incumbent choose Withdraw given the belief that the challenger has selected Ready. Just as subgame perfect and weak sequential equilibria eliminate certain unreasonable Nash equilibria, stronger equilibrium concepts have been defined that eliminate the former unreasonable assessment, while retaining the latter, more reasonable, assessment for the Primary game. For more information on these stronger equilibrium concepts see [69].

We close this section by stating theorems that ensure the existence of weak sequential equilibria and interconnect game models and equilibrium concepts. For strategic games, the primary solution concept has been Nash equilibrium. For sequential games, we have used subgame perfect equilibrium. For extensive games, we have used weak sequential equilibrium. In the following theorems, we provide a sufficient condition for the existence of a weak sequential equilibrium and observe that strategic and sequential games are special cases of extensive games with the equilibrium concepts being equivalent in these special cases.

Theorem 6.1.1 (Weak Sequential Equilibrium Existence). If an extensive game has perfect recall and a finite number of histories, then a weak sequential equilibrium exists.

Theorem 6.1.2. Given a sequential game, the corresponding extensive game puts each non-terminal history in its own information set. Each subgame perfect equilibrium of the sequential game becomes a weak sequential equilibrium for the corresponding extensive game by adding a belief system achieving consistency of beliefs, and the strategy profile of each weak sequential equilibrium of the corresponding extensive game is a subgame perfect equilibrium of the sequential game.

Theorem 6.1.3. Given a strategic game with strategy sets $S_1, S_2, \ldots, S_n$ the corresponding extensive game consists of the set of terminal histories $S_1 \times S_2 \times \cdots \times S_n$ and an information set $I_i = S_1 \times \cdots \times S_{i-1}$ for each player i. Each Nash equilibrium of the strategic game becomes a weak sequential equilibrium for the corresponding extensive game by adding a belief system achieving consistency of beliefs, and the strategy profile of each weak sequential equilibrium of the corresponding extensive game is a subgame perfect equilibrium of the sequential game.

6.2 ROMANS AND GERMANS

In this section, we will construct a model for an unfortunately common scenario in human history, that of war. We'll use a historical scenario, but our model and analysis can apply to many modern-day situations.

Warfare. The Roman Empire and the Germanic tribes of northern Europe fought a series of wars throughout the early part of the common era. Many of these wars consisted of short campaigns by the Romans raiding villages to capture or destroy tribal military supplies. In a typical campaign, the Roman legions can advance on the village either through a forest or across a small lake. The defending tribal soldiers are sufficient to defend one or the other approaches, and if they choose correctly they will win the resulting battle. If they choose incorrectly, the Romans will capture the village and secure the weapons and food cache. The tribal soldiers are then faced with a second choice: attack the Romans in the village, or wait in ambush. The success of either option depends on a decision by the Romans about whether they will return to their base immediately, or wait until nightfall. If the tribe is waiting in ambush and the Romans return by day, the tribe wins the resulting battle. If the tribe is waiting in ambush, but the Romans withdraw at night, the Romans successfully survive the resulting battle. On the other hand, if the tribe attacks and Romans have decided to wait to withdraw there is a vicious battle in which both sides lose. If the tribe attacks, but the Romans had withdrawn during the day, the Romans are successful in their mission. Of course, the Romans could decide not to attack at all, but this results in a loss of morale among the soldiers and possibly the removal of the commander.

Barron [8] describes a method for modeling a fictitious battle in the Greco-Persian wars. We present an adaptation of his model for the Greek and Roman battle described above.

Using RE and GT to name the players representing the Roman Empire and the Germanic tribe, the extensive game tree is given in Figure 6.8. To more easily discern the reasoning behind the payoffs, a short description of each outcome and the corresponding payoffs are rank ordered in Tables 6.3 and 6.4. Observe that payoffs were assigned so that positive numbers correspond to an overall "win" for that player and negative numbers correspond to an overall "loss" for that player, and the numbers are to be interpreted as vNM utilities. We encourage the reader to consider whether different numbers might be more realistic.

TABLE 6.3   Roman Empire Preferences

Terminal History	Outcome	Utility
(Forest, Lake, Ambush, Night)	RE raids village & leaves without casualties	2
(Forest, Lake, Attack, Day)	"	2
(Lake, Forest, Ambush, Night)	"	2
(Lake, Forest, Attack, Day)	"	2
(Forest, Lake, Attack, Night)	RE raids village and both lose a battle	1
(Lake, Forest, Attack, Night)	"	1
(Forest, Lake, Ambush, Day)	RE raids village but loses a battle afterwards	0
(Lake, Forest, Ambush, Day)	"	0
(Don't Attack)	RE has no casualties but loses morale	−1
(Forest, Forest)	RE loses a battle to raid the village	−2
(Lake, Lake)	"	−2

This model is different from the three games analyzed in the first section because both players lack information. This lack of information is again indicated by the dashed line boxes around three pairs of the decision nodes. Within each of these boxes, the player who is making the decision does not know which history has occurred. We have also labeled each decision point so that they can be easily referenced. For example, within the GT1 box, when the German tribe is making its decision whether to defend the forest or lake, it does not know the Roman Empire legion's decision whether to advance through the forest or over the lake, however, at the GT2 node (which may also be considered as an information set containing

TABLE 6.4   German Tribe Preferences

Terminal History	Outcome	Utility
(Forest, Forest)	RE loses a battle to raid the village	2
(Lake, Lake)	"	2
(Don't Attack)	Status quo for GT; RE a future threat	1
(Forest, Lake, Ambush, Day)	RE raids village but GT wins a battle	1
(Lake, Forest, Ambush, Day)	"	1
(Forest, Lake, Attack, Night)	RE raids village and both lose a battle	0
(Lake, Forest, Attack, Night)	"	0
(Forest, Lake, Ambush, Night)	RE raids village & leaves without casualties	−2
(Forest, Lake, Attack, Day)	"	−2
(Lake, Forest, Ambush, Night)	"	−2
(Lake, Forest, Attack, Day)	"	−2

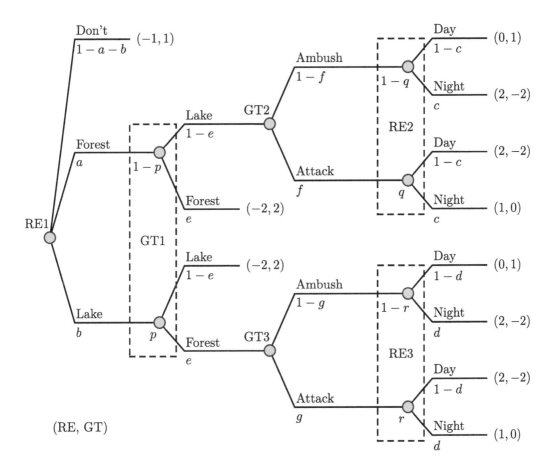

FIGURE 6.8   Warfare Game Tree.

just this node), the German tribe knows that they have defended an advance across the lake but the Roman empire legion has attacked the village after advancing through the forest.

The Warfare game has three subgames: First, the entire game has the empty history as its root. Second, the subgame with root (Forest, Lake) which corresponds in Figure 6.8 to the node labeled GT2 and everything to the right of GT2. Third, the subgame with root (Lake, Forest) which corresponds in Figure 6.8 to the node labeled GT3 and everything to the right of GT3. The effect of requiring the consistency condition to hold on every subgame rather than just the entire game is to ensure players choose an equilibrium even on subgames that have zero probability of being reached.

We now find the weak sequential equilibria. Figure 6.8 includes variable names below edges and nodes corresponding to the probabilities associated with an arbitrary assessment. Assume that the assessment is a weak sequential equilibrium.

Similar to the backward induction used for sequential games, we will start by analyzing the right-most action choices (under the assumption that they are reached). At the information set RE3 by choosing Day or Night, RE expects payoffs of $(1-r)(0) + r(2) = 2r$ or $(1-r)(2) + r(1) = 2-r$, respectively. If $r > 2/3$, then Day is RE's unique best response, and so GE's unique best response at GT3 would be Ambush (a payoff of 1 is larger than a payoff of -2). Now to achieve consistency of beliefs, it follows that $r = 0$. Since $r > 2/3$ and $r = 0$ are mutually contradictory, a weak sequential equilibrium cannot have $r > 2/3$. Similarly, if $r < 2/3$, then Night is RE's unique best response, and so GT's unique best response at GT3 would be Attack (a payoff of 0 is larger than a payoff of -2). Again, to achieve consistency of beliefs, it follows that $r = 1$. Since $r < 2/3$ and $r = 1$ are mutually contradictory, a weak sequential equilibrium cannot have $r < 2/3$.

Therefore, $r = 2/3$, and to achieve consistency of beliefs, $g = 2/3$. For this to be part of GT's best response strategy, GT's expected payoffs from choosing at GT3 to Ambush or Attack, $(1-d)(1) + d(-2) = 1 - 3d$ and $(1-d)(-2) + d(0) = 2d - 2$, must be equal. This implies that $d = 3/5$, and the expected payoffs at GT3 must be

$$(1-2/3)((1-3/5)(0,1) + (3/5)(2,-2)) + (2/3)((1-3/5)(2,-2) + (3/5)(1,0)) = (4/3, -4/5).$$

Since the structure and payoffs in the subgames rooted at GT2 and GT3 are identical, in any sequential equilibrium, $f = g = 2/3$, $q = r = 2/3$, $c = d = 3/5$, and the expected payoffs at GT2 are the same as at GT3, $(4/3, -4/5)$. Our analysis so far has resulted in the truncated game tree shown in Figure 6.9.

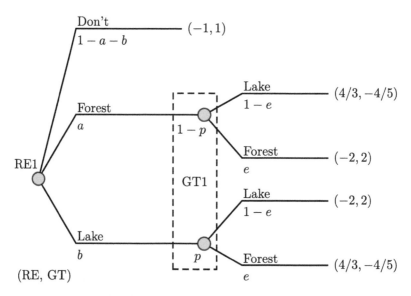

FIGURE 6.9 Warfare Truncated Game Tree.

Now we can analyze the GT1 information set. The expected payoff to GT when they choose Lake ($e = 0$) is $(1-p)(-4/5) + p(2) = 2.8p - 0.8$ and when they choose Forest

$(e = 1)$ is $(1 - p)(2) + p(-4/5) = 2 - 2.8p$. If $p > 1/2$, then Lake is the GT's unique best response, and so RE's best response at RE1 would be Forest (since the resulting payoff $4/3$ is larger than the payoffs of -1 and -2 by choosing Don't and Lake, respectively) which forces, by consistency of beliefs, $p = 0$. Since $p > 1/2$ and $p = 0$ are mutually contradictory, a weak sequential equilibrium cannot have $p > 1/2$. A similar argument shows that a weak sequential equilibrium cannot have $p < 1/2$. Thus, $p = 1/2$. To achieve consistency of beliefs, $a = b$.

If $a = b = 0$, then RE's expected payoff by choosing Don't Attack, -1, must be at least as big as RE's expected payoffs by choosing Forest and Lake, $(1 - e)(4/3) + e(-2)$ and $(1 - e)(-2) + e(4/3)$. By summing the two inequalities $-1 \geq (1 - e)(4/3) + e(-2)$ and $-1 \geq (1 - e)(-2) + e(4/3)$, we obtain $-2 \geq -2/3$, a contradiction showing that a weak sequential equilibrium cannot have $a = b = 0$.

Since $a = b > 0$, RE's expected payoffs by choosing Forest and Lake, $(1-e)(4/3)+e(-2)$ and $(1 - e)(-2) + e(4/3)$, must be equal. Thus, $e = 1/2$ and RE's expected payoffs at RE1 for choosing Forest or Lake is $-1/3$. Since $-1/3 > -1$, RE's unique best response at RE1 is $a = b = 1/2$.

In conclusion, we have shown that the assessment satisfying $a = b = 1/2$, $c = d = 3/5$, $f = g = 2/3$, $p = 1/2$, and $q = r = 2/3$ is the only possible weak sequential equilibrium. Since we found the best responses at each information set and chose the belief system to achieve consistency of beliefs, this assessment is the unique weak sequential equilibrium.

Observe that our analysis depends crucially on nontrivial mixed strategies, probabilistic beliefs, and expected payoffs. Expected payoffs are only meaningful if the payoffs are interpreted as vNM, rather than only ordinal, utilities.

6.3 INCOMPLETE INFORMATION

With imperfect information games, players may not always know where they are in a game. In a game with incomplete information, some players may be wondering even what game they are playing. Most commonly, games with incomplete information leave players not knowing the utility of the various outcomes to the other players. The following scenario demonstrates this information gap.

College Choices. Karl and Juanita are seniors attending the same high school and are in the midst of making decisions about which college to attend. Both could attend either a big state university or a small liberal arts college. Karl has a preference for the big state university because many of his friends are attending there, and Juanita has a preference for the small liberal arts college because it has a highly ranked marine biology program. They will make independent decisions about which college to attend because of financial and academic reasons, however, information about their relationship impacts their school preference. If they want to be a couple, they would prefer to be at the same institution even if it is someone's second choice. But, alas, as with many high school couples, neither really knows if the other wants the relationship to continue beyond the end of the school year.

In this scenario, both players know their own preferences (and hence payoffs) for the various outcomes, but they do not know the payoff function of the other player. To model Karl's uncertainty about Juanita's preferences, we will say that Juanita can be one of many possible types, with each type associated with different preferences. We'll assign a probability distribution to the types of Juanita based on Karl's beliefs. Similarly, we will model Juanita's uncertainty about Karl's preferences with a probability distribution over the possible types of Karl.

For Karl to make a wise choice of actions, he needs to know how each type of Juanita would play the game. For Juanita to make a wise choice of actions, she needs to know how each type of Karl would play the game. Thus, it is important to consider each pairing of a Karl type with each Juanita type. This suggests that before Karl and Juanita choose their actions, each is randomly assigned a type, each one knows their own type, but neither knows the type of the other player. This can be visualized with a game tree in which the first move belongs to an impartial nonplayer, called *chance*, which assigns to each player their type. Formally, we have the following definition wherein we also allow for the possibility that different types of Karl may have different beliefs about which types of Juanita are more likely or that different types of Juanita may have different beliefs about which types of Karl are more likely.

Definition 6.3.1. A *Bayesian game* consists of the following:

1. A set $N = \{1, 2, \ldots, n\}$ of at least two players.

2. A set T_i of types for each player i. We let $T = T_1 \times T_2 \times \cdots \times T_n$ be the set of type profiles, which we call the *type space*.

3. A set A_i of actions available to player i for each $i \in N$.

4. The set of outcomes $O = A = A_1 \times \cdots \times A_n$ that occur when typed players choose their actions according to the type profiles.

5. Belief functions ϕ_i, where $\phi_i(t_{-i}|t_i)$ is the probability that the other player types come from the type profile t_{-i} given that player i has type t_i. The function ϕ_i is a probability distribution on the truncated type space $T_{-i} = T_1 \times T_2 \times \cdot T_{i-1} \times T_{i+1} \times \cdots T_n$. (The notation $t_{-i} \in T_{-i}$ indicates a type profile with t_i removed.)

6. A rule stating that players simultaneously choose actions after privately learning their own types.

7. Utility functions u_i, where $u_i(a|t_i)$ is the cardinal utility player i of type t_i ascribes to outcome a, for each player i.

Under a certain condition, we can model Bayesian games of incomplete information as extensive games. To understand this condition, we consider a probability distribution φ on the entire type space T. For $t \in T$, consider the conditional probability $\varphi(t|t_i$ is player i's type in $t)$. When for each i this conditional probability equals the probability $\phi_i(t_{-i}|t_i)$, player i's belief that profile t occurs given that player i knows they are type t_i, then we say the belief functions ϕ_i *arise from a common prior*. All of the examples in this section will arise from a common prior, and so we can model any of them as extensive games.

For Bayesian games of incomplete information modeled as extensive games, the terminal histories are of the form $(t_1, \ldots, t_n, a_1, \ldots, a_n)$ where chance chooses the types $t_1, \ldots, t_n$ according to the common prior probability distribution φ on the type space T and each player i in turn chooses an action a_i in accordance with their type t_i. An information set for player i consists of all histories that contain a fixed type for player i, an arbitrary type for each of the other players, and arbitrary actions for all preceding players j in the tree.

Returning to our College Choice scenario, we see Karl and Juanita each have two actions available: either select the big state university (B) or the small liberal arts college (S), that is, $A_K = A_J = \{B, S\}$. There are two types of Karl: one who prefers to be at the same school as Juanita (K_a) and one who does not (K_b). Similarly, there are two corresponding types of Juanita, J_a and J_b. The outcomes are given in the set $\{BB, BS, SB, SS\}$, where the first element of each pair is Karl's action and the second is Juanita's action. Table 6.5 gives the utility functions for each type of Karl and Juanita. Both type a players place primary importance on being at the same school as the other person, but Karl secondarily prefers going to the big state university while Juanita secondarily prefers going to the small liberal arts college. Karl type b places primary importance on going to the big state university and secondary importance on avoiding Juanita, while Juanita type b places primary importance on avoiding Karl and secondary importance on going to the small liberal arts college.

TABLE 6.5 Utilities $u_i(\alpha|t_i)$ for Karl and Juanita in College Choices

α	K_a	K_b	J_a	J_b
BB	8	6	6	0
BS	2	8	2	8
SB	0	2	0	6
SS	6	0	8	2

For simplicity, we will assume that both types of Karl assign a probability of $1/2$ to each of Juanita's types. We assume that each of Juanita's types assign the same probability distribution to Karl's types, but in this case it is $2/3$ to type K_a and $1/3$ to type K_b. That is, Juanita is more optimistic about Karl's interest in being with her than Karl is about her wanting to be with him. Formally,

$$\phi_K(J_a|K_a) = 1/2 \quad \phi_K(J_b|K_a) = 1/2 \quad \text{and} \quad \phi_J(K_a|J_a) = 2/3 \quad \phi_J(K_a|J_b) = 2/3$$
$$\phi_K(J_a|K_b) = 1/2 \quad \phi_K(J_b|K_b) = 1/2 \qquad \qquad \phi_J(K_b|J_a) = 1/3 \quad \phi_J(K_b|J_b) = 1/3.$$

Observe that these belief functions arise from the common prior $\varphi(K_a, J_a) = \varphi(K_a, J_b) = \frac{1}{3}$ and $\varphi(K_b, J_a) = \varphi(K_b, J_b) = 1/6$ because based on this common prior, for example, the probability that Juanita of type J_b thinks Karl is of type K_a is

$$\frac{\varphi(K_a, J_b)}{\varphi(K_a, J_b) + \varphi(K_b, J_b)} = \frac{1/3}{1/3 + 1/6} = \frac{2}{3} = \phi_J(K_a|J_b).$$

All of this information is displayed (1) in a manner that emphasizes the individual strategic form games that types of players play in Table 6.6, and (2) in an extensive game tree representation in Figure 6.10.

TABLE 6.6 Preferences for Karl and Juanita in College Choices

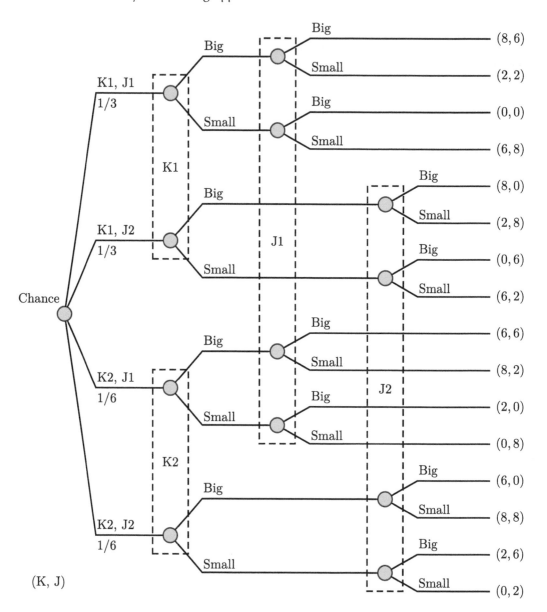

FIGURE 6.10 College Choices Game Tree.

To play this game, Karl and Juanita each must first receive a private signal (from chance) indicating which type of player they are; this models Karl and Juanita figuring out (for themselves individually) whether or not they want the relationship to continue. Then, they proceed to act accordingly. Determining an equilibrium will require an *a-priori* balancing of the actions and the expected payoffs before receiving the signals. Thus, we are led to the following definition.

Definition 6.3.2. In a Bayesian game, a *strategy* for player $i \in N$ is a function s_i from actions A_i and types T_i to probabilities, that is, $s_i(a_i|t_i)$ is the probability that player i of type t_i will choose action a_i. Given a strategy profile $s = (s_1, s_2, \ldots, s_n)$, we define the payoff to player i of type t_i of s to be the expected utility

$$v_i(s|t_i) = \sum_{t_{-i} \in T_{-i}} \sum_{a \in A} \phi_i(t_{-i}|t_i) s_1(a_1|t_1) \ldots s_n(a_n|t_n) u_i(a|t_i).$$

Recall that (r_i, s_{-i}) is the strategy profile that replaces s_i with r_i in the strategy profile $s = (s_1, s_2, \ldots, s_n)$. A strategy profile s is a *Bayesian equilibrium* if

$$v_i(s|t_i) \geq v_i((r_i, s_{-i})|t_i).$$

for every strategy r_i, for every type $t_i \in T_i$, and for every player $i \in N$.

If the belief functions arise from a common prior that assigns a positive probability to each element of the type space, then there is a one-to-one correspondence between the Bayesian equilibria in the Bayesian game and the weak sequential equilibria in the corresponding extensive game. To find the Bayesian equilibrium for the College Choices Bayesian game, we consider the game from the perspective of each type of each of the players. Table 6.7 shows the relevant calculations for Karl, and Table 6.8 shows the relevant calculations for Juanita.

TABLE 6.7   Karl's Preferences by Type

		Juanita's Strategy (J_a, J_b)			
		(B, B)	(B, S)	(S, B)	(S, S)
Karl K_a	B	$\frac{1}{2}8 + \frac{1}{2}8 = 8$	$\frac{1}{2}8 + \frac{1}{2}2 = 5$	$\frac{1}{2}2 + \frac{1}{2}8 = 5$	$\frac{1}{2}2 + \frac{1}{2}2 = 2$
	S	$\frac{1}{2}0 + \frac{1}{2}0 = 0$	$\frac{1}{2}0 + \frac{1}{2}6 = 3$	$\frac{1}{2}6 + \frac{1}{2}0 = 3$	$\frac{1}{2}6 + \frac{1}{2}6 = 6$
Karl K_b	B	$\frac{1}{2}6 + \frac{1}{2}6 = 6$	$\frac{1}{2}6 + \frac{1}{2}8 = 7$	$\frac{1}{2}8 + \frac{1}{2}6 = 7$	$\frac{1}{2}8 + \frac{1}{2}8 = 8$
	S	$\frac{1}{2}2 + \frac{1}{2}2 = 2$	$\frac{1}{2}2 + \frac{1}{2}0 = 1$	$\frac{1}{2}0 + \frac{1}{2}2 = 1$	$\frac{1}{2}0 + \frac{1}{2}0 = 0$

TABLE 6.8   Juanita's Preferences by Type

		Karl's Strategy (K_a, K_b)			
		(B, B)	(B, S)	(S, B)	(S, S)
Juanita J_a	B	$\frac{2}{3}6 + \frac{1}{3}6 = 6$	$\frac{2}{3}6 + \frac{1}{3}0 = 4$	$\frac{2}{3}0 + \frac{1}{3}6 = 2$	$\frac{2}{3}0 + \frac{1}{3}0 = 0$
	S	$\frac{2}{3}2 + \frac{1}{3}2 = 2$	$\frac{2}{3}2 + \frac{1}{3}8 = 4$	$\frac{2}{3}8 + \frac{1}{3}2 = 6$	$\frac{2}{3}8 + \frac{1}{3}8 = 8$
Juanita J_b	B	$\frac{2}{3}0 + \frac{1}{3}0 = 0$	$\frac{2}{3}0 + \frac{1}{3}6 = 2$	$\frac{2}{3}6 + \frac{1}{3}0 = 4$	$\frac{2}{3}6 + \frac{1}{3}6 = 6$
	S	$\frac{2}{3}8 + \frac{1}{3}8 = 8$	$\frac{2}{3}8 + \frac{1}{3}2 = 6$	$\frac{2}{3}2 + \frac{1}{3}8 = 4$	$\frac{2}{3}2 + \frac{1}{3}2 = 2$

With this information, we can now find the Bayesian equilibria. Action B strongly dominates action S for Karl type K_b; hence, in any Bayesian equilibrium, Karl's strategy must be of the form

$$((1 - p)B + pS, B) \text{ where } 0 \leq p \leq 1.$$

We consider three cases. First, if $p = 0$, then Karl's strategy is (B, B), and Juanita's unique best response is (B, S) because $6 > 2$ and $0 < 8$. Karl's best response to (B, S) is (B, B) because $5 > 3$ and $7 > 1$. Thus, one Bayesian equilibrium is

$$((B, B), (B, S))$$

with payoffs of 5, 7, 6, and 8 to K_a, K_b, J_a, and J_b, respectively.

Second, if $p = 1$, then Karl's strategy is (S, B), and Juanita's best responses are $(S, (1 - q)B + qS)$ for $0 \leq q \leq 1$ because $2 < 6$ and $4 = 4$. Karl's best response to $(S, (1 - q)B + qS)$ will be (S, B) if and only if $5(1 - q) + 2q \leq 3(1 - q) + 6q$, which is equivalent to $q \geq (1/3)$. Thus, a second set of Bayesian equilibria are

$$((S, B), (S, (1 - q)B + qS)) \text{ for } \tfrac{1}{3} \leq q \leq 1$$

with payoffs of $3(1 - q) + 6q$, $7(1 - q) + 8q$, 6, and 4 to K_a, K_b, J_a, and J_b, respectively.

Third, if $0 < p < 1$, then Juanita's best responses must be of the form $((1 - q)B + qS, S)$ for some $0 \leq q \leq 1$, and Karl's best response for type K_a must include both actions. This implies that $5(1 - q) + 2q = 3(1 - q) + 6q$ which is equivalent to $q = 1/3$. This implies that Juanita's best response for type J_a must include both actions. This implies that $6(1 - p) + 2p = 2(1 - p) + 6p$ which is equivalent to $p = 1/2$.

Thus, a last Bayesian equilibrium is

$$((\tfrac{1}{2}B + \tfrac{1}{2}S, B), (\tfrac{2}{3}B + \tfrac{1}{3}S, S))$$

with payoffs of 4, 22/3, 4, and 6 to K_a, K_b, J_a, and J_b, respectively.

The first Bayesian equilibrium has a natural interpretation: either type of Karl should go to the big university because, in comparison with Juanita, he attaches more importance to his college environment than the presence or absence of the other person, and the Juanita type who wants to be with Karl should go to the big university while the Juanita type who does not want to be with Karl should go to the liberal arts college. Juanita obtains the largest payoffs for this Bayesian equilibrium: 6 vs. 6 vs. 4 for Juanita type J_a and 8 vs. 4 vs. 6 for Juanita type J_b. The second set of Bayesian equilibria contains one pure-strategy Bayesian equilibrium which has a similar interpretation with the roles of Juanita and Karl interchanged. The remaining mixed-strategy Bayesian equilibria seem less likely both because of the non-efficient payoff pairs and the human psychological difficulty in implementing a mixed strategy. Type b players should go to the institution they most favor independent of the other player's choice. Type a players should go to the same institution, but the presence of two pure-strategy Bayesian equilibria makes the choice a Battle of the Sexes game.

In our first model for the College Choices scenario, we posited two preference types for each player. We will now examine a model in which each player can have any weighted average of the two types previously assumed, as summarized in Table 6.9. Juanita believes

TABLE 6.9   Utilities for Karl and Juanita in College Choices

α	$u_K(\alpha\|t)$	$u_J(\alpha\|s)$
BB	$8(1 - t) + 6t = 8 - 2t$	$6(1 - s) + 0s = 6 - 6s$
BS	$2(1 - t) + 8t = 2 + 6t$	$2(1 - s) + 8s = 2 + 6s$
SB	$0(1 - t) + 2t = 2t$	$0(1 - s) + 6s = 6s$
SS	$6(1 - t) + 0t = 6 - 6t$	$8(1 - s) + 2s = 8 - 6s$

Karl is of some type t in the closed interval $[0, 1]$ with probability density $\phi_J(t|s) = g(t) = t/2$. Karl believes Juanita is of some type s in the closed interval $[0, 1]$ with probability density $\phi_K(s|t) = f(s) = 1$. As was the case in the first model, each player's beliefs do not depend upon their own type. For each player, the new type 0 corresponds to the previous type a and the new type 1 corresponds to the previous type b. As was the case in the first model, Juanita's types are distributed uniformly while Karl's type are more heavily concentrated towards b.

Suppose $s = (s_K, s_J)$ is a Bayesian equilibrium. Then

$$p = \int_0^1 s_J(S|t) f(t) \, dt$$

is the probability that Karl is faced with a Juanita who chooses the action S, and

$$q = \int_0^1 s_K(S|t) g(t) \, dt$$

is the probability that Juanita is faced with a Karl who chooses the action S. Under these conditions, Karl's payoffs for his two pure action choices are

$$u_K(B) = (1-p)(8 - 2t) + p(2 + 6t) = 8 - 6p - 2t + 8pt$$
$$u_K(S) = (1-p)(2t) + p(6 - 6t) = 6p + 2t - 8pt,$$

and Juanita's payoffs for her two pure action choices are

$$u_J(B) = (1-q)(6 - 6s) + q(6s) = 6 - 6q - 6s + 12qs$$
$$u_J(S) = (1-q)(2 + 6s) + q(8 - 6s) = 2 + 6q + 6s - 12qs$$

Thus, Karl's best response is S if and only if $u_K(B) \leq u_K(S)$ if and only if $p \geq 2/3$ and $t \leq (3p - 2)/(4p - 1)$ as shown on the left half of Figure 6.11. Also, Juanita's best response

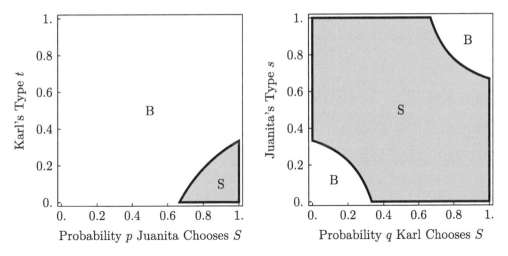

FIGURE 6.11   College Choices Best Responses.

is B if and only if $u_J(B) \geq u_J(S)$ if and only if (1) $p \leq 1/3$ and $s \leq (1 - 3q)/(3 - 6q)$, or (2) $p \geq 2/3$ and $s \geq (3q - 1)/(6q - 3)$ as shown on the right half of Figure 6.11.

Based upon these best responses, the probability that Karl will choose S is 0 if $p \leq 2/3$ and

$$\int_0^{(3p-2)/(4p-1)} t/2 \, dt = (3p - 2)^2/(4p - 1)^2$$

if $p > 2/3$, and the probability that Juanita will choose S is 1 if $1/3 \leq q \leq 2/3$,

$$\int_{(1-3q)/(3-6q)}^1 dt = 1 - (1 - 3q)/(3 - 6q)$$

if $q \leq 1/3$, and

$$\int_0^{(3q-1)/(6q-3)} dt = (3q-1)/(6q-3)$$

if $q \geq 2/3$. In Figure 6.12, the probability that Karl will choose S is the thick line, and the probability that Juanita will choose S is the dashed line. The Bayesian equilibrium occurs

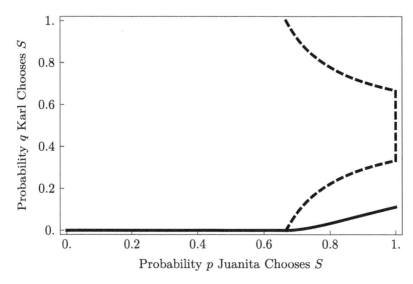

FIGURE 6.12 College Choices Best Responses.

at the unique intersection of the two curves: $(p,q) = (2/3, 0)$. This corresponds to Karl choosing B regardless of his type while Juanita chooses B if of type $s < 1/3$, chooses S if of type $s > 1/3$, and chooses either if of type $s = 1/3$.

Its reasonable to assume that this latter model of the College Choices scenario has high fidelity because it accounts for the entire range of potential feelings of Karl and Juanita towards their relationship with each other. On the other hand, it is more costly in terms of the mathematical effort required for its execution. While both the discrete and continuous models are flexible in the sense that the various parameters can be easily changed, neither model is flexible in the sense that they can be easily generalized.

6.4 BARGAINING GAMES WITH PRIVATE INFORMATION

The bargaining games studied in Chapter 4 included complete information about the various players' preferences about outcomes. The essential work of the chapter was in exploring the various solution concepts and their fairness properties.

In this section, we explore the problem of bargaining with incomplete information. At least one of the players has private information that they may or may not wish to reveal to the other player. However, because the players are allowed to communicate to each other as they make proposals and counter-proposals, this information may end up being revealed in the process.

Takeover Bid. Bisco Corp, a major food corporation, is considering buying Urchin Laboratories, a company that makes food processing machinery. The Urchin owners value the company at either $20 million or $80 million, depending on whether a new, more efficient machine is successful. Bisco, which has a reputation for making take-it-or-leave-it offers to other companies, is aware that Urchin is developing this machine but does not know if they have been successful although Bisco believes Urchin has a 1/2 probability of being successful. Regardless, Bisco does believe that they can add 50% to the value of the company if they purchase it.

We first model this as a discrete game of private information, that is one in which Bisco will pay one of two prices: $100 million or $25 million. After the offer, Urchin can choose to accept or reject. While Bisco's reputation suggests that they would not follow a rejection with a second offer, in order to determine whether building such a reputation is rational, we will assume that if a $25 million offer is rejected, Bisco could follow with a $100 million offer, which we will assume Urchin would accept. Chance first determines whether the new piece of machinery has been successfully built (Succeed) or not (Fail), each with probability 1/2. Without knowledge of chance's choice, Bisco then chooses a price: $100 million or $25 million. Urchin knows both its success status and the value of Bisco's offer when it chooses to either Accept or Reject the offer. Since we assume that both companies are self-interested and risk-neutral, for an accepted offer Bisco's payoff is 150% of the value of the company less the purchase price, and Urchin's payoff is the purchase price less the value of the company. This gives us the game tree in Figure 6.13.

Let us now find all weak sequential equilibria. At nodes U1 and U4, Accept is Urchin's best response since it yields a payoff that is greater than 0, the payoff for Reject. This means at information set B1, Bisco's payoff for choosing Offer $100 is $(1/2)(20)+(1/2)(-70) = -25$. If Bisco chooses Offer $25 at B1 and Quit at B2, then their payoff will be nonnegative. Thus, Bisco must choose Offer $25 at B1.

At node U2, Reject is Urchin's best response since it yields a payoff between 0 and 20 while Accept yields a payoff of -55. We now only need to determine Urchin's choice at U3 and Bisco's beliefs and choice at B2 as shown in Figure 6.14.

If at B2 Bisco chooses Quit, they get a payoff of 0, and if they choose Offer $100, they get a payoff of $(1-q)(20) + q(-70) = 20 - 90q$. Since $0 = 20 - 90q$ if and only if $q = 2/9$, there are three cases to consider.

Case 1. Suppose $q < 2/9$. Since $20 - 90q > 0$, Bisco will choose Offer $100 at B2. Then Urchin will Reject at U3 $(80 > 5)$. Now the probability the bottom node will be reached if B2 is reached is 1/2. In order for Bisco's beliefs to be consistent, $q = 1/2$. This contradiction to $q < 2/9$ shows that there is no weak sequential equilibrium in this case.

Case 2. Suppose $q = 2/9$. For there to be consistency of beliefs, $(1/2)c/(1/2+(1/2)c) = q$ which implies $c = q/(1-q) = 2/7$. For Urchin's completely mixed action choice at U3 to be a best response, their payoff for Reject, $(1-e)(80)+e(0) = 80-80e$, must equal their payoff for Accept, 5. Hence, $e = 75/80$. We have found one weak sequential equilibrium with payoff pair $(1/2)(1 - 75/80)(20,20) + (1/2)(75/80)(0,0) + (1/2)(1 - 2/7)(1 - 75/80)(-70,80) + (1/2)(1 - 2/7)(75/80)(0,0) + (1/2)(2/7)(5,5) = (-25/112, 25/8) \approx (-0.223, 3.125)$.

Case 3. Suppose $q > 2/9$. Since $20 - 90q < 0$, Bisco will choose Quit at B2. Then Urchin will Accept at U3 $(5 > 0)$. Consistency of beliefs implies $q = 1$. We have found one weak sequential equilibrium with payoff pair $(1/2)(0,0) + (1/2)(5,5) = (2.5, 2.5)$.

We can now see the value in Bisco building a reputation for giving take-it-or-leave-it offers. By doing so, the weak sequential equilibrium in which they Quit gives them a payoff of 2.5, whereas the other equilibrium gives them a negative expected payoff.

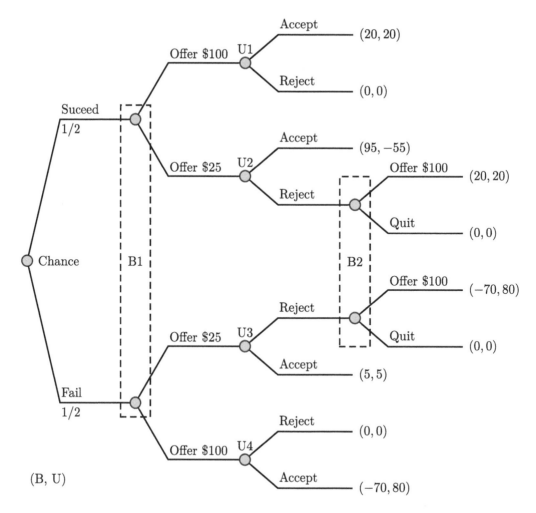

FIGURE 6.13 Takeover Bid Game Tree.

In the model above, we assumed Bisco had to choose between a high bid of $100 million or a low bid of $25 million. It's more realistic to assume that Bisco can actually choose from a continuous range of price options.

In a continuous model, we again have Chance first determining whether the new piece of machinery has been successfully built (Succeed) or not (Fail), each with probability $1/2$. Without knowledge of Chance's choice, Bisco then chooses a price $P \geq 0$ to offer Urchin; the sectors labeled P indicate the infinite number of edges corresponding to possible offers. Now knowing whether it has been successful and the value of Bisco's offer, Urchin chooses to either Accept or Reject the offer. We still assume that both companies are self-interested and risk-neutral, so for an accepted offer Bisco's payoff is 150% of the value of the company less the purchase price, and Urchin's payoff is the purchase price less the value of the company. Bisco has to determine a price P to offer Urchin, and Urchin needs to decide whether or not to accept the offer. Figure 6.15 illustrates the situation.

To find weak sequential equilibria which are equivalent to Bayesian equilibria given the common prior, we start at the right end of the game tree. Given an offer P by Bisco, Urchin's best response is Accept if (1) Succeed and $80 < P$ or (2) Fail and $20 < P$; is Reject if (1) Succeed and $0 \leq P < 80$ or (2) Fail and $0 \leq P < 20$; and is any probabilistic

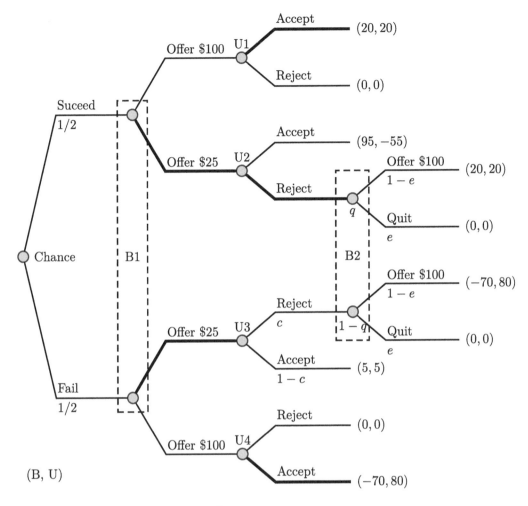

FIGURE 6.14 Takeover Bid Game Tree.

mixture of Reject and Accept if (1) Succeed and $P = 80$ or (2) Fail and $P = 20$. Given Urchin's strategy, Bisco's payoff is 0 if $P < 20$, $(1/2)(0) + (1/2)(30 - P) = 15 - (1/2)P$ if $20 < P < 80$, and $(1/2)(120 - P) + (1/2)(30 - P) = 75 - P$ if $80 < P$. If $P = 20$, Bisco's payoff is somewhere between 0 and $15 - (1/2)(20) = 5$ depending upon Urchin's strategy choice. If $P = 80$, Bisco's payoff is somewhere between $15 - (1/2)(80) = -25$ and $75 - 80 = -5$ depending upon Urchin's strategy choice. Bisco's payoff is graphed in Figure 6.16

It is clear from the graph that Bisco's unique best response is $P = 20$, but only if Urchin accepts an offer of 20 when it has failed, that is, the unique Bayesian equilibrium has Bisco offer 20 and Urchin chooses Accept if (1) Succeed and $80 < P$ or (2) Fail and $20 \leq P$; chooses Reject if (1) Succeed and $0 \leq P < 80$ or (2) Fail and $0 \leq P < 20$; and chooses any probabilistic mixture of Reject and Accept if Succeed and $P = 80$.

It is also interesting to realize that the act of rejecting offers larger than $20 million reveals that Urchin was in fact successful in building the new machine. However, unlike in the Tourist Bazaar in which there was a second round of bargaining, Bisco needs to maintain its reputation of take-it-or-leave-it bargaining and so must walk away from the

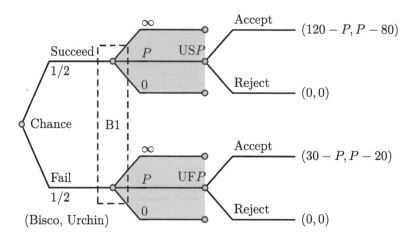

FIGURE 6.15 Takeover Game Tree.

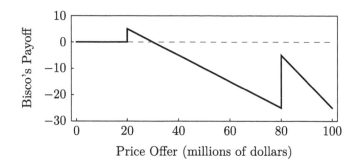

FIGURE 6.16 Bisco Payoff in Takeover Game If Urchin Is Using a Best Response.

opportunity. Although the equilibrium reveals Urchin's private information to Bisco, it does so after Bisco can take advantage of the revealed information.

In games with multiple equilibria, some equilibria can distinguish between types of players, and these are known as *separating equilibria*. Other equilibria mask the types of the players and are called *pooling equilibria*. We see this phenomenon in bargaining games and more generally in signaling games. Banks [6] provides more information on signaling games, but we briefly discuss one type below.

> **College Education**. In recent years there have been many editorials written about and debates over the value of a college education (See for example [113].) Some argue that it provides individuals with the skills needed to succeed in the employment market, while others argue that a college education provides no real advantage. Game theoretic models can shed some light on this subject.

The goal of a model of the above situation is to determine what education can signal about a person, especially to an employer. The education signaling games model, attributed to Michael Spence [105], and analyzed by many authors including [78], [112], and [12], gives us the tools to do this. The players in such a game are often an individual who decides what level of education to receive, and employers who use that information as a signal to determine the wages to offer.

6.5 INTERNATIONAL COLLABORATION

> **International Collaboration.** The International Telegraph Union was organized in 1865 to coordinate international telegraph communications. Two of the issues to be resolved included establishing a common code system and a common billing system. In fact, by 1875, the latter became a critical issue as the use of the telegraph system began to extend between continents. Within Europe, billing was done in twenty-word increments, and countries were using non-European communication to subsidize their European business. On the other hand, non-European members of the ITU wanted billing to be in single-word increments to increase the amount of traffic. There were many solutions available and initially the compromise was to have a high billing rate within Europe and a low rate outside of Europe [24], summarized in [73].

Morrow [73] discusses the ITU scenario and the more general topic of international collaboration across complex issues. In the ITU scenario, there were many solutions possible, and it was advantageous for everyone to achieve a coordinated solution. Different nations had different preferences among the solutions and had some information about the solutions. Yet no single nation could clearly see the consequences of implementing any particular solution. He further argues that these distribution and information problems often inhibit countries reaching agreements. The distribution problem reflects the fact that countries may have different preferences over the solution choices, while the information problem reflects the fact that countries may not know the value of the solution choices. These two problems are intertwined in that if the countries share information about the solutions, it can change their preferences among the solutions. While in the reverse direction, their preferences among the solutions may affect their willingness to share information. A game theoretic model can offer some insight on managing the tension between these two problems.

In this section, we present and discuss the model developed by Morrow and find as he did that different equilibria illustrate different ways that the problems of international collaboration can be resolved. At times we have been concerned about the predictive power of a model which has multiple equilibria. Here we will see that having multiple equilibria is reflective of reality, provides us with guidelines for how to achieve best outcomes, and so exhibits the flexibility of the model. While we have restructured the presentation of Morrow's model and introduced some new terminology and additional details, all results including the theorems in this section are due to him.

The basic framework for the International Collaboration extensive game begins with the Battle of the Sexes (BOS) strategic game, introduced in Chapter 3, and reproduced here in Table 6.10. The parameter $a > 1$ indicates the payoff for the "winner" of the "battle."

TABLE 6.10   Botswana or Switzerland (BOS)

(Botswana, Switzerland)	H	L
H	$(a, 1)$	$(0, 0)$
L	$(0, 0)$	$(1, a)$

We name the two countries Botswana and Switzerland to represent the non-European and European countries generally. Each has two actions available, H and L, which represent high and low pricing options, respectively. The two countries prefer to agree on an action, but have different preferences between the two available actions. As a strategic game, BOS has three Nash equilibria: (H, H), (L, L), and

$$\left(\frac{a}{a+1} H + \frac{1}{a+1} L, \frac{1}{a+1} H + \frac{a}{a+1} L \right).$$

As a bargaining game the first two equilibria seem to be reasonable outcomes, as might the correlated strategy $(1/2)(H, H) + (1/2)(L, L)$.

We now extend the BOS game model to allow the possibility that the countries do not know the value of the outcome (i.e., the consequences of their actions). In this more general framework, the countries are playing one of three games: either the BOS Game, a variation Coordinate H (COH) in which the countries prefer to coordinate on (H, H), or the analogous variation Coordinate L (COL) in which they prefer to coordinate on (L, L). These latter two games are shown in Table 6.11 and Table 6.12. COH and COL have the same three Nash equilibria as BOS.

TABLE 6.11   Coordinate H (COH)

(Botswana, Switzerland)	H	L
H	(a, a)	$(0, 0)$
L	$(0, 0)$	$(1, 1)$

TABLE 6.12   Coordinate L (COL)

(Botswana, Switzerland)	H	L
H	$(1, 1)$	$(0, 0)$
L	$(0, 0)$	(a, a)

We can now describe in detail the International Collaboration extensive game. There will be four stages as summarized in Table 6.13.

TABLE 6.13   International Collaboration Extensive Game Stages

0.	Parameters $a > 1$ and $1 \geq p > 0$ are set.
1.	Chance chooses the strategic game.
2.	Chance sends private signals to the countries.
3.	Countries exchange messages simultaneously in rounds.
4.	Countries choose actions simultaneously.

To introduce uncertainty, chance will choose the strategic game to be played during stage 1. The game selected by chance is unknown to either country. For a given probability p, they are playing the BOS Game with probability $1 - p$ and they are playing either COH or COL with equal probabilities of $p/2$. Symbolically,

$$\Pr(\text{BOS}) = 1 - p \qquad \Pr(\text{COH}) = \Pr(\text{COL}) = p/2.$$

The countries know p, but do not know the random action of chance.

In stage 2, each country is given a private signal of either the number 1 or the number 2 representing the best advice of its technical support staff. If the BOS Game is being played, each country's private signal will be determined by separate flips of a fair coin. If COH (resp., COL) is being played, then both countries receive the same signal 1 (resp., 2).

Symbolically, we use s_i to denote that country i receives signal s, and so

$$\begin{array}{ll}
\Pr(1_B | \text{BOS}) = \Pr(1_S | \text{BOS}) = 1/2 & \Pr(2_B | \text{BOS}) = \Pr(2_S | \text{BOS}) = 1/2 \\
\Pr(1_B | \text{COH}) = \Pr(1_S | \text{COH}) = 1 & \Pr(2_B | \text{COH}) = \Pr(2_S | \text{COH}) = 0 \\
\Pr(1_B | \text{COL}) = \Pr(1_S | \text{COL}) = 0 & \Pr(2_B | \text{COL}) = \Pr(2_S | \text{COL}) = 1.
\end{array}$$

We note that the unconditional probabilities for receipt of each signal are

$$\Pr(1_B) = \Pr(1_S) = (1/2)(1-p) + (1/2)(p) = 1/2$$
$$\Pr(2_B) = \Pr(2_S) = 1 - (1/2) \qquad\qquad = 1/2.$$

Having received a signal 1, Botswana realizes the probability it is playing BOS is (where $\Pr(EF)$ means the probability that events E and F occur)

$$\Pr(\mathrm{BOS}|1_B) = \frac{\Pr(1_B\mathrm{BOS})}{\Pr(1_B)} = \frac{\Pr(1_B|\mathrm{BOS})\Pr(\mathrm{BOS})}{\Pr(1_B)} = \frac{(1/2)(1-p)}{1/2} = 1 - p.$$

Obtained in a similar manner, the probabilities of playing each of the strategic games given the signal received are

$$\Pr(\mathrm{BOS}|1_B) = \Pr(\mathrm{BOS}|2_S) = \Pr(\mathrm{BOS}|2_B) = \Pr(\mathrm{BOS}|1_S) = 1 - p$$
$$\Pr(\mathrm{COH}|1_B) = \Pr(\mathrm{COL}|2_S) = \Pr(\mathrm{COL}|2_B) = \Pr(\mathrm{COH}|1_S) = p$$
$$\Pr(\mathrm{COL}|1_B) = \Pr(\mathrm{COH}|2_S) = \Pr(\mathrm{COH}|2_B) = \Pr(\mathrm{COL}|1_S) = 0.$$

With signal 1, Botswana knows it is in BOS or COH, and in either of those games, (H, H) is the Nash equilibrium that gives it the highest payoff. On the other hand, signal 2 is problematic for Botswana because the Nash equilibrium that gives it the highest payoff is either (H, H) if playing BOS or (L, L) if playing COL. Since Botswana's strategy goal is clear when it receives signal 1, we use Morrow's terminology for signal 1, calling it Botswana's *favorable signal*. Since Botswana's strategy goal is unclear when it receives signal 2, this is Botswana's *unfavorable signal*. By symmetry, signal 2 is Switzerland's favorable signal and signal 1 is Switzerland's unfavorable signal.

We can begin to see advantages to sharing information by each country sending a message to the other about their own signal. If both countries accurately report to the other that they received different signals, then they both know that they are playing BOS. If they accurately report the same signal, their confidence rises that they are playing the corresponding COH or COL game. The possibility that they are playing BOS creates a distribution problem, while the fact that neither country knows which game they are playing creates an information problem. Sharing information by sending messages may change the distribution problem. Further, as p increases, there is a larger chance the game is COL or COH and therefore sharing information becomes even more valuable in determining which game is likely being played.

In stage 3, each country sends a message m_{i1}, $i = \{S, B\}$ consisting of the number 1 or number 2. These messages are sent simultaneously, and what message is sent can depend upon the signal s_i the country received in stage 2. After receipt of the message from the other country, each country sends a second message m_{i2} consisting of the number 1 or number 2. These messages are sent simultaneously, and what message is sent can depend upon the signal s_i the country received in stage 2 and the first-round messages m_{B1} and m_{S1}.

In stage 4, each country chooses an action a_i that is either H or L. This action choice can depend upon the signal s_i the country received in stage 2, the first-round messages m_{B1} and m_{S1}, and the second-round messages m_{B2} and m_{S2}. Countries receive the payoff corresponding to the strategic game chance selected in stage 1 and the actions chosen by the countries in stage 4.

A strategy specifies the messages sent by a country as a function of its signal and the previous messages sent and received, and its action as a function of its signal and messages sent and received. Since chance first selects one of three strategic games and then sends one of two possible signals to each country, then each country sends one of two possible messages in two rounds, and finally each country chooses one of two possible actions, there are $(3)(4)(2)(2)(2)(2)(2)(2) = 768$ terminal histories. Each country bases its first message on the signal it received (2 possibilities), bases its second message on the signal it received and the first-round messages (2^3 possibilities), and bases its action choice on the signal it received and the two rounds of messages (2^5 possibilities); hence, each country has $2 + 2^3 + 2^5 = 42$ information sets. Since at each information set, a country has two options, there are $2^{42} = 4,398,046,511,104$ pure strategies. A complete analysis of this game seems daunting. Instead we consider some reasonable strategy choices, determine when those choices lead to an equilibrium, compare the resulting payoffs with maximum possible payoffs with any pair of strategies, and interpret the results in the context of the original scenario.

Strategies can be very simple or quite complex. We will identify and explore different types of strategies with natural interpretations. A strategy is *unaware* if the messages sent and actions chosen are independent of the signal received; otherwise, the strategy is *aware*. A strategy is said to *send uninformative messages* if the messages sent are independent of the signal received and the action choice is independent of the messages sent. A strategy is *babbling* if it sends uninformative messages and the second message sent is independent of the first-round messages.

TABLE 6.14   Important Strategy Types and Strategies

	Message Sent Depends On			Action Chosen Depends On		
	Signal	Previous Messages Sent	Previous Messages Received	Signal	Previous Messages Sent	Previous Messages Received
Unaware	No			No		
Babbling	No		No		No	No
Choose H and Choose L	No		No	No	No	No
Coordinate on Message	No		No	No	Yes	Yes
Follow the Signal	No		No	Yes	No	No
Communicate	Yes	Yes	Yes	Yes	Yes	Yes
Lead	Yes		No	Yes		No
Follow	No		No	No	No	Yes

These definitions, and the characteristics of strategies to be introduced later, are summarized in Table 6.14. Note that if a strategy type may or may not have a particular strategy dependency, then the corresponding cell is left blank.

Unaware strategies ignore the information provided by the signal, strategies that send uninformative messages provide no information to the other country, and babbling strategies neither provide or use information provided by messages. Notice that we make no restriction on whether messages sent are independent of previous messages sent. Messages based on independent coin flips and messages that are repeated based on a single coin flip are equally uninformative if sent independent of the signal and messages received and the future action chosen.

Strategies that are unaware and babbling model situations in which there is no technical expertise available and no international forum for discussion or negotiation. One such strategy is *Choose H*: send uninformative messages and choose the action H. Likewise, *Choose L*: send uninformative messages and choose the action L is unaware and babbling. They are the only such strategies that are part of an equilibrium.

Theorem 6.5.1. The strategy profiles (Choose H, Choose H) and (Choose L, Choose L) are weak sequential equilibria in the International Collaboration extensive game. Furthermore, they are the only equilibria using unaware and babbling strategies, and no other pair of unaware strategies yields a higher sum of expected payoffs.

Proof. Suppose Switzerland uses an unaware and babbling strategy. Then it ignores its signal and messages received, sends uninformative messages, and chooses action H with probability $1 - r$ and action L with probability r. Since Switzerland sends uninformative messages and ignores messages received, Botswana's best response can also send uninformative messages and ignore messages received because to do otherwise will not change the outcome. On the other hand, it may be beneficial for Botswana to choose its actions based on its signal. Given the assumed strategy choice for Switzerland, the expected payoff to Botswana for choosing action a_B given the signal s_B it receives is

$$
\begin{aligned}
u_B(a_B|s_B) = {} & \Pr(\text{BOS}|s_B)\Pr(H_S)u_B(a_B H_S|\text{BOS}) \\
& + \Pr(\text{BOS}|s_B)\Pr(L_S)u_B(a_B L_S|\text{BOS}) \\
& + \Pr(\text{COH}|s_B)\Pr(H_S)u_B(a_B H_S|\text{COH}) \\
& + \Pr(\text{COH}|s_B)\Pr(L_S)u_B(a_B L_S|\text{COH}) \\
& + \Pr(\text{COL}|s_B)\Pr(H_S)u_B(a_B H_S|\text{COH}) \\
& + \Pr(\text{COL}|s_B)\Pr(L_S)u_B(a_B L_S|\text{COH}).
\end{aligned}
$$

Since the payoff is zero whenever the action choices are not the same, three of the above terms are zero. An additional term will be zero because given the signal received by Botswana, one of COH or COL is impossible.

The following calculations start by including only those terms that are non-zero, plug in the appropriate values for the probabilities and payoffs, and finish with a simplified form.

$$
\begin{aligned}
u_B(H_B|1_B) &= \Pr(\text{BOS}|1_B)\Pr(H_S)u_B(H_B H_S|\text{BOS}) \\
&\quad + \Pr(\text{COH}|1_B)\Pr(H_S)u_B(H_B H_S|\text{COH}) \\
&= (1-p)(1-r)(a) + (p)(1-r)(a) \\
&= a(1-r) \\
u_B(L_B|1_B) &= \Pr(\text{BOS}|1_B)\Pr(L_S)u_B(L_B L_S|\text{BOS}) \\
&\quad + \Pr(\text{COH}|1_B)\Pr(L_S)u_B(L_B L_S|\text{COH}) \\
&= (1-p)(r)(1) + (p)(r)(1) \\
&= r \\
u_B(H_B|2_B) &= \Pr(\text{BOS}|2_B)\Pr(H_S)u_B(H_B H_S|\text{BOS}) \\
&\quad + \Pr(\text{COL}|2_B)\Pr(H_S)u_B(H_B H_S|\text{COL}) \\
&= (1-p)(1-r)(a) + (p)(1-r)(1) \\
&= (a + p - ap)(1-r) \\
u_B(L_B|2_B) &= \Pr(\text{BOS}|2_B)\Pr(L_S)u_B(L_B L_S|\text{BOS}) \\
&\quad + \Pr(\text{COL}|2_B)\Pr(L_S)u_B(L_B L_S|\text{COL}) \\
&= (1-p)(r)(1) + (p)(r)(a) \\
&= (1 - p + ap)r
\end{aligned}
$$

We can now see that $u_B(H_B|1_B) \geq u_B(L_B|1_B)$ if and only if $r \leq a/(a+1)$, and $u_B(H_B|2_B) \geq u_B(L_B|2_B)$ if and only if $r \leq (a+p-ap)/(a+1+2p)$. Observe that the right-hand side quantity in each inequality is strictly between 0 and 1. Hence, if Switzerland chooses H with certainty, then $r = 0$ and Botswana's best response is to choose H with certainty. Similarly, if Switzerland chooses L with certainty, then $r = 1$ and Botswana's best response is to choose L with certainty. Because of the symmetry of the two countries, this shows that (Choose H, Choose H) and (Choose L, Choose L) are equilibria.

For Botswana to have an unaware and babbling strategy as a best response when $0 < q < 1$, it must be indifferent between its two actions after receiving either signal, which implies $r = a/(a+1)$ and $r \leq (a+p-ap)/(a+1+2p)$. After some algebra, we obtain $p(a^2 + 2a - 1) = 0$. Since $p > 0$ and $a > 1$ implies $a^2 + 2a - 1 > 2$, we obtain a contradiction. Thus, no other equilibria using unaware and babbling strategies exist.

For the (Choose H, Choose H) equilibrium ($r = 0$), Botswana's expected payoff is

$$\Pr(1_B)u_B(H_B|1_B) + \Pr(2_B)u_B(H_B|2_B)$$
$$= (1/2)a(1-0) + (1/2)(a+p-ap)(1-0)$$
$$= a - (1/2)(a-1)p.$$

For the (Choose L, Choose L) equilibrium ($r = 1$), Botswana's expected payoff is

$$\Pr(1_B)u_B(L_B|1_B) + \Pr(2_B)u_B(L_B|2_B)$$
$$= (1/2)(1) + (1/2)(1-p+ap)(1)$$
$$= 1 + (1/2)(a-1)p.$$

By symmetry, Switzerland's expected payoff is $1 + (1/2)(a-1)p$ in the (Choose H, Choose H) equilibrium and $a - (1/2)(a-1)p$ in the (Choose L, Choose L) equilibrium. The sum of the country payoffs in each equilibrium is $a - (1/2)(a-1)p + 1 + (1/2)(a-1)p = a + 1$.

The expected payoff to each country is zero for the strategy pairs (Choose H, Choose L) and (Choose L, Choose H). Given a pair of unaware strategies, there is a resulting probability distribution over the four action pairs that is independent of the strategic game chosen. So, the expected payoff pair for a pair of unaware strategies is a weighted average of the payoff pairs for (Choose H, Choose H), (Choose H, Choose L), (Choose L, Choose H), and (Choose L, Choose L). Since the sum of these payoffs are $a + 1$, 0, 0, and $a + 1$, the sum of the expected payoffs for a pair of unaware strategies is no greater than $a + 1$, the sum obtained for the strategy for the two equilibria described in the theorem statement. □

The theorem tells us that if there is no technical expertise available and no international forum for discussion or negotiation, the countries would each like to adopt the same pricing option; however, since there are two equilibria with each favored by one of the countries, the distribution problem shows itself in full force. Without an international forum, it is unclear how the countries will settle into one of the two equilibria, and there is a clear danger that no equilibrium strategies will be chosen, resulting in zero payoffs.

We can model the presence of an international forum by having the countries make use of informative messages. We will first consider a strategy that is not babbling but is still unaware. *Coordinate on Message* has the country flip a fair coin to determine which message to send in all rounds and then choose action H if the last round messages sent and received match and choose action L otherwise. This strategy addresses the distribution problem superbly.

Theorem 6.5.2. The strategy profile (Coordinate on Message, Coordinate on Message) is a weak sequential equilibrium in the International Collaboration extensive game. Furthermore, no other pair of unaware strategies yields a higher sum of expected payoffs.

It is somewhat surprising that Botswana cannot in some way make use of its signal to better its response to Switzerland's use of Coordinate on Message. This happens because Switzerland's fair coin flip determination of its message makes it impossible for Botswana to change the probability of messages matching. Had Switzerland used an unfair coin, then Botswana's message choice would have affected the matching probability, and so Botswana's signal could have been put to productive use.

In the world of negotiations, the (Coordinate on Message, Coordinate on Message) equilibrium might be called "splitting the difference." This equilibrium solves the distribution problem by having the countries choose a correlated strategy. Unfortunately, this equilibrium may lead to non-efficient solutions if COH or COL is actually being played, which is an indicator that it is not addressing the information problem.

We now solve the information problem while ignoring the distribution problem by using an aware but babbling strategy. Such strategies model countries having and using technical expertise but having no international forum for discussion and negotiation. *Follow the Signal* sends uninformative messages and chooses the action that corresponds to the signal received (i.e., choose action H if signal 1 is received and choose action L if signal 2 is received). The intent of this strategy is to choose the action that is most likely part of the better equilibrium for the country. Follow the Signal will do this if p is sufficiently large and one of COH or COL is the strategic game being played.

Theorem 6.5.3. The strategy profile (Follow the Signal, Follow the Signal) is a weak sequential equilibrium in the International Collaboration extensive game if and only if $p \geq (a-1)/(3a-1)$. Furthermore, no pair of babbling strategies yields a higher sum of expected payoffs if and only if $p \geq (a-1)/(3a-1)$.

Proof. Suppose Switzerland uses Follow the Signal, summarized in Table 6.15.

TABLE 6.15   Switzerland Uses Follow the Signal

signal → message		signal → action	
1_S	→ babble	1_S	→ H_S
2_S	→ babble	2_S	→ L_S

We will show that Follow the Signal is a best response for Botswana when $p \geq (a-1)/(3a-1)$. Since Switzerland sends uninformative messages and ignores messages received, Botswana's best response can also send uninformative messages and ignore messages received because to do otherwise will not change the outcome. On the other hand, it may be beneficial for Botswana to choose its actions based on its signal. Our approach to calculating Botswana's expected payoff for an action choice a_B will be to first partition Switzerland's action choices into the possible events $\text{BOS1}_S H_S$, $\text{BOS2}_S L_S$, $\text{COH1}_S H_S$, and $\text{COL2}_S L_S$. Hence, Botswana's payoff for choosing a_B after receiving signal s_B is

$$
\begin{aligned}
u_B(a_B|s_B) = {} & \Pr(\text{BOS1}_S H_S|s_B)u_B(a_B H_S|\text{BOS}) \\
& + \Pr(\text{BOS2}_S L_S|s_B)u_B(a_B L_S|\text{BOS}) \\
& + \Pr(\text{COH1}_S H_S|s_B)u_B(a_B H_S|\text{COH}) \\
& + \Pr(\text{COL2}_S L_S|s_B)u_B(a_B L_S|\text{COL}).
\end{aligned}
$$

Two of the terms will be zero because the payoff is zero whenever the action choices of the countries do not match, and a third term will be zero when the strategic game is incompatible with the signal received by Botswana. The remaining probabilities can be calculated using standard conditional probability rules, e.g.,

$$\Pr(\text{BOS}1_S H_S | 1_B) = \Pr(\text{BOS}|1_B)\Pr(1_S|\text{BOS}1_B)\Pr(H_S|1_S\text{BOS}1_B)$$
$$= (1-p)(1/2)(1).$$

In the following calculations, we first include only the nonzero terms, then state the relevant conditional probabilities and payoffs, and finally state a simplified form.

$$u_B(H_B|1_B) = \Pr(\text{BOS}1_S H_S|1_B)u_B(H_B H_S|\text{BOS})$$
$$+ \Pr(\text{COH}1_S H_S|1_B)u_B(H_B H_S|\text{COH})$$
$$= (1-p)(1/2)(1)(a)$$
$$+ (p)(1)(1)(a)$$
$$= (1/2)a(1+p)$$
$$u_B(L_B|1_B) = \Pr(\text{BOS}2_S L_S|1_B)u_B(L_B L_S|\text{BOS})$$
$$= (1-p)(1/2)(1)(1)$$
$$= (1/2)(1-p)$$
$$u_B(H_B|2_B) = \Pr(\text{BOS}1_S H_S|2_B)u_B(H_B H_S|\text{BOS})$$
$$= (1-p)(1/2)(1)(a)$$
$$= (1/2)a(1-p)$$
$$u_B(L_B|2_B) = \Pr(\text{BOS}2_S L_S|2_B)u_B(L_B L_S|\text{BOS})$$
$$+ \Pr(\text{COL}2_S L_S|2_B)u_B(L_B L_S|\text{COL})$$
$$= (1-p)(1/2)(1)(1)$$
$$+ (p)(1)(1)(a)$$
$$= (1/2)(1-p) + ap.$$

Since $u_B(H_B|1_B) = (1/2)a(1+p) > (1/2)(1-p) = u_B(L_B|1_B)$ for all probabilities p, Botswana's best response is to choose action H_B when it receives signal 1_B. Since $u_B(L_B|2_B) \geq u_B(H_B|2_B)$ if and only if $(1/2)(1-p) + ap \geq (1/2)a(1-p)$ if and only if $p \geq (a-1)/(3a-1)$, Follow the Signal is a best response for Botswana if and only if $p \geq (a-1)/(3a-1)$.

The expected payoff to each country when both use Follow the Signal is

$$\Pr(1_B)u_B(H_B|1_B) + \Pr(2_B)u_B(L_B|2_B)$$
$$= (1/2)(1/2)a(1+p) + (1/2)((1/2)(1-p) + ap)$$
$$= (1/4)((1+a) + (3a-1)p).$$

Consider a pair of babbling strategies. Since payoffs are directly dependent only on the strategic games being played and the action choices, and with babbling strategies the action choices are dependent only on the private signals received, the possible payoff pairs obtained from babbling strategies are weighted combinations of the payoff pairs obtained by using strategies in which actions are pure functions of the signal received. Each such strategy pair can be denoted by $(a_{B1}, a_{B2}, a_{S1}, a_{S2})$ where a_{is} is the action choice by country i when it receives the signal s. If $\bar{u}(a_B a_S|g)$ denotes the sum of the two country payoffs when they are playing the strategic game g and their actions are a_B and a_S, then the sum of expected payoffs to the two countries when using the strategy denoted by $(a_{B1}, a_{B2}, a_{S1}, a_{S2})$ is

$$\begin{aligned}
\bar{u}(a_{B1}, a_{B2}, a_{S1}, a_{S2}) &= \Pr(\text{BOS}1_B 1_S)\bar{u}(a_{B1}a_{S1}|\text{BOS}) \\
&+ \Pr(\text{BOS}1_B 2_S)\bar{u}(a_{B1}a_{S2}|\text{BOS}) \\
&+ \Pr(\text{BOS}2_B 1_S)\bar{u}(a_{B2}a_{S1}|\text{BOS}) \\
&+ \Pr(\text{BOS}2_B 2_S)\bar{u}(a_{B2}a_{S2}|\text{BOS}) \\
&+ \Pr(\text{COH}1_B 1_S)\bar{u}(a_{B1}a_{S1}|\text{COH}) \\
&+ \Pr(\text{COL}2_B 2_S)\bar{u}(a_{B2}a_{S2}|\text{COL}) \\
&= (1/4)(1-p)\bar{u}(a_{B1}a_{S1}|\text{BOS}) \\
&+ (1/4)(1-p)\bar{u}(a_{B1}a_{S2}|\text{BOS}) \\
&+ (1/4)(1-p)\bar{u}(a_{B2}a_{S1}|\text{BOS}) \\
&+ (1/4)(1-p)\bar{u}(a_{B2}a_{S2}|\text{BOS}) \\
&+ (1/2)(p)\bar{u}(a_{B1}a_{S1}|\text{COH}) \\
&+ (1/2)(p)\bar{u}(a_{B2}a_{S2}|\text{COL}).
\end{aligned}$$

Given the symmetries in the game definition, the sixteen pure strategies of the form $(a_{B1}, a_{B2}, a_{S1}, a_{S2})$ have the following expected payoff sums:

$$\begin{aligned}
\bar{u}(H,H,H,H) &= (4/4)(1-p)(a+1) + (1/2)p(2a) + (1/2)p(2) &&= a+1 \\
\bar{u}(H,H,H,L) &= (2/4)(1-p)(a+1) + (1/2)p(2a) + (1/2)p(0) &&< a+1 \\
\bar{u}(H,H,L,H) &= (2/4)(1-p)(a+1) + (1/2)p(0) + (1/2)p(2) &&< a+1 \\
\bar{u}(H,H,L,L) &= (0/4)(1-p)(a+1) + (1/2)p(0) + (1/2)p(0) &&= 0 \\
\bar{u}(H,L,H,L) &= (2/4)(1-p)(a+1) + (1/2)p(2a) + (1/2)p(2a) && \\
\bar{u}(H,L,L,H) &= (2/4)(1-p)(a+1) + (1/2)p(0) + (1/2)p(0) &&< a+1.
\end{aligned}$$

Therefore, the largest sum is $(1/2)((a+1) + (3a-1)p)$ if $p \geq (a+1)/(3a-1)$ and $a+1$ if $p \leq (a+1)/(3a-1)$. The sum of expected payoffs for (Follow the Signal, Follow the Signal) is equal to this largest sum if $p \geq (a+1)/(3a-1)$. □

If p is sufficiently small, BOS is the most likely strategic game being played and it may be advantageous to not follow an unfavorable signal. Another aware but babbling strategy type that captures this idea is *Follow the Signal Mostly*: send uninformative messages, and choose the favorable action if the signal is favorable and choose the favorable action with probability q if the signal is unfavorable.

Theorem 6.5.4. The strategy profile (Follow the Signal Mostly, Follow the Signal Mostly) is a weak sequential equilibrium in the International Collaboration extensive game if and only if $p \leq (a-1)/(3a-1)$ and the probability of choosing the favorable action given receipt of an unfavorable signal is $q = (a - 1 - (3a - 1)p)/((a+1)(1-p))$.

While we have found the equilibrium using Follow the Signal for high values of p is the best possible among babbling strategies, the equilibrium using Follow the Signal Mostly for low values of p does not yield an expected payoff sum equal to the upper bound we found for babbling strategy pairs in the proof to a previous theorem. The equilibrium strategy pairs (Choose H, Choose H) and (Choose L, Choose L) achieve that upper bound for low values of p. The trade-off is the asymmetry of the individual payoffs. We suspect, but have not proved, that (Follow the Signal Mostly, Follow the Signal Mostly) yields the highest possible expected payoff sum among equilibrium payoff pairs in babbling strategies in which the expected payoffs are the same. If our suspicion is correct, then we have completely solved the distribution and information problems separately.

To address both the distribution and information problems simultaneously, we require that the messages convey meaningful information about the signals received and the action choices make use of both the signals and messages. It seems reasonable to consider what conditions might lead the countries to honestly send messages revealing their signals.

In the *Communicate Honestly* strategy, the country sends its signal as its first message. If the first messages sent and received are 1, send 1 again and choose action H. If the first messages sent and received are 2, send 2 again and choose action L. If the first messages sent and received do not match, use Coordinate on Message, that is, flip a fair coin to choose the second message to send, and then choose H if the second-round messages sent and received match and choose L otherwise. If both countries were to use Communicate Honestly, then their sum of payoffs would be the highest possible. Unfortunately, these strategies are not always in equilibrium.

Theorem 6.5.5. The strategy profile (Communicate Honestly, Communicate Honestly) is a weak sequential equilibrium in the International Collaboration extensive game if and only if $p \geq 1/2$. Furthermore, no pair of strategies yields a higher sum of expected payoffs for any value of p.

For $p < 1/2$, the countries will want to send dishonest messages about the signals that they have received. In particular, when the signal is unfavorable, the countries have an incentive and opportunity to misinform the other country. Therefore, each country must discount the messages that they received based on their interpretation of the size of the distribution problem (i.e., the value of a) and the likelihood that the countries have common interests (i.e., the value of p).

In order to dissuade dishonesty, a mechanism must be put into place that punishes countries who may be lying, leading to the *Communicate Dishonestly* strategy which consists of the following instructions:

1. If the country receives a favorable signal, send that as its initial message.

2. If the country receives an unfavorable signal, send a favorable initial message with probability r.

3. If both countries send message 1 initially, send message 1 again and then choose H.

4. If both countries send message 2 initially, send message 2 again and then choose L.

5. If Botswana sends message 2 and Switzerland sends message 1 initially, then use Coordinate on Message.

6. If Botswana sends message 1 and Switzerland sends message 2 initially, then use Follow the Signal.

The second statement formalizes acting on the incentive to be dishonest, while the fifth statement rewards the countries with coordination when they both send honest messages of unfavorable signals received and the sixth statement protects countries against receiving dishonest messages of favorable signals.

If both send message 1, Switzerland is telling the truth about receiving an unfavorable signal and both countries know that they are playing BOS or COH. Therefore, choosing H is optimal for both countries. Similarly, if both send message 2, Botswana is telling the truth that they are playing BOS or COL, in which case choosing L is optimal. If both countries truthfully send messages of unfavorable signals, then they are likely playing BOS, in which case their optimal strategy is to coordinate their play. Conversely, if one or both countries are potentially sending untruthful messages, their best responses depend on the probability of deceit (r), and an equilibrium is achieved when this probability is chosen so that the message is effectively uninformative, and hence a country ignores the message and relies exclusively on their signal. Hence for the right choice of r the strategy profile described above is an equilibrium.

Theorem 6.5.6. The strategy profile (Communicate Dishonestly, Communicate Dishonestly) is a weak sequential equilibrium in the International Collaboration extensive game if and only if the probability of acting dishonestly is $r = (a-1)/(a+1)$ and $p \geq (a-1)/(3a+1)$.

Since $(a-1)/(3a+1) < 1/2$, the strategy profile (Communicate Dishonestly, Communicate Dishonestly) is an equilibrium over a larger interval of p than the interval over which (Communicate Honestly, Communicate Honestly) is an equilibrium. Unfortunately, (Communicate Dishonestly, Communicate Dishonestly) yields smaller payoffs than (Communicate Honestly, Communicate Honestly). We suspect, but have not proved, that there are no symmetric equilibria that yield payoffs approaching the upper bound when $p < 1/2$.

In international relationships, countries are not always treated identically and often there is a lead country and following country. This insight yields yet another equilibrium profile. In a *Lead* strategy, the eventual action choice depends on the signal in some manner and the messages sent match the pre-determined action (i.e., if action H is chosen and message 1 is sent and if action L is chosen, message 2 is sent). *Follow* sends uninformative messages and chooses the action corresponding to the last message received.

Theorem 6.5.7. The strategy profiles (Lead, Follow) and (Follow, Lead) are weak sequential equilibria in the International Collaboration extensive game if and only if the Lead country chooses its action in the following manner: if $p > 1/2$ follow its signal, if $p < 1/2$, choose the favorable action, and if $p = 1/2$, choose the favorable action upon receipt of the favorable signal and choose either action with indifference upon receipt of an unfavorable signal. Furthermore, no pair of strategies yields a higher sum of expected payoffs for $p > 1/2$.

Proof. Suppose Botswana uses Lead. Then there is no message that Switzerland can send that will change Botswana's action choice. Since Botswana has announced its planned action choice through its messages, Switzerland's best response is to choose the same action announced by Botswana, achieving a positive payoff versus obtaining a zero payoff by not choosing the same action.

Suppose Switzerland uses Follow. Then Switzerland's action choice will be Botswana's action choice, and the payoff to Botswana for its action choices given the signal received are

$$
\begin{aligned}
u_B(H_B|1_B) &= \Pr(\text{BOS}|1_B)u_B(H_BH_S|\text{BOS}) \\
&\quad + \Pr(\text{COH}|1_B)u_B(H_BH_S|\text{COH}) \\
&= (1-p)(a) + (p)(a) = a \\
u_B(L_B|1_B) &= \Pr(\text{BOS}|1_B)u_B(L_BL_S|\text{BOS}) \\
&\quad + \Pr(\text{COH}|1_B)u_B(L_BL_S|\text{COH}) \\
&= (1-p)(1) + (p)(1) = 1 < a \\
u_B(H_B|2_B) &= \Pr(\text{BOS}|2_B)u_B(H_BH_S|\text{BOS}) \\
&\quad + \Pr(\text{COL}|2_B)u_B(H_BH_S|\text{COL}) \\
&= (1-p)(a) + (p)(1) \\
u_B(L_B|2_B) &= \Pr(\text{BOS}|2_B)u_B(L_BL_S|\text{BOS}) \\
&\quad + \Pr(\text{COL}|2_B)u_B(L_BL_S|\text{COL}) \\
&= (1-p)(1) + (p)(a).
\end{aligned}
$$

Since $u_B(H_B|1_B) > u_B(L_B|1_B)$, Botswana's best response having received a favorable signal is to choose its favorable action. If $p > 1/2$, then $u_B(H_B|2_B) < u_B(L_B|2_B)$ and Botswana should follow its signal. If $p < 1/2$, then $u_B(H_B|2_B) > u_B(L_B|2_B)$ and Botswana should always choose its favorable action. If $p = 1/2$, then $u_B(H_B|2_B) = u_B(L_B|2_B)$ and Botswana can choose any action upon receipt of an unfavorable signal.

If Botswana uses the described Lead strategy and Switzerland uses Follow, then Botswana's expected payoff is

$$
\begin{array}{ll}
a - (1/2)(a-1)p & \text{if } p < 1/2 \\
(1/4)(3a+1) & \text{if } p = 1/2 \\
(1/2)((a+1) + (a-1)p) & \text{if } p > 1/2,
\end{array}
$$

Switzerland's expected payoff is

$$
\begin{array}{ll}
1 + (1/2)(a-1)p & \text{if } p < 1/2 \\
(1/4)[a+3, 3a+1] & \text{if } p = 1/2 \\
(1/2)((a+1) + (a-1)p) & \text{if } p > 1/2,
\end{array}
$$

the sum of the expected payoffs is

$$
\begin{array}{ll}
a + 1 & \text{if } p < 1/2 \\
[a+1, (3a+1)]/2 & \text{if } p = 1/2 \\
(a+1) + (a-1)p & \text{if } p > 1/2.
\end{array}
$$

Notice the brackets in the middle of the display of Switzerland's expected payoff. This is not an typographic error. When $p = 1/2$, Switzerland's payoff does in fact vary over an interval because Botswana has a continuum of best response strategies. Also observe that if $p > 1/2$, the formula for the sum of expected payoffs matches the upper bound obtained in the proof of an earlier theorem. □

Figure 6.17 shows the payoffs to the two countries, in terms of the degree of distributional conflict a as the value of informational conflict p varies.

We clearly see that when $p > 1/2$, indicating higher probabilities of playing COH or COL, the payoffs to the two countries increase by sharing their information and coordinating their action choices. For $p < 1/2$, the situation is more complicated. Clearly, Lead is best if the countries fall into leadership and follower roles. Absent those roles, and for very small values of p, the best that the two countries can hope for is obtained by playing their Coordinate on Message strategies.

The International Telegraph Union process of holding periodic international conferences to discuss and re-negotiate arrangements is but one example of a scenario to which this normative model might be applied. The fact is that there are many scenarios that might be modeled by this process of repeated negotiations in the presence of a distributional problem (i.e., the value of a) and/or an information problem (i.e., the value of p). This suggests that one could test the fidelity of the framework and each of the models. In situations in which the countries clearly have different interests and most of the information is known, then one would expect to see pure coordination equilibria or leadership equilibria emerge. Conversely, in situations where the distribution problem is low, but the information problem is high, communicative equilibria should be more likely to occur.

Our construction and organization of this model highlights another feature of the modeling process that is frequently overlooked, that is, its ability to communicate the resulting analysis and interpretations effectively. In particular, the fact that we used a series of theorems to draw attention to each of the equilibria and the differences between them assists

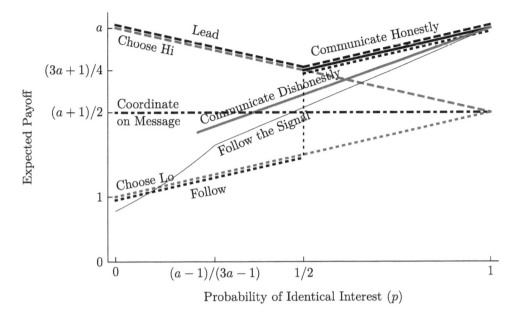

FIGURE 6.17 Countries' Expected Payoffs from Various Equilibria. Reproduced from James D. Morrow, "Modeling the Forms of International Cooperation: Distribution versus Information," *International Organization*, 48:3 (Summer, 1994), pp. 387-423. 1994 by the World Peace Foundation and the Massachusetts Institute of Technology. Reprinted by permission of The MIT Press.

the reader in seeing the overall structure as well as the specific results. Putting the detailed arguments in proofs allows a reader to skip over them when they are trying to understand the overall structure of the model and its interpretation. At the same time, reading the proofs later assures the reader's confidence in the results. This approach will be used more extensively in the final chapters of this book.

6.6 AUCTIONS

Auctions are used extensively in commerce, including livestock sales, offshore oil exploration leases, and internet marketplaces. In Chapter 3 we briefly examined Second Price Auction and Ascending Bid Auction scenarios. Here we explore an auction that happens in nature.

Biological Auction. Two animals vie for a prize: territory, food, or a mate. They display their prowess and may even fight. Eventually a winner emerges and the loser sulks away. Because both players have exerted effort but only one wins the prize, this is analogous to an auction where the higher of the two bidders wins but both bidders pay. It is rare in such intraspecies contests for an animal to die or be severely injured, and so it is evident that in most contests, the loser could have continued to vie for the prize. How long and with what ferocity should an animal fight? [70]

We will model this scenario as a two-player game. The players represent the two animals competing for a prize. Each will choose a maximal bid, representing the effort each is willing to expend to win the prize. The player choosing the higher bid wins the prize and both players pay the lower of the two bids. This payment of the lower bid represents the fact that continuing the fight requires increasing effort and escalates the final price. The auction ends when one animal reaches its maximum threshold and walks away. We also assume that if both players place the same bid, i.e., exert the same effort, each player has a 50% chance of winning the prize. We will analyze this game by finding the Nash equilibrium. Biologically speaking, players using best responses in a Nash equilibrium corresponds to the concept that only the fittest survive to produce offspring and propagate their behavior.

If the value of the prize to each animal was the same and commonly known (say v), then we would have the following strategic game [102]: players 1 and 2 choose bids $b_1 \geq 0$ and $b_2 \geq 0$, respectively, and the expected payoffs are

$$(u_1(b_1, b_2), u_2(b_1, b_2)) = \begin{cases} (-b_1, v - b_1) & \text{if } b_1 < b_2 \\ (\frac{1}{2}v - b_1, \frac{1}{2}v - b_2) & \text{if } b_1 = b_2 \\ (v - b_2, -b_2) & \text{if } b_1 > b_2. \end{cases}$$

Given player 2's bid b_2, player 1's payoff for a bid of b_1 is graphed in Figure 6.18 for a generic $b_2 < v$ and $b_2 > v$. Thus, player 1's best response to player 2 bidding b_2 is any

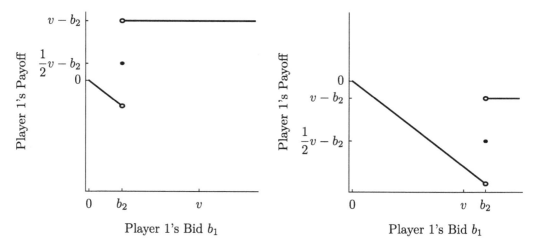

FIGURE 6.18 Player 1's Payoff versus Bid.

amount larger than b_2 if $b_2 < v$ and 0 if $b_2 \geq v$. By symmetry, player 2's best response to player 1 bidding b_1 is any amount larger than b_1 if $b_1 < v$ and 0 if $b_1 \geq v$. Therefore, the only pure-strategy Nash equilibria are $(0, b_2)$ for $b_2 \geq v$ and $(b_1, 0)$ for $b_1 \geq v$. This suggests that the animals should flip a coin, without taking any effort, to determine who wins the prize. Clearly this is not what happens in nature, but the Nash equilibrium is in fact the regret-free choice because of the threat that the winning player is willing to place at least as much effort in the fight as the prize is worth.

We next examine whether there is a symmetric mixed-strategy Nash equilibrium. Biologically speaking, all players using the same strategy says that the players have fully adapted to their environment. Suppose that player 2 uses the mixed strategy where their bid is a random variable with probability density of g, and that $g > 0$ only in some interval $(\underline{b}, \overline{b})$. Against this strategy, player 1's payoff using a bid b in the interval $(\underline{b}, \overline{b})$ is given by the function

$$w(b) = \int_{\underline{b}}^{b} (1-t)g(t)\, dt - b \int_{b}^{\overline{b}} g(t)\, dt$$

where the first term corresponds to player 1 winning the prize worth v and paying player 2's lower bid t and the second term corresponds to player 1 losing the prize and paying its own lower bid b. Observe that

$$w(\underline{b}) = \int_{\underline{b}}^{\underline{b}} (1-t)g(t)\, dt - \underline{b} \int_{\underline{b}}^{\overline{b}} g(t)\, dt$$
$$= 0 - \underline{b}(1) = -\underline{b};$$

hence, if $\underline{b} > 0$, player 1's best response would be to bid zero. Thus, in any symmetric Nash equilibrium, we must have

$$\underline{b} = 0.$$

For player 2's strategy g to be a best response for player 1, then every b in the interval $(0, \overline{b})$ must be a best response, and so we must have $w(b) = c$ for some constant c on this interval.

Taking the derivative of both sides of $w(b) = c$ with respect to b, we obtain

$$(1-b)g(b) - \int_{b}^{\infty} g(t)\, dt + bg(b) = 0$$

where we have used the Fundamental Theorem of Calculus and the product rule. Combining the first and third terms and rewriting the second term, we obtain

$$g(b) - \left(1 - \int_{0}^{b} g(t)\, dt\right) = 0,$$

which can be written

$$G'(b) + G(b) = 1$$

where

$$G(b) = \int_{0}^{b} g(t)\, dt.$$

This is a first-order linear differential equation whose general solution is

$$G(b) = 1 + Ae^{-b}$$

for an arbitrary constant A. Since g is a probability density function and $g(b) > 0$ if and only if $0 < b < \overline{b}$, it follows that $G(0) = 0$ and $G(\overline{b}) = 1$. This gives $A = -1$, and so

$$G(b) = 1 - e^{-b}$$

and

$$g(b) = G'(b) = e^{-b}.$$

Substituting $G(\overline{b}) = 1$ into our current formula for G, we obtain $1 = 1 - e^{-\overline{b}}$, which implies $\overline{b} = \infty$.

Now substituting this back into Player 1's payoff function, we obtain

$$w(b) = \int_{0}^{b} (1-t)e^{-t}\, dt - b \int_{b}^{\infty} e^{-t}\, dt = 0$$

after employing integration by parts. Therefore, there is a symmetric mixed-strategy Nash equilibrium where each player chooses its bid in accordance with the probability density

$$g(b) = e^{-b}, \text{ for } b \geq 0.$$

This seems more realistic in the sense that it predicts some level of posturing or fighting when two animals both desire a limited resource; however, it is still unclear why such behavior is evolutionarily reasonable since the expected payoffs are zero.

We will build a slightly more sophisticated model that leads to more evolutionarily reasonable results [13]. The key assumption will be that the prize may be valued differently by the two players, and while each player knows its own valuation of the prize, each player does not know the valuation of the other player. To model this incomplete information, we will assume that each player's valuation for the prize is chosen in accordance with the same probability density f, that is, $f(v)$ is the probability density that the prize is worth v to a player. (The density f also gives each player's beliefs about how much the prize is worth to the other player.) We assume that the prize is truly a benefit, that is, $f(v) > 0$ only if $v \geq 0$. After each player learns their type (how much they themselves value the prize), the game proceeds as before: each player simultaneously chooses a bid, the player with the larger bid wins the prize, and both players pay the lesser bid. A strategy is a function β that tells a player to bid $\beta(v)$ if the prize is worth v to the player.

Suppose that (β, β) is a Bayesian equilibrium. Suppose player 2 adopts the strategy to bid $\beta(t)$ if the prize is worth t to player 2. Against this strategy, if v is how much the prize is worth to player 1 and player 1 bids b, then player 1's payoff is

$$w(b) = \int_{\beta(t)<b} (v - \beta(t))f(t)\,dt - b\int_{\beta(t)\geq b} f(t)\,dt$$

where the first term corresponds to player 1 winning the prize worth v and paying player 2's lower bid $\beta(t)$ and the second term corresponds to player 1 losing the prize and paying its lower bid b.

Assuming that β is an increasing function (the more the prize is worth, the higher the bid), it has an (increasing) inverse function which will be denoted by β^{-1}, that is, $b = \beta(t)$ if and only if $t = \beta^{-1}(b)$. Since β is increasing, the inequality $\beta(t) < b$ is equivalent to the inequality $t < \beta^{-1}(b)$, and so

$$w(b) = \int_0^{\beta^{-1}(b)} (v - \beta(t))f(t)\,dt - b\int_{\beta^{-1}(b)}^{\infty} f(t)\,dt.$$

Defining

$$F(v) = \int_0^v f(t)\,dt$$

to be the probability the prize is worth no more than v and recognizing that $F(\infty) = 1$, we obtain

$$w(b) = \int_0^{\beta^{-1}(b)} (v - \beta(t))f(t)\,dt - b(1 - F(\beta^{-1}(b))). \tag{6.4}$$

Player 1 should choose $b \geq 0$ to maximize $w(b)$. Since we have assumed that (β, β) is a Bayesian equilibrium, player 1's best response is $\beta(v)$, that is, $w(\beta(v)) \geq w(b)$ for all $b \geq 0$. If $\beta(v) > 0$, then it must be the case that $w'(\beta(v)) = 0$.

Now we calculate the derivative of w using the Fundamental Theorem of Calculus, chain and product rules, and the derivative of an inverse function formula:

$$w'(b) = \frac{(v - \beta(\beta^{-1}(b)))f(\beta^{-1}(b))}{\beta'(\beta^{-1}(b))} - (1 - F(\beta^{-1}(b))) + \frac{bf(\beta^{-1}(b))}{\beta'(\beta^{-1}(b))}.$$

Part of the first term cancels the third term yielding

$$w'(b) = \frac{vf(\beta^{-1}(b))}{\beta'(\beta^{-1}(b))} - (1 - F(\beta^{-1}(b))) \tag{6.5}$$

where the first-order necessary condition $w'(\beta(v)) = 0$ now is

$$\frac{vf(\beta^{-1}(\beta(v)))}{\beta'(\beta^{-1}(\beta(v)))} - \left(1 - F\left(\beta^{-1}(\beta(v))\right)\right) = 0$$

which simplifies to

$$\frac{vf(v)}{\beta'(v)} - (1 - F(v)) = 0.$$

Solving for $\beta'(v)$, we obtain

$$\beta'(v) = \frac{vf(v)}{1 - F(v)}. \tag{6.6}$$

Integrating, we obtain

$$\beta(v) = \int_0^v \frac{tf(t)}{1 - F(t)}\, dt. \tag{6.7}$$

Observe that the integrand is positive whenever $f(t) > 0$ and so our assumptions that β is increasing and differentiable hold. To verify that we have indeed found a best (rather than worst) response, substitute Equation 6.6 into Equation 6.5 to obtain

$$w'(b) = \frac{v - \beta^{-1}(b))}{\beta^{-1}(b))}(1 - F(\beta^{-1}(b)),$$

which is positive if $b < \beta(v)$ and negative if $b > \beta(v)$. Therefore, (β, β), where β is defined by Equation 6.7, is the unique symmetric Bayesian equilibrium. The payoff in this equilibrium to a player who values the prize at v is

$$w(\beta(v)) = \int_0^v (v - \beta(t))f(t)\, dt - \beta(v)(1 - F(v)). \tag{6.8}$$

For the special case when prize values are chosen uniformly on the interval $[0, 1]$, we have

$$f(t) = \begin{cases} 0 & \text{if } t \leq 0 \\ 1 & \text{if } 0 < t < 1 \\ 0 & \text{if } 1 \leq t \end{cases} \quad \text{and} \quad F(t) = \begin{cases} 0 & \text{if } t \leq 0 \\ t & \text{if } 0 < t < 1 \\ 1 & \text{if } 1 \leq t \end{cases}$$

Substituting into Equation 6.7, we obtain

$$\beta(v) = \int_0^v \frac{t}{1 - t}\, dt = -v - \ln(1 - v),$$

for $0 \leq v < 1$, and substituting into Equation 6.8, we obtain

$$w(\beta(v)) = \int_0^v (v - (-t - \ln(1 - t)))\, dt - (-v - \ln(1 - v))(1 - v) = \frac{1}{2}v^2.$$

These functions of value v are shown in Figure 6.19. Observe that the payoff is positive (for $v > 0$) and strictly increasing with the value of the prize (for $0 < v < 1$). This provides an evolutionarily reasonable justification for the attrition behavior observed in nature. Also observe that for low prize values, the optimal bid is smaller than the value of the prize,

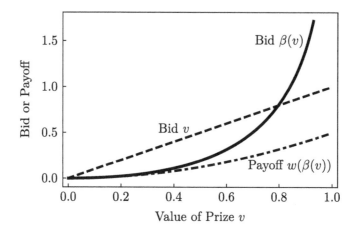

FIGURE 6.19 Biological Auction Extensive Game Special Case Equilibrium Results.

but for large prize values the optimal bid become larger than the prize value, approaching infinity as the prize value approaches 1. Of course, the probability that an animal will actually pay more than the prize value is small (since otherwise the expected payoff would not be positive).

The two observations made in the special case turn out to be true in general. First, by Equations 6.7 and 6.8, $\beta(0) = 0$ and $w(\beta(0)) = 0$. Now taking the derivative of Equation 6.8 (being careful to use the product rule on both terms) and using Equation 6.6, we obtain

$$\frac{d}{dv}w(\beta(v)) = (v - \beta(v))f(v) + F(v) - \beta'(v)(1 - F(v)) + \beta(v)f(v)$$

$$= vf(v) + F(v) - \frac{vf(v)}{1 - F(v)}(1 - F(v))$$

$$= F(v) > 0$$

and so $w(\beta(v))$ is an increasing function of v. Use Equation 6.7 to find the average bid

$$\int_0^\infty \beta(v)f(v)\,dv = \int_0^\infty \int_0^v \frac{tf(t)}{1 - F(t)}dt\,f(v)\,dv.$$

Interchanging integrals ($0 \leq u \leq v < \infty$), we obtain

$$\int_0^\infty \beta(v)f(v)\,dv = \int_0^\infty \frac{tf(t)}{1 - F(t)}\int_t^\infty f(v)\,dv\,dt.$$

Since the inner integral is $1 - F(t)$,

$$\int_0^\infty \beta(v)f(v)\,dv = \int_0^\infty tf(t)\,dt.$$

Thus, the average bid equals the average value. This means that if for some prize values the bid is less, then there must be other prize values for which the bid is actually greater than the prize value! The biological auction, like other scenarios discussed in this chapter, shows that surprising things can happen when players have incomplete information.

EXERCISES

6.1 Give a rationale for the utilities assigned to each player and outcome in the FoodPro scenario.

6.2 For the FoodPro game, give all of the pure strategies for Mark and Ben.

6.3 In FoodPro, one pair of behavior strategies is

$$s_{\text{Mark}}(\text{Mark1}) = (1/3)\text{No} + (1/3)\text{Lo} + (1/3)\text{Hi}$$

and

$$s_{\text{Ben}}(\text{Ben1}) = (1/2)\text{No} + (1/2)\text{Hi}$$
$$s_{\text{Ben}}(\text{Ben2}) = (1/3)\text{No} + (2/3)\text{Hi}.$$

These behavior strategies indicate that Mark will choose randomly among his three actions, each with the same probability, and Ben will randomly choose No or Hi with equal probabilities at the information set Ben1 but will be twice as likely to choose Hi over No at the information set Ben2. The probabilities for behavior strategies are displayed below the edges in Figure 6.20. If the players choose the strategy profile s,

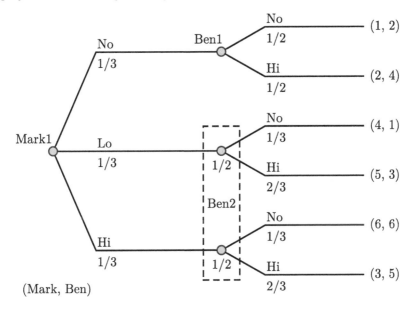

FIGURE 6.20 FoodPro Game Tree with Assessment (s, β).

find Mark and Ben's expected payoffs.

6.4 For the FoodPro extensive game, create the payoff matrix for the corresponding strategic game and find its Nash equilibria.

6.5 For the FoodPro extensive game, a short argument is given to verify that (s', β') is a weak sequential equilibrium. Provide similar arguments to verify that (s'', β'') and (s''', β''') below are weak sequential equilibria.

$$s''(\text{Mark1}) = \text{Hi}$$
$$s''(\text{Ben1}) = \text{Hi}$$
$$s''(\text{Ben2}) = \text{No}$$
$$\beta''(\text{Ben2}) = (\text{Hi})$$

and

$$s'''(\text{Mark1}) = (1/3)\text{Lo} + (2/3)\text{Hi}$$
$$s'''(\text{Ben1}) \quad = \text{Hi}$$
$$s'''(\text{Ben2}) \quad = (1/2)\text{No} + (1/2)\text{Hi}$$
$$\beta'''(\text{Ben2}) = (1/3)(\text{Lo}) + (2/3)(\text{Hi})$$

*6.6 Verify that $(s', \beta'), (s'', \beta'')$ and (s''', β''') are the only weak sequential equilibria for the FoodPro scenario.

*6.7 For the FoodPro extensive game with unspecified payoffs as displayed in Figure 6.21, state the minimal restrictions on the utilities that are required for (s', β') to be a weak sequential equilibrium. Describe what these restrictions mean in terms of the original scenario.

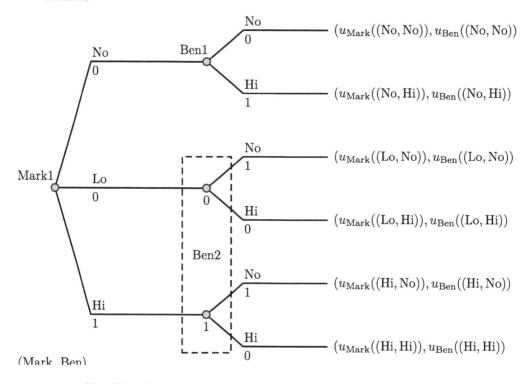

FIGURE 6.21 FoodPro Game Tree with Unspecified Payoffs.

6.8 Describe a scenario which can be modeled as a two-person game with missing informa-tion, where the missing information is something other than the other player's utilities. Sketch a game tree, indicating how the missing information leads to information sets with more than one node.

6.9 Show that the consistency condition, Equation (6.1), need not be checked for one of the histories in an information set. Specifically, suppose (s, β) is an assessment and $I = \{h_1, h_2, \ldots, h_m\}$ is an information set with $m \geq 2$ and $\Pr(h_k|G', s) > 0$ for some k. Let

$$\Pr{}_{G'}(I|s) = \sum_{k=1}^{m} \Pr{}_{G'}(h_k|s).$$

Show that if

$$\beta(h_k|I) = \frac{\Pr_{G'}(h_k|s)}{\Pr_{G'}(I|s)}$$

for $k = 1, 2, \ldots, m-1$, then the equality also holds for $k = m$.

6.10 Verify the weak sequential equilibrium in One Card Poker that satisfies the necessary conditions.

6.11 Consider the One Card Poker game in which the number of tokens won is unchanged but Colin's payoffs for winning or losing more than one token are changed as shown in Figure 6.22. Note that the payoff pair $(2, -2.4)$ shows Rose winning two tokens, so Colin is losing two tokens, but here this outcome now has a utility of -2.4 for Colin.

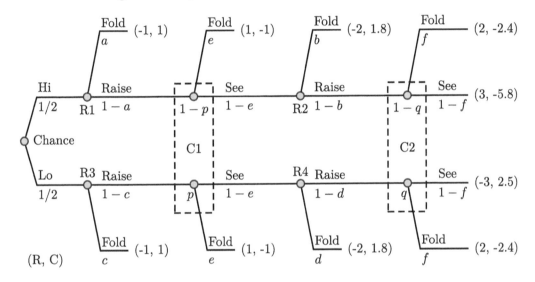

FIGURE 6.22 Modified One Card Poker.

a. Find Colin's payoffs for winning $-3, -2, -1, 1, 2$, and 3 tokens and plot his payoff versus number of tokens won.

b. Sketch a graph of Colin's payoff as a function of tokens won that includes the points plotted in (a) and clearly shows that Colin is risk-adverse with respect to the number of tokens won.

*6.12 Continuing with the modified One Card Poker given in Figure 6.22, we now explore weak sequential equilibria.

a. Show that the assessments satisfying $a = b = c = d = 0$, $e = f = 1$, $p = 1/2$, and $q < 34/83$ are weak sequential equilibria.

b. Show that the assessments satisfying $a = b = c = d = 0$, $e = 1$, $f > 1/5$, $p = 1/2$, and $q = 34/83$ are weak sequential equilibria.

c. Show that the only weak sequential equilibria are those described above.

d. Predict what will happen in this game and compare with what happens in the original One Card Poker game.

*6.13 Modify the One Card Poker game by changing each payoff pair (x, y) to $(v_R(x), v_C(y))$ where v_R and v_C are strictly increasing functions from $\{-3, -2, -1, 1, 2, 3\}$ to real numbers, that is, the payoffs are ordered by the number of tokens won but the intensities may be different. Prove that if (s, β) is an assessment satisfying $s(\text{R3}) = \text{Fold}$ (that is, Rose does not bluff), then (s, β) is not a weak sequential equilibrium.

*6.14 In this chapter, we described and modeled as an extensive game a highly simplified game of poker. In so doing, we were able to highlight the rationality of a central feature of poker, bluffing. Choose a different card or board game (e.g., Rummy, Monopoly, or Risk) and describe and model as an extensive game a highly simplified version of the game.

6.15 Consider the Primary game depicted in Figure 6.23.

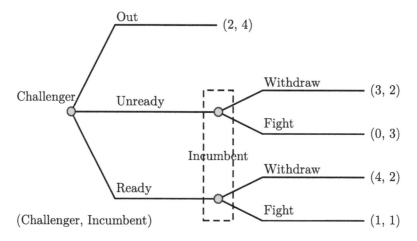

FIGURE 6.23 Primary Game Tree.

    a. Consider the assessment in which the challenger selects the strategy Out and the incumbent selects the strategy Fight given a belief that the challenger has selected Unready. Verify this assessment is a weak sequential equilibrium.

    b. Consider the assessment in which the challenger selects the strategy Ready and the incumbent selects the strategy Withdraw given the belief that the challenger has selected Ready. Verify this assessment is a weak sequential equilibrium.

    c. Find all weak sequential equilibria.

    d. For which weak sequential equilibria can the payoffs be interpreted simply as ordinal utilities?

    e. Comment on the reasonableness of each weak sequential equilibrium.

6.16 Consider modifying the payoffs for the Primary game as shown in Figure 6.24.

    a. Describe the viewpoints of the two players that would result in these preferences.

    b. Find the unique weak sequential equilibrium.

    c. Explain why the payoffs must be interpreted as vNM utilities.

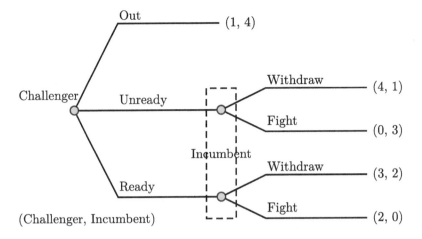

FIGURE 6.24 Modified Primary Game Tree.

d. One way to interpret a non-pure strategy is that a player randomly (in accordance with the given probabilities) chooses a pure strategy. What could be a different interpretation in the Primary scenario?

*6.17 Consider modifying the payoffs for the Primary game. Figure 6.25 provides abbreviated notation for the various utilities, e.g., $u_C(UW)$ is the abbreviation for $u_{Challenger}((Unready, Withdraw))$.

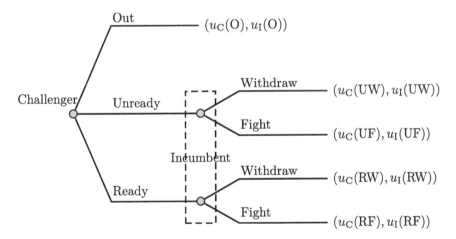

FIGURE 6.25 Modified Primary Game Tree.

a. Describe conditions on the payoffs so that the following assessment is a weak sequential equilibrium: The challenger selects the strategy Out and the incumbent selects the strategy Fight given a belief that the challenger is Unready.

b. Describe conditions on the payoffs so that the following assessment is a weak sequential equilibrium: The challenger selects the strategy Ready and the incumbent selects the strategy Withdraw given the belief that the challenger has selected Ready.

6.18 *Selten's Horse* In the three-player game displayed in Figure 6.26, find the weak sequential equilibria in which each player's strategy is pure [95].

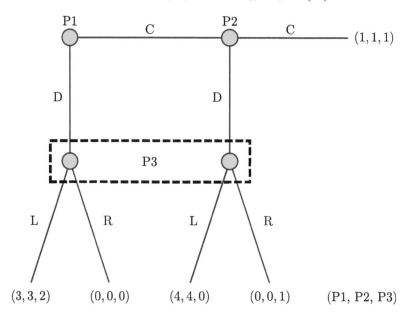

FIGURE 6.26 Selten's Horse Game Tree.

*6.19 **Hike**. Rose and Colin travel separately to a park where they plan to do some hiking and meet at an observation tower in order to continue their hike together. Colin reaches the tower first and watches for Rose. From the trail she is on, Rose knows that she must turn at the second crossing to get to the tower. Unfortunately, Rose is somewhat distracted by the flora, and when she comes to a crossing, she cannot remember whether she had previously hiked by a crossing. So, at each crossing, Rose can either Turn or Continue, but she does not know which crossing she is at. Fortunately, if Rose turns at the first crossing, Colin will be able to eventually see that Rose has taken the wrong trail and could either catch up with her (Seek) so as to spend the rest of the day hiking with her, although not where he had planned, or he could hike without her (Continue). Figure 6.27 summarizes an extensive game model for this scenario and an arbitrary assessment.

a. Explain why the payoffs are reasonable based on the scenario description.

b. Explain why the probability of Rose choosing each Turn is the same number $1 - a$.

c. Model this extensive game as a strategic game. Notice that not every outcome in the game tree is an outcome in the strategic game. Find all Nash equilibria.

d. Show that the assessment with $a = 1/9$, $b = 0$, and $p = 1/10$ is a weak sequential equilibrium.

e. Conclude that for some extensive games, a weak sequential equilibrium need not

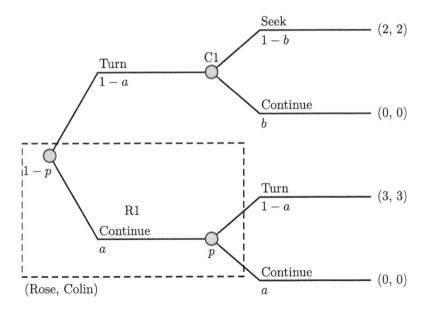

FIGURE 6.27 Hike Game Tree.

be a Nash equilibrium for the corresponding strategic game. Explain why this does not violate Theorem 6.1.3.

f. Use Gambit to find Nash equilibria for this extensive game and its corresponding strategic game. Compare with previous results.

*6.20 **Trivial Bridge**. Consider the extremely simple partner card game modeled in Figure 6.28. Initially, a winning card is dealt to either Jim or Bob. Whoever receives the winning card can either show it and obtain 2 tokens from the other player or continue the game. If the game is continued, Jim's partner Jan, without knowing who had the winning card, can either end the game resulting in no exchange of tokens or can double the stakes with the holder of the winning card obtaining 4 tokens from the other player. Note that the partners Jim and Jan are considered a single player called "J." We have assumed that the players are self-interested and risk-neutral in the number of tokens obtained. The behavior strategy for J is specified by a and d, the behavior strategy for Bob is specified by b, and the belief system is specified by p.

a. Show that the assessment defined by $a = b = 0$, $d = 1$, and $p = 1/2$ achieves consistency of beliefs, J's strategy is a best response at information sets Jim and Jan, and Bob's strategy is a best response at all information sets, but J's strategy is not a best response at information sets Chance and Bob.

b. Directly show that the extensive game has no weak sequential equilibrium. Hints: Consider the cases $d < 1/2$, $d > 1/2$, and $d = 1/2$, and remember to consider the Chance information set when checking for sequential rationality.

c. Explain why the result of part (a) does not contradict Theorem 6.1.1.

d. Find the strategic game corresponding to this extensive game. Note that J's pure strategies can be denoted by (x, y) where x is the action choice by Jim and y is

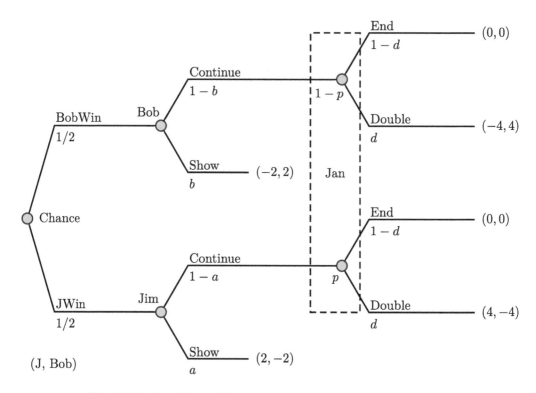

FIGURE 6.28 Trivial Bridge Game Tree and Assessment.

the action choice by Jan. Show that the unique Nash equilibrium is

$$((1/2)(\text{Continue, Double}) + (1/2)(\text{Show, End}), (1/2)\text{Continue} + (1/2)\text{Show}).$$

e. Show that J's arbitrary behavior strategy summarized in Figure 6.28 is equivalent to the mixed strategy

$$(1-a)(1-d)(\text{Continue, End}) + (1-a)d(\text{Continue, Double})$$
$$+ a(1-d)(\text{Show, End}) + ad(\text{Show, Double}).$$

f. Using the results of the previous two parts, show that there is no behavior strategy for J that is equivalent to J's mixed strategy that is part of the unique Nash equilibrium.

g. Explain why the result of the previous part provides a second, indirect proof that the extensive game has no weak sequential equilibrium.

h. Change the extensive game so that Jim and Jan act as separate players having the same utilities as J. Find all weak sequential equilibria.

i. Explain why each Nash equilibrium of the strategic form game corresponding to the changed extensive game is associated with a weak sequential equilibrium for the changed extensive game.

j. Predict what will happen in the card game given the results obtained for the two extensive games and corresponding strategic games.

k. Use Gambit to find Nash equilibria for this extensive game and its corresponding strategic game. Compare with the previous results.

6.21 Describe a different reasonable set of preferences for the Warfare game. Use Gambit to determine the weak sequential equilibria for the modified game and explain in what ways they are different from the original weak sequential equilibrium.

6.22 For the Warfare extensive game, consider the assessment (s, β) where

$$s(\text{RE1}) = \text{Don't}$$
$$s(\text{RE2}) = \text{Day}$$
$$s(\text{RE3}) = \text{Day}$$
$$s(\text{GT1}) = (1/2)\text{Lake} + (1/2)\text{Forest}$$
$$s(\text{GT2}) = \text{Ambush}$$
$$s(\text{GT3}) = \text{Ambush}$$
$$\beta(\text{RE1}) = (\text{Forest}, \text{Lake}, \text{Attack})$$
$$\beta(\text{RE2}) = (\text{Lake}, \text{Forest}, \text{Attack})$$
$$\beta(\text{GT1}) = (1/2)(\text{Forest}) + (1/2)(\text{Lake}).$$

a. Show that (s, β) is sequentially rational.

b. Show that (s, β) satisfies

$$\beta(h|I) = \frac{\Pr(h \text{ reached given } s \text{ used})}{\Pr(I \text{ reached given } s \text{ used})}$$

for each h in each information set I satisfying $\Pr(I$ reached given s used$) > 0$ in the original game G. Explain how this differs from the consistency of beliefs condition.

c. Explain why (s, β) is not a weak sequential equilibrium.

d. Discuss whether (s, β) is a reasonable solution.

6.23 Complete the calculations necessary to show that $((B, B), (B, S))$ is a Bayesian equilibrium for Karl and Juanita.

6.24 Modify the College Choices Bayesian game by interchanging the values of $\phi_J(K_a|J_b)$ and $\phi_J(K_b|J_b)$.

a. Describe the difference between the beliefs of the two types of Juanita.

b. Show that the new belief functions cannot arise from a common prior.

c. Find the Bayesian equilibria. Hint: Very little changes in the analysis done on the original College Choices Bayesian game.

6.25 Consider the Tourist Bazaar scenario in which there are two types of buyers and only one type of seller. One type of buyer avoids negotiation and places a value of 12 on the necklace. The other enjoys negotiation and puts a value of 6 on the necklace. Assume that the seller wishes to avoid negotiation and values the necklace at 0. There is a common discount factor of 2/3, and the probability that the buyer enjoys negotiation is 1/3. Construct a game tree to model this scenario and then find a sequential equilibrium.

6.26 Modify the Takeover Bid scenario by assuming both players know whether Urchin is successful with its new machine. Draw the new game tree and find the Bayesian equilibrium.

6.27 Generalize the Takeover Bid game by assuming that the probability that Urchin's machine works is some number q that may be different from $1/2$. For what values of q is there an equilibrium in which Bisco offers the maximum of \$80?

6.28 Does the Takeover Bid game have a separating equilibrium?

*6.29 Modify the Takeover Bid scenario by having Urchin first make the offer to sell to Bisco and then have Bisco decide whether to accept or reject the offer. Keep the same information conditions, that is, Urchin knows whether it has been successful before it makes its offer, but Bisco does not know whether Urchin has been successful when deciding to accept or reject.

    a. Draw the game tree for this modified scenario.

    b. Find all pooling Bayesian equilibria.

    c. Find all separating Bayesian equilibria in which Bisco Accepts offers of $P \leq P^*$ and Rejects offers of $P > P^*$ for some fixed value P^*.

    d. Are there separating Bayesian equilibria that do not satisfy the condition given in part (c)?

6.30 Suppose you inadvertently overhear two people in a restroom discuss a plan to blow up a public building in Washington. You consider calling the FBI and offering to sell this information to it. Assuming that the FBI gets calls like this every day, is there any way for you to actually sell his information? Suppose, on the other hand, that you had heard, from a knowledgeable source, that a large public company would be declaring bankruptcy. Is it possible for you to sell this information? How is this second scenario different from the first scenario?

6.31 Show that the strategy profile (Coordinate on Message, Coordinate on Message) is a weak sequential equilibrium in the International Collaboration extensive game. Furthermore, no other pair of unaware strategies yields a higher sum of expected payoffs.

*6.32 Prove one or more of the theorems not proved in the International Collaboration section.

6.33 The last observation for the Biological Auction Bayesian game claimed that the optimal bid was sometimes less and sometimes more than the value of the prize. Prove this by setting the optimal bid equal to the prize value, i.e., $\beta(v) = v$, in the equation defining the equilibrium strategy β and derive a contradiction.

6.34 **First-Price Sealed Bid Auction.** In a first-price sealed bid auction, a resource is awarded to the bidder submitting the highest bid at a cost of their bid. Only the winning bidder pays anything. Assume that the resource's value to each bidder is always nonnegative and is chosen randomly from a known probability density f. Each bidder knows their own valuation but not their opponent's valuation. Model this as a Bayesian game, following the biological auction model with different rules regarding what each payer plays. Find a symmetric Bayesian equilibrium.

6.35 **Second-Price Sealed Bid Auction**. In a second-price sealed bid auction, a resource is awarded to the bidder submitting the highest bid at a cost of the second highest bid. Only the winning bidder pays anything. Assume that the resource's value to each bidder is always nonnegative and is chosen randomly from a known probability density f. Each bidder knows their own valuation but not their opponent's valuation. Model this as a Bayesian game, following the biological auction model with different rules regarding what each payer plays. Find a symmetric Bayesian equilibrium.

CHAPTER 7

Repetitious Play

The strategic, extensive, and bargaining games we have discussed so far assume that the players' interactions are independent of any previous or future interaction. More realistically, players interact with each other multiple times in a variety of ways, and the actions they choose in one interaction are likely to depend upon past experience and predictions about future behavior. We examine models of repetitious play and features unique to these models. In particular, we examine the multiplicity of Nash equilibria and the robustness of qualitative behavior in different models.

Up to this point, we have examined various models of games that may have multiple stages but ultimately are assumed to be played once. In this chapter, we will introduce repeated play games as a new mathematical modeling tool. Mathematically, all of these models are extensive games once we resolve how the payoffs in individual games are to be combined into a payoff for the repeated game. Nonetheless, the special properties of repeated play lead to special considerations (such as errors in perception and implementation) and interesting results (such as the ubiquity of equilibria). The dynamics of strategy choice that are passed on in a species also become a focus. We begin with the simplest of repeated games.

Definition 7.0.1. A *finite repeated game* consists of the following:

1. A game G with all of its embedded players, outcomes, rules, and utilities.

2. An integer $m \geq 2$.

3. In stage $k = 1, 2, \ldots, m$, the game G is played with all players knowing the actions chosen in the previous stage.

4. Utilities for each player i are defined to be the mean of the utilities player i obtained in each stage of the repeated game.

A common example of a game that can be extended to a finite repeated game is the Prisoner's Dilemma game. When the Prisoner's Dilemma game is played only once, players choose strategies with little information about how their opponent might behave, leading them to the Nash equilibrium which is not efficient. However, if we imagine partners in crime that find themselves in this dilemma often, we can see that over time they might develop a more informed strategy because they know how their partner has behaved in the past. The Prisoner's Dilemma is just one of many scenarios we call Social Dilemmas.

7.1 REPEATED SOCIAL DILEMMAS

A *social dilemma* is an interaction among players for which one action, *defect*, dominates another action, *cooperate*; however, all players prefer the outcome when all choose to cooperate over the outcome when all choose to defect. It is possible to listen to National Public Radio (NPR) without donating (defect), but programming is of higher quality if everyone donates (cooperate) in comparison with no one donating. A unilateral imposition of import tariffs (defect) is economically beneficial to a country, but all countries benefit from adhering to a well-crafted free trade agreement (cooperate) in comparison with all countries imposing stiff import tariffs.

Although defect may be the dominant choice in these and many other social dilemmas, we often observe real-world players cooperating. One possible reason for this observed cooperation is that we modeled the preferences incorrectly. Stewardship, integrity, and fidelity may be so morally compelling to the players that cooperation dominates defection. With such preferences, a social dilemma no longer exists; the corrected model matches observation. A second possibility is that the outcomes have not been described correctly. A donation announcement could engender beneficial goodwill with an acquaintance or employer, while flaunting a trade agreement could result in a trade embargo.

A proper evaluation of these outcomes may again change the modeled preferences so that defection no longer dominates cooperation.

A final possibility, explored here, is that the players see a short-term advantage to defecting but will play the game more than once and can choose their subsequent actions based on knowledge of previous players' choices. Donation and programming decisions are made at least once each year. The countries involved in trade agreements may have interests in other issues of trade, migration, environmental protection, culture, etc.

Intuitively, cooperation begets cooperation. NPR often reminds its listeners of the many who donate and the resulting quality programming. Successful trade negotiators use confidence-building and face-saving measures.

Our goal will be to understand mathematical conditions that result in cooperative behavior.

Repeated Social Dilemmas. The same set of agents (people, businesses, animals, etc.) interact in a specific social dilemma multiple times. Each time, the agents have knowledge of how the others have acted in this social dilemma on previous occasions and can base their own subsequent choices on this knowledge. Agents consider their gains or losses over the entire time that they are interacting with others.

We will model the social dilemmas as Prisoner's Dilemma games using the standard payoffs of S for cooperating when their partner defects, P for both defecting, R for both cooperating, and T for defecting when their partner cooperates. These are shown in Table 7.1 where $S < P < R < T$ and $S + T < 2R$.

TABLE 7.1   Payoffs for Prisoner's Dilemma

(Rose, Colin)	Cooperate	Defect
Cooperate	(R, R)	(S, T)
Defect	(T, S)	(P, P)

The players will be called Rose and Colin, and we will use the specific payoffs $S = 0$, $P = 1$, $R = 3$, and $T = 5$ from [5] which was an early study of repeated play. When

Prisoner's Dilemma is played only once in isolation, we referred to Cooperate and Defect as strategies, but now we will refer to them as actions taken within a single stage of the repeated game. In the one-stage game, Defect dominates Cooperate, (Defect, Defect) is the unique Nash equilibrium, and the efficient strategy pairs involve at least one player cooperating with certainty. Figure 7.1 shows the payoff pairs for the four pure action pairs and all correlated action pairs (recall that a correlated action is a probability distribution over action pairs, as defined in Chapter 4).

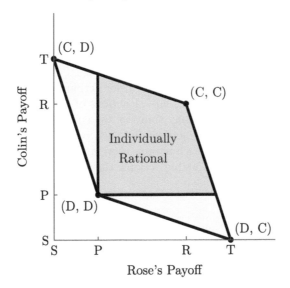

FIGURE 7.1 Prisoner's Dilemma Feasible and Individually Rational Payoff Pairs.

If Rose and Colin could negotiate and make a binding agreement, the resulting bargaining game would be symmetric, and all efficient and unbiased solution methods would choose the payoff pair (R, R) and the outcome in which both players Cooperate. The dilemma is in finding a way for both players to Cooperate when no binding agreement is made. We will first model this scenario as a finite repeated game. That is, we assume that there is a fixed and known number m of stages and the overall payoff is the mean of the payoffs in the individual stages.

In this model, G is the Prisoner's Dilemma strategic game, and there will be $m = 6$ stages. Table 7.2 shows the actions (C = Cooperate and D = Defect) and payoffs associated with one possible play of the finite repeated game.

TABLE 7.2   A Finite Repeated Game Play

Stage	1	2	3	4	5	6	Mean
Rose's Action	C	C	D	D	D	C	
Colin's Action	C	D	C	C	D	C	
Rose's Payoff	3	0	5	5	1	3	2.83
Colin's Payoff	3	5	0	0	1	3	2.00

The table does not tell us the strategies Rose and Colin used to determine their actions. Perhaps Colin has decided to alternate C and D but restart the pattern the stage after Rose first chooses D. If Colin is using this strategy, Table 7.3 shows how Colin's actions and payoffs would change if Rose chooses D in each stage except the last.

TABLE 7.3 A Finite Repeated Game Play

Stage	1	2	3	4	5	6	Mean
Rose's Action	D	D	D	D	D	C	
Colin's Action	C	C	D	C	D	C	
Rose's Payoff	5	5	1	5	1	3	3.33
Colin's Payoff	0	0	1	0	1	3	0.83

A strategy for a finite repeated game specifies what to do in stage 1 (C or D), what to do in stage 2 given the stage 1 actions of both players (e.g., C if both players' choices were the same in stage 1 and D otherwise), what to do in stage 3 given the stage 1 and 2 actions of both players (e.g., C if and only if 3 of the 4 prior actions were C), and in general, what to do in stage m given the stage 1, 2, ..., $m-1$ actions of both players (e.g., C if and only if the opponent chose C at least as many times as the current player chose C). In stage k, a player knows their own $k-1$ prior actions and their opponent's $k-1$ prior actions, and since there are two possibilities for each action, there are $2^{2(k-1)} = 4^{k-1}$ possible histories. Since for each history there are 2 possible actions, there are

$$2^{4^0} 2^{4^1} 2^{4^2} \cdots 2^{4^{m-1}} = 2^{4^0 + 4^1 + 4^2 + \cdots + 4^{m-1}} = 2^{(4^m - 1)/3}$$

possible strategies. For $m = 2$, there are 32 strategies. By the time $m = 6$, there are more than 10^{410} strategies, a number larger than the number of particles in the universe. While a strategy can be very complex and the number of possible strategies is enormous, we will be interested in only a few including the following.

1. *AllDefect*: Defect in each stage.

2. *RandomCooperate(ρ)*: Cooperate with probability ρ in each stage.

3. *TitForTat*: Cooperate in the first stage. In subsequent stages, act as the other player did in the previous stage.

4. *GrimTrigger*: Cooperate in the first and subsequent stages until the other player defects, and then defect in all subsequent stages.

5. *Pavlov*: Cooperate in the first stage and whenever the player's actions in the previous stage matched each other.

6. *TwoForOne*: Cooperate in the first stage. In subsequent stages, defect whenever the other player defected in either of the two most recent previous stages, and otherwise cooperate.

7. *Flow*: Cooperate in the first stage. In subsequent stages, cooperate with probability equal to the fraction of the time the other player has cooperated in previous stages.

Observe that strategies AllDefect and RandomCooperate(ρ) require no memory because the action chosen in a particular stage does not depend on actions taken in previous stages. The difference between these two, however, is that one is deterministic and the other is probabilistic. The strategies TitforTat, GrimTrigger, and Pavlov require a one-stage memory because the action chosen in a particular stage depends on the actions taken in the previous, but no other, stage. The strategy TwoforOne requires a two-stage memory and the strategy Flow requires a memory bounded only by the number of stages because the action chosen requires knowing the other player's actions in all previous stages.

For a probabilistic strategy like RandomCooperate(ρ), there are many possibilities for the sequence of actions a player using this strategy will have. Table 7.4 shows one instance of a six-stage repeated game with Rose using RandomCooperate(0.6) paired with Colin using TitForTat. For clarity in the table we refer to the players by their strategy names.

TABLE 7.4   A Finite Repeated Game Play

Stage	1	2	3	4	5	6	Mean
RandomCooperate(0.6)'s Action	D	C	C	D	D	C	
TitForTat's Action	C	D	C	C	D	D	
RandomCooperate(0.6)'s Payoff	5	0	3	5	1	0	2.33
TitForTat's Payoff	0	5	3	0	1	5	2.33

Since Rose using RandomCooperate(0.6) chooses C with probability 0.6 and D with probability 0.4, this instance of the game occurs with probability $(0.4)(0.6)(0.6)(0.4)(0.4)(0.6)$ ≈ 0.0138. A player deciding whether to use this strategy or a player playing against it would need to consider all possible instances and the mean expected payoffs. Calculating mean expected payoffs requires some work, but one approach would be to construct all 64 possible instances of the game, calculate their probabilities of occurring, and calculate the mean expected payoffs to each player over all 64 instances with the given probabilities.

A different approach is to determine the expected payoffs in each stage of the repeated game and calculate their means. To do this, we first need to calculate the probability of a sequence of action pairs occurring. For example, in stage one of the game, the action pair CC occurs with probability 0.6 and DC with probability 0.4 since Colin, playing second and using TitForTat, will always choose C in stage 1.

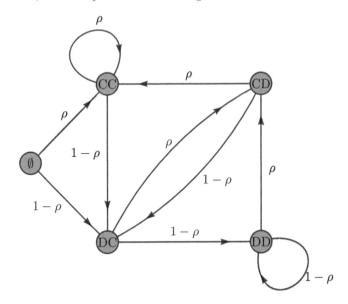

FIGURE 7.2   RandomCooperate(ρ) versus TitForTat State Transition Diagram.

The probability of the first two stages having the sequence of action pairs CC, CD is zero since TitForTat will never follow the action pair CC with an action of D. In contrast, Colin using TitForTat will follow CC with C with probability 1 so the probability of the first two stages having the sequence of action pairs CC, DC is $(0.6)(0.4)$, the probability of Rose choosing C and then D. The state transition diagram Figure 7.2 shows all possible action pairs and the probability of moving from one action pair to another for the general strategy RandomCooperate(ρ) against TitForTat. We can calculate the probability of any sequence of action pairs by multiplying the probabilities given on the arrows.

The information encoded in Figure 7.2 can be encoded in the following state transition matrix, $\mathbf{A}$, and initial state probability distribution vector, $\mathbf{x}$.

$$
\mathbf{A} = \begin{pmatrix} \rho & \rho & 0 & 0 \\ 0 & 0 & \rho & \rho \\ 1-\rho & 1-\rho & 0 & 0 \\ 0 & 0 & 1-\rho & 1-\rho \end{pmatrix} \begin{matrix} CC \\ CD \\ DC \\ DD \end{matrix} \quad , \quad \mathbf{x} = \begin{pmatrix} \rho \\ 0 \\ 1-\rho \\ 0 \end{pmatrix} \begin{matrix} CC \\ CD \\ DC \\ DD \end{matrix} \tag{7.1}
$$

with column headers $CC \quad CD \quad DC \quad DD$ From/To for $\mathbf{A}$ and $\emptyset$ From/To for $\mathbf{x}$.

The row i column j entry of the state transition matrix $\mathbf{A}$ is the probability of transitioning from state j to state i in the next stage, and row i of the initial state probability distribution vector $\mathbf{x} = \mathbf{x}^1$ is the probability of being in state i in stage 1. The probability distribution across states in subsequent stages can be found using matrix multiplication. Specifically, the state probability distribution vector in stage $k+1$ is defined to be

$$
\mathbf{x}^{k+1} = \mathbf{A}\mathbf{x}^k = \mathbf{A}^k\mathbf{x}.
$$

For the given matrix $\mathbf{A}$ and vector $\mathbf{x}$, we find that

$$
\mathbf{x}^k = \begin{pmatrix} \rho^2 \\ \rho(1-\rho) \\ (1-\rho)\rho \\ (1-\rho)^2 \end{pmatrix}, \text{ for } k = 2, 3, \ldots, 6.
$$

This shows that by stage 2, the probability distribution over states becomes stationary. Using the state transition diagram and matrix then gives a method for directly computing the probability of any sequence of action pairs in this game. Our goal is to find the expected payoffs at each stage using these probabilities.

Returning to the game RandomCooperate(0.6) versus TitforTat, the expected payoff for Rose using RandomCooperate(0.6) in stage 1 is $3(0.6) + 5(0.4) = 3.8$. In stage two, the probability of the sequences of action pairs CC, DC is $(0.6)(0.4) = 0.24$. The mean payoff for Rose for this sequence is $(3 + 5)/2 = 4$. To find the expected payoff for Rose we would need the probabilities of all other two-stage sequences of action pairs and their mean payoffs. In general, we can use the state probability distribution to payoff matrix Γ to compute these expected payoffs. For the Prisoner's Dilemma game,

$$
\Gamma = \begin{pmatrix} R & S & T & P \\ R & T & S & P \end{pmatrix} \begin{matrix} \text{Rose} \\ \text{Colin} \end{matrix} .
$$

with column headers $CC \quad CD \quad DC \quad DD$ State/Player.

Specifically, the expected payoffs in stage k are given in the vector

$$
\mathbf{v}^k = \Gamma \mathbf{x}^k.
$$

The mean expected payoffs for an m-stage game are then given in

$$
\mathbf{v} = \frac{1}{m}(\mathbf{v}^1 + \mathbf{v}^2 + \cdots + \mathbf{v}^m).
$$

For our original 6-stage finite repeated Prisoner's Dilemma game with Rose using RandomCooperate(0.6) versus Colin using TitForTat, we had $S = 0$, $P = 1$, $R = 3$, and $T = 5$. We therefore obtain

$$\mathbf{v}^1 = \begin{pmatrix} 3 & 0 & 5 & 1 \\ 3 & 5 & 0 & 1 \end{pmatrix} \begin{pmatrix} 0.6 \\ 0 \\ 1 - 0.6 \\ 0 \end{pmatrix} = \begin{pmatrix} 3.8 \\ 1.8 \end{pmatrix},$$

$$\mathbf{v}^k = \begin{pmatrix} 3 & 0 & 5 & 1 \\ 3 & 5 & 0 & 1 \end{pmatrix} \begin{pmatrix} 0.6^2 \\ (0.6)(1 - 0.6) \\ (1 - 0.6)(0.6) \\ (1 - 0.6)^2 \end{pmatrix} = \begin{pmatrix} 2.44 \\ 2.44 \end{pmatrix}$$

for $k = 2, 3, 4, 5, 6$, and

$$\mathbf{v} = \frac{1}{6} \left(\begin{pmatrix} 3.8 \\ 1.8 \end{pmatrix} + 5 \begin{pmatrix} 2.44 \\ 2.44 \end{pmatrix} \right) = \begin{pmatrix} 2.67 \\ 2.33 \end{pmatrix}.$$

This shows that Rose using RandomCooperate(0.6) has an expected payoff per stage of 2.67, and Colin using TitForTat has an expected payoff per stage of 2.33 in this 6-stage repeated game. It turns out that neither strategy is a best response to the other strategy in this game. If Rose cooperated in the first five stages but defected in stage 6 while Colin continued to use TitForTat, the sequence of action choices would be (CC, CC, CC, CC, CC, DC) and Rose's mean expected payoff would increase from 2.67 to $((5)(3) + 5)/6 = 3.33$. If Colin defected in all six stages while Rose continued to use RandomCooperate(0.6), the action choices in each stage would be CD with probability 0.6 and DD with probability $1 - 0.6 = 0.4$, and Colin's mean expected payoff would increase from 2.33 to $(0.6)(5) + (1 - 0.6)(1) = 3.4$. Clearly, (RandomCooperate(0.6), TitForTat) is not a Nash equilibrium in the 6-stage repeated game. The next theorem characterizes all Nash equilibria for a finite repeated Prisoner's Dilemma game.

Theorem 7.1.1. In a finite repeated Prisoner's Dilemma game, (AllDefect, AllDefect) is a Nash equilibrium. Furthermore, any Nash equilibrium results in each player defecting at every stage.

Proof. Suppose Colin is using the AllDefect strategy. If Rose chooses a different strategy from AllDefect that changes her payoff, she will necessarily cooperate in at least one stage with positive probability. Her payoff in that stage will decrease, because Colin does not change his actions and defection is unilaterally better than cooperation. Thus her expected mean payoff for the repeated game will decrease. Hence, AllDefect is a best response to AllDefect.

In order to prove the second part of the theorem, suppose to the contrary we have a Nash equilibrium such that in at least one stage there is a positive probability that some player will cooperate. Let stage k be the last such stage, and without loss of generality, let Rose be the player that has a positive probability of cooperating. Consider what would happen if Rose changes her strategy to unconditionally defect in stages k and later. This change in Rose's strategy will not change either player's actions before stage k, in stage k it will change Rose's action to defect with probability 1 but not change Colin's action, and after stage k will leave Rose's defections unchanged but may change Colin's defections into some positive probability for cooperation. Hence, Rose's expected payoff from the first $k - 1$ states is unchanged, expected payoff from stage k increases (since defection within a single stage is always better than cooperation), and expected payoff from the later stages is unchanged or increased (since increased cooperation by an opponent is always better). Thus, Rose's original strategy was not a best response to Colin's strategy, a contradiction to having a Nash equilibrium. □

Therefore, cooperation does not result from repeated play when the number of stages is known. However, we still observe cooperation in many real-life scenarios. Donors give to NPR and it remains funded and countries enter trade agreements all the time. The reason for this observed cooperation is that in real life the number of stages of the game is rarely known. To capture the idea that the number of stages is not known beforehand, we determine the number of stages randomly using a probability distribution. For simplicity, we will assume that the next stage of the game happens with a fixed probability δ. Thus, the first stage happens with probability 1, a second stage happens with probability δ, a third stage happens with probability δ^2, and a kth stage happens with probability δ^{k-1}. These assumptions lead to the following definition.

Definition 7.1.1. A *random repeated game* consists of the following:

1. A game G with all of its embedded players, outcomes, rules, and utilities.

2. A continuation probability $\delta < 1$.

3. In each stage, the game G is played with all players knowing the actions chosen in the previous stages. The next stage is then played with probability δ.

4. Utilities for each player i are defined to be the mean of the utilities player i obtained in each stage of the repeated game.

Here we pause to discuss the fidelity of this model for repeated social dilemmas. It is often the case, that while players may have multiple opportunities to play a game, those opportunities decrease over time as modeled by our probability distribution. For example, listeners to an NPR station eventually move away from the broadcast region and while countries have multiple opportunities to negotiate trade treaties, these opportunities decrease as more agreements are reached or new governments are installed.

The other assumptions match our finite repeated game model and have worked well in investigating such social dilemmas.

Returning to our Prisoner's Dilemma game example, Table 7.5 gives a random repeated game that begins with the action pair DD and continues by alternating the action pairs DC, CC in a two-stage cycle.

TABLE 7.5   A Random Repeated Game Play

Stage	1	2	3	4	5	...	$2k$	$2k + 1$	...
Rose's Action	D	D	C	D	C	...	D	C	...
Colin's Action	D	C	C	C	C	...	C	C	...
Rose's Payoff	P	T	R	T	R	...	T	R	...
Colin's Payoff	P	S	R	S	R	...	S	R	...

To calculate Rose's expected payoff we must first calculate the expected number of stages

$$1 + \delta + \delta^2 + \cdots + \delta^{k-1} + \cdots = 1/(1 - \delta).$$

We find her payoffs at each stage from the table above and thus Rose's mean expected payoff is

$$
\begin{aligned}
v_{\text{Rose}} &= (1 - \delta)(P + T\delta + R\delta^2 + T\delta^3 + R\delta^4 + \cdots) \\
&= (1 - \delta)(P + (T + R\delta)\delta(1 + \delta^2 + \cdots)) \\
&= (1 - \delta)(P + (T + R\delta)\delta/(1 - \delta^2)) \\
&= P(1 - \delta) + (T + R\delta)\delta/(1 + \delta).
\end{aligned}
$$

Similarly, Colin's expected payoff per stage is

$$
v_{\text{Colin}} = P(1 - \delta) + (S + R\delta)\delta/(1 + \delta).
$$

If two strategies are deterministic, such as AllDefect, then the actions can be determined in each stage and expected payoffs can be calculated as above. If one or both of the strategies has a random aspect and both strategies have a bounded memory, then a state transition matrix $\mathbf{A}$, an initial state probability distribution vector $\mathbf{x}$, and state probability distribution to payoffs matrix $\mathbf{\Gamma}$ can be used. The probability that stage k will be reached is δ^{k-1}, and if it is reached, the expected payoffs in stage k are in the vector $\mathbf{\Gamma}\mathbf{A}^{k-1}\mathbf{x}$. Therefore, the mean expected payoffs are

$$
\begin{aligned}
v &= (1 - \delta)(\mathbf{\Gamma}\mathbf{x} + \delta\mathbf{\Gamma}\mathbf{A}\mathbf{x} + \delta^2\mathbf{\Gamma}\mathbf{A}^2\mathbf{x} + \cdots + \delta^{k-1}\mathbf{\Gamma}\mathbf{A}^{k-1}\mathbf{x} + \cdots) \\
&= (1 - \delta)\mathbf{\Gamma}(\mathbf{I} + \delta\mathbf{A} + \delta^2\mathbf{A}^2 + \cdots + \delta^{k-1}\mathbf{A}^{k-1} + \cdots)\mathbf{x} \\
&= (1 - \delta)\mathbf{\Gamma}(\mathbf{I} - \delta\mathbf{A})^{-1}\mathbf{x}
\end{aligned}
\tag{7.2}
$$

where it can be shown that the matrix $\mathbf{I} - \delta\mathbf{A}$ must have an inverse if $0 \leq \delta < 1$.

For example, for RandomCooperate(ρ) versus TitForTat in the random repeated game with probability δ of continuation, we obtain (using Mathematica or similar software to obtain the second line)

$$
\begin{aligned}
v &= (1 - \delta)\mathbf{\Gamma} \left(\begin{pmatrix} 1 & 0 & 0 & 0 \\ 0 & 1 & 0 & 0 \\ 0 & 0 & 1 & 0 \\ 0 & 0 & 0 & 1 \end{pmatrix} - \delta \begin{pmatrix} \rho & \rho & 0 & 0 \\ 0 & 0 & \rho & \rho \\ 1-\rho & 1-\rho & 0 & 0 \\ 0 & 0 & 1-\rho & 1-\rho \end{pmatrix} \right)^{-1} \begin{pmatrix} \rho \\ 0 \\ 1-\rho \\ 0 \end{pmatrix} \\
&= \begin{pmatrix} R & S & T & P \\ R & T & S & P \end{pmatrix} \begin{pmatrix} \rho(1 - \delta(1 - \rho)) \\ \delta\rho(1 - \rho) \\ (1 - \rho)(1 - \delta(1 - \rho)) \\ \delta(1 - \rho)^2 \end{pmatrix} \\
&= \begin{pmatrix} R\rho(1 - \delta(1 - \rho)) + S\delta\rho(1 - \rho) + T(1 - \rho)(1 - \delta(1 - \rho)) + P\delta(1 - \rho)^2 \\ R\rho(1 - \delta(1 - \rho)) + T\delta\rho(1 - \rho) + S(1 - \rho)(1 - \delta(1 - \rho)) + P\delta(1 - \rho)^2 \end{pmatrix}.
\end{aligned}
$$

If $S = 0$, $P = 1$, $R = 3$, $T = 5$, $\rho = 0.6$, and $\delta = 5/6$, then the mean expected payoffs for Rose and Colin are 2.67 and 2.33. These mean expected payoffs happen to be the same as for the finite repeated Prisoner's Dilemma game with 6 stages because the expected number of stages in the random repeated game $1/(1 - \rho) = 6$ equals the number of stages in the finite repeated game and by stage 2 the probability distribution over states becomes stationary.

Once again, we have chosen a pair of strategies that are not best responses to each other. If Rose switches to AllCooperate while Colin sticks with TitForTat, Rose's expected payoff per stage increases from 2.67 to 3. If Colin switches to AllDefect while Rose sticks with RandomCooperate(0.6), Colin's expected payoff per stage increases from 2.33 to 3.4.

Unlike in the finite repeated game, the following theorem shows that in the random repeated game there may be a Nash equilibrium in which players cooperate in every stage.

Theorem 7.1.2. In the random repeated Prisoner's Dilemma game, (GrimTrigger, GrimTrigger) is a Nash equilibrium with mean expected payoffs of (R, R) if the continuation probability δ is sufficiently large.

Proof. If both Rose and Colin use GrimTrigger, then both will cooperate in every stage resulting in mean expected payoffs per stage of

$$(1 - \delta)(R + R\delta + R\delta^2 + \cdots) = R$$

for each player. For Rose to obtain a different mean expected payoff, Rose must defect in some stage. Once Rose defects, Colin will defect in all subsequent stages regardless of Rose's subsequent actions. Hence, Rose will obtain the largest expected payoff by continuing to defect after her first defection. If it is advantageous for Rose to start defecting in some stage, it will be advantageous for Rose to start defecting in the first stage. In this case, Rose's mean expected payoff per stage is

$$(1 - \delta)(T + P\delta + P\delta^2 + \cdots) = T(1 - \delta) + P\delta.$$

Thus, GrimTrigger will be a best response to GrimTrigger whenever

$$T(1 - \delta) + P\delta \leq R,$$

which is equivalent to

$$\frac{T - R}{T - P} \leq \delta.$$

Since $P < R$, it follows that $(T - R)/(T - P) < 1$, hence, δ can be appropriately chosen. $\qquad\square$

Cooperation can result from repeated play if the continuation probability is sufficiently high, that is, cooperation can emerge when the likelihood of continued interaction is sufficiently high. Recall that in a single-stage Prisoner's Dilemma game, Defect is the dominant strategy for each player, and in every Nash equilibrium in the finite repeated game, players always choose Defect. Thus, our decision to model potentially infinite repeated games was the correct way to show the possibility of the cooperation that we observe in the real world, but we need a high enough continuation probability for this to happen.

Although (AllDefect, AllDefect) with payoff pair (P, P) continues to be a Nash equilibrium for all continuation probabilities, high continuation probabilities allow many other Nash equilibria such as (TitForTat, TitForTat) and (Pavlov, Pavlov) which yield the payoff pair (R, R). Other non-punitive outcomes are possible. In Figure 7.1 these non-punitive outcomes are the individually rational payoff pairs excluding the left and bottom borders. The following theorem shows that for any such payoff pair, there is a Nash equilibrium of the random repeated game whose mean expected payoff per stage pair is arbitrarily close when there is a large enough chance of the game continuing.

The origin of this result is not clear, so it has become known in the literature as the Folk Theorem.

Theorem 7.1.3 (Folk). Suppose (a, b) is an individually rational payoff pair in Prisoner's Dilemma with $a > P$ and $b > P$, and let $\epsilon > 0$. In the random repeated Prisoner's Dilemma game with a sufficiently large continuation probability, there is a Nash equilibrium with mean expected payoff pair (c, d) satisfying $|a - c| < \epsilon$ and $|b - d| < \epsilon$ [4].

We illustrate the idea embedded in the proof of the Folk Theorem with an example. Suppose our desired payoff is

$$(a, b) = \tfrac{2}{3}(R, R) + \tfrac{1}{3}(T, S).$$

Let Rose's strategy be to repeat the three-stage pattern Cooperate, Cooperate, Defect as long as Colin always Cooperates, but Defect in all subsequent stages if Colin Defects in any stage. Let Colin's strategy be to Cooperate in every stage as long as Rose follows the three-stage pattern Cooperate, Cooperate, Defect, but Defect in all subsequent stages if Rose does not follow this pattern in some stage. With these strategies, Rose will repeat the three-stage pattern Cooperate, Cooperate, Defect, and Colin will Cooperate in every stage. Rose's expected payoff per stage is

$$c = (1 - \delta)(R + R\delta + T\delta^2)/(1 - \delta^3) = (R + R\delta + T\delta^2)/(1 + \delta + \delta^2)$$

which converges upward to a as δ approaches 1. Similarly, Colin's expected payoff per stage is

$$d = (1 - \delta)(R + R\delta + S\delta^2)/(1 - \delta^3) = (R + R\delta + S\delta^2)/(1 + \delta + \delta^2)$$

which converges downward to b as δ approaches 1. So, for sufficiently large $\delta < 1$, it follows that $|a - c| < \epsilon$ and $|b - d| < \epsilon$. Finally, we show that the given strategies form a Nash equilibrium. If Colin were to deviate in the first stage, his expected payoff per stage would be $T(1 - \delta) + P\delta$ which is less than b for sufficiently large δ (by our Theorem 7.1.3 assumption that $b > P$) which is less than d, hence, the deviation is disadvantageous. Similar arguments can be used to show that deviations by Rose or Colin in any stage would be disadvantageous.

Theorem 7.1.3 demonstrates that Nash equilibria make very unspecific predictions about player behavior in the repeated Prisoner's Dilemma. Indeed, we observe players adopting a plethora of strategies. We see individuals who donate to NPR each year, while others donate randomly from year to year. Likewise, national positions on trade may change as administrations change

Perhaps the theory qualitatively matches reality. On the other hand, it seems unlikely that real agents adopt the increasingly complex and sensitive strategies suggested by the proof of Theorem 7.1.3, and so much of the research has focused on relatively simple strategies such as the 1-stage memory strategies that have been the primary focus of this section. Good sources for additional information are [5], [104], and [67].

7.2   MODELING NOISE

Sometimes players believe their opponent has defected although the opponent actually cooperated. Other times players intend to cooperate but inadvertently defect. The former is a *perception error*, and the latter is an *implementation error*. The potential for such errors introduces noise into the situation, and the resulting repeated game will have different Nash equilibria.

Definition 7.2.1. A *noisy repeated game* consists of the following:

1. A strategic game G.

2. A continuation probability $\delta < 1$, a perception error probability ξ, and an implementation error probability η.

3. In stage $k = 1, 2, \ldots$, the game G is played with player i's chosen action actually being used with probability $1 - \eta$ and a different action being used with probability η. With probability $1 - \xi$, player i will perceive the actual outcome and with probability ξ will perceive a different outcome has occurred. The conditional probability that stage $k = 2, 3, \ldots$ is played given that stage $k - 1$ was played is δ.

4. The payoff received by player i is the player's expected payoff per stage.

In each stage that is played, there are intended, implemented, and perceived actions. The actions chosen by the players are their intended actions. With probability $1 - \eta$, a player's intended action is implemented correctly, and with probability η, their intended action is implemented incorrectly, Finally, with probability $1 - \xi$, a player's implemented action is perceived correctly, and with probability ξ, their implemented action is perceived incorrectly. Each player receives a stage payoff based on the implemented strategies and the payoff table given above. With probability $1 - \delta$, the game ends, and with probability δ, a new stage is played.

In general, a (behavioral) strategy consists of a function that returns a probability distribution over actions given the history of past actions. To faithfully model perception errors, we will assume that players do not know the intended or implemented actions of their opponent nor their own perceived actions. Hence, a player's strategy can only depend on the player's own intended and implemented actions and their opponent's perceived actions.

To illustrate, we again use Prisoner's Dilemma as the underlying game. Suppose Rose and Colin each use TitForTat, where in the current stage each player copies the perceived action of their opponent in the most recent stage. As in the previous section, we can describe the state of the repeated game by the implemented actions of Rose and Colin in the most recent stage of the game. If there is no noise, then the probability distribution across states is determined by the following state transition matrix $\mathbf{A}$ and initial state probability distribution vector $\mathbf{x}$.

$$
\mathbf{A} = \begin{array}{cccc c}
CC & CD & DC & DD & From/To \\
\begin{pmatrix} 1 & 0 & 0 & 0 \\ 0 & 0 & 1 & 0 \\ 0 & 1 & 0 & 0 \\ 0 & 0 & 0 & 1 \end{pmatrix} & & & & \begin{array}{c} CC \\ CD \\ DC \\ DD \end{array}
\end{array}
\quad , \quad
\mathbf{x} = \begin{array}{c c}
\emptyset & From/To \\
\begin{pmatrix} 1 \\ 0 \\ 0 \\ 0 \end{pmatrix} & \begin{array}{c} CC \\ CD \\ DC \\ DD \end{array}
\end{array}
$$

If there is a positive probability ξ of a perception error (and no probability of an implementation error), then the state transition matrix and initial state probability distribution vector will change to the following.

$$
\mathbf{A} = \begin{array}{cccc c}
CC & CD & DC & DD & From/To \\
\begin{pmatrix} (1-\xi)^2 & \xi(1-\xi) & (1-\xi)\xi & \xi^2 \\ (1-\xi)\xi & \xi^2 & (1-\xi)^2 & \xi(1-\xi) \\ \xi(1-\xi) & (1-\xi)^2 & \xi^2 & (1-\xi)\xi \\ \xi^2 & (1-\xi)\xi & \xi(1-\xi) & (1-\xi)^2 \end{pmatrix} & & & & \begin{array}{c} CC \\ CD \\ DC \\ DD \end{array}
\end{array}
\quad , \quad
\mathbf{x} = \begin{array}{c c}
\emptyset & From/To \\
\begin{pmatrix} 1 \\ 0 \\ 0 \\ 0 \end{pmatrix} & \begin{array}{c} CC \\ CD \\ DC \\ DD \end{array}
\end{array}
$$

For example, consider the first column of the transition matrix, which corresponds to each player's intended and implemented actions being Cooperate in the most recent stage of the game. Each player has probability $1 - \xi$ of perceiving the correct action (Cooperate) of his opponent and probability ξ of perceiving the incorrect action (Defect) of his opponent. Hence, $(1 - \xi)^2$ is the probability that each player perceives correctly that the opponent previously chose Cooperate, and so each player should Cooperate in the next stage. Next, $(1 - \xi)\xi$ is the probability that Rose perceives correctly and Colin perceives incorrectly, which implies that Rose should Cooperate and Colin should Defect in the next stage. Next, $\xi(1-\xi)$ is the probability that Rose perceives incorrectly and Colin perceives correctly, which implies that Rose should Defect and Colin should Cooperate in the next stage. Finally, ξ^2 is the probability that each player perceives incorrectly and so each player should Defect in the next stage. The other entries are determined in a similar manner. Since there is nothing to perceive before players choose their first stage actions, both players will Cooperate in stage 1 with probability 1.

For a fixed $\xi > 0$, we can approximate the expected payoff per stage when δ is near 1 by calculating (preferably using Mathematica or similar software)

$$\mathbf{v} = \lim_{\delta \to 1}(1 - \delta)\mathbf{\Gamma}(\mathbf{I} - \delta\mathbf{A})^{-1}\mathbf{x} = \frac{1}{4}(S + P + R + T)\begin{pmatrix}1\\1\end{pmatrix}.$$

Intuitively, two TitForTat players will both cooperate until one player makes a perception error. At that point, the players will alternate between cooperation and defection until the next perception error. At that point, the players will either engage in mutual defection or mutual cooperation. In the long run, each payoff pair will be equally likely.

If Rose instead uses AllCooperate while Colin continues to use TitForTat, a similar calculation shows that Rose's expected payoff per stage as δ approaches 1 is $(1 - \xi)R + \xi S$. Intuitively, AllCooperate and TitForTat will both Cooperate most of the time (with probability $1 - \xi$) except when a perception error is made by Colin (with probability ξ). When there is no perception error, Rose obtains R, and when Colin makes a perception error, Rose obtains S in the next stage. Observe that our assumptions that $P < R$ and $S + T < 2R$ imply that $(1/4)(S + P + R + T) < R$. Thus, for sufficiently small ξ, Rose obtains a higher payoff using AllCooperate instead of TitForTat when Colin is using TitForTat. Although (TitForTat, TitForTat) is a Nash equilibrium for the random repeated game with no noise, we have just shown that (TitForTat, TitForTat) is not a Nash equilibrium for the random repeated game with noise arising from perception errors.

Let us now consider a noisy repeated game in which there is a positive probability η of an implementation error (and no probability of a perception error). If Rose uses AllDefect and Colin uses Pavlov, then the state transition matrix and initial state probability distribution vector will be as follows.

$$\mathbf{A} = \begin{array}{c} \\ \\ \begin{pmatrix} \eta(1-\eta) & \eta^2 & \eta^2 & \eta(1-\eta) \\ \eta^2 & \eta(1-\eta) & \eta(1-\eta) & \eta^2 \\ (1-\eta)^2 & (1-\eta)\eta & (1-\eta)\eta & (1-\eta)^2 \\ (1-\eta)\eta & (1-\eta)^2 & (1-\eta)^2 & (1-\eta)\eta \end{pmatrix} \end{array} \begin{array}{l} CC \\ CD \\ DC \\ DD \end{array}, \quad \mathbf{x} = \begin{pmatrix} \eta(1-\eta) \\ \eta^2 \\ (1-\eta)^2 \\ (1-\eta)\eta \end{pmatrix} \begin{array}{l} CC \\ CD \\ DC \\ DD \end{array}$$

with column headings $CC \quad CD \quad DC \quad DD \quad From/To$ and $\emptyset \quad From/To$.

For example, in the second and third columns of the transition matrix, CD and DC are the implemented action pairs in the most recent stage. In either case, both Rose/AllDefect and Colin/Pavlov will intend to Defect in the next stage. If neither make an implementation error, DD will be the next stage implemented action pair, and the probability of neither player making an implementation error is $(1 - \eta)^2$. If both make an implementation error, CC will be the next stage implemented action pair, and this will occur with probability η^2. Finally, if exactly one player makes an implementation error, then CD or DC will be the next stage implemented action pair, and each will occur with probability $\eta(1 - \eta)$. In the initial state probability distribution vector and the first and fourth columns of the transition matrix, the intended action pair by the players is DC, and the probabilities of the different possible action pairs that are actually implemented are derived in a similar fashion.

For a fixed $\eta > 0$, we can approximate the expected payoff per stage when δ is near 1 by calculating

$$\mathbf{v} = \lim_{\delta \to 1}(1 - \delta)\mathbf{\Gamma}(\mathbf{I} - \delta\mathbf{A})^{-1}\mathbf{x} = \begin{pmatrix} \frac{1}{2}(T + P) - \frac{\eta}{2}(T - R + P - S) \\ \frac{1}{2}(S + P) + \frac{\eta}{2}(T - P + R - S) \end{pmatrix}.$$

Intuitively, the player using AllDefect will usually Defect, and in those stages, the player using Pavlov will alternate between Defect and Cooperate. It can be shown that the expected payoff per stage of each player when both are using Pavlov in a noisy repeated game in which the probability of continuation δ is near 1 and the probability of an implementation error is near 0 is approximately R. Thus, for at least some choices for the Prisoner's Dilemma payoff parameters (specifically, when $2R < T + P$), the strategy pair (Pavlov, Pavlov) is not a Nash equilibrium in a noisy repeated game.

Although many of the Nash equilibria when there is no noise are no longer Nash equilibria when there is noise, essentially the same set of payoff pairs can be obtained via Nash equilibria of the noisy random repeated game as was obtained from the (no noise) random repeated game.

Theorem 7.2.1 (Another Folk). Suppose (a, b) is an expected payoff pair that can be obtained with mixed strategies in Prisoner's Dilemma, $a > P$ and $b > P$, and $\epsilon > 0$. In the noisy repeated Prisoner's Dilemma game with a sufficiently large continuation probability and sufficiently small but positive perception and/or implementation errors, there is a Nash equilibrium with expected payoff pair per stage (c, d) satisfying $|a - c| < \epsilon$ and $|b - d| < \epsilon$.

Since the conclusions of the two Folk Theorems are the same, we can make the same remarks here as we did at the end of the previous section. Nash equilibria for these repeated game models provide very unspecific predictions about player behavior, but that may correspond well will our qualitative and anecdotal observations about people, companies, and nations interacting in repeated social dilemmas. In the next section, we will examine one way players might select among the plethora of Nash equilibria.

7.3 EVOLUTIONARY GAME THEORY

Natural Evolution. In Chapter 3 we discussed animal behavior related to competition and the decisions animals make about engaging in fights. We treated animals as players making rational decisions and seeking optimal outcomes. In reality, animals may not have the cognitive abilities to make such reasoned choices, but evidence suggests they do learn from past interactions and adapt their behaviors, and we can study the behavioral patterns that evolve.

While the motivation for the field of evolutionary game theory was biological, the methods and concepts have widespread applications. Businesses and governments have limited computational resources and may not always be able to make the investment required to find an optimal strategy. They can, however, learn from past interactions and adapt their strategies, especially when they face repeated interactions with the same opponent.

To model this adaptive behavior across a population, we'll assume that the population of players plays the game pairwise, i.e., one player from each population plays according to a strategy. We'll model the distribution of strategies across each population using probability density functions over the set of available strategies, and we'll model the pairwise interactions of the players using two-player strategic games. The likelihood that a strategy will continue to be used will increase with the total payoff received by a player using that strategy because either (1) players with low total scores will learn about and change to better performing strategies, or (2) better performing players will have more progeny that make use of the better performing strategies.

We also include in our model the possibility that some players may adopt a strategy somewhat at random and that some players may make sophisticated predictions about the behaviors of others and choose a strategy to do well based on those predictions. The goal of our model is to predict the time evolution of the probability distributions of strategies used by the players. We want to see if one strategy will dominate and drive the others to extinction if a stable distribution of two or more strategies will arise, if there will be continual oscillation in a regular or sporadic manner. The model we use is known as a *strategy evolution system*. Formally,

Definition 7.3.1. A *strategy evolution system* consists of the following:

1. A two-player symmetric strategic game with m strategies $\sigma_1, \sigma_2, \ldots, \sigma_m$ and payoffs $V_{i,j} \geq 0$ for a player using strategy σ_i against a player using strategy σ_j. Note that the mean payoff to a player using strategy σ_i against each player in a population consisting of proportions $p_1, p_2, \ldots, p_m$ using strategies $\sigma_1, \sigma_2, \ldots, \sigma_m$, respectively, is

$$v(\sigma_i, \mathbf{p}) = p_1 V_{i,1} + p_2 V_{i,2} + \cdots + p_m V_{i,m} = (\mathbf{Vp})_i.$$

2. An *evolution mechanism* $\mathbf{F}$, which is a function that given a strategy distribution $\mathbf{p}$ (i.e., $\mathbf{p} = (p_1, p_2, \ldots, p_m)$ satisfies $p_1 \geq 0, p_2 \geq 0, \ldots, p_m \geq 0$, and $p_1 + p_2 + \cdots + p_m = 1$) and the payoff matrix $\mathbf{V}$ outputs the next-generation strategy distribution $\mathbf{F}(\mathbf{p}, \mathbf{V})$, which will be abbreviated $\mathbf{F}(\mathbf{p})$ when the payoff matrix $\mathbf{V}$ is clear from the context.

The *evolution* of a strategy distribution $\mathbf{p}$ is the sequence $\mathbf{p}^0, \mathbf{p}^1, \mathbf{p}^2, \ldots$ of strategy distributions satisfying $\mathbf{p}^0 = \mathbf{p}$ and $\mathbf{p}^{t+1} = \mathbf{F}(\mathbf{p}^t)$ for generations $t = 0, 1, 2, \ldots$.

Although it includes a game, a strategy evolution system is not a game. There are no true players making independent decisions about what to do. Instead, there is a strategy distribution that changes in accordance with the evolution mechanism. We are interested in the behavior of these evolutions. In particular, a strategy distribution $\mathbf{p}$ is *fixed* if $\mathbf{p} = \mathbf{F}(\mathbf{p}, \mathbf{V})$. This formalizes a population being in stasis. We are most interested in populations that are likely to arise and persist. A strategy distribution $\mathbf{p}$ is *dynamically stable* if the evolution for every strategy distribution $\mathbf{q}$ sufficiently close to $\mathbf{p}$ converges to $\mathbf{p}$. Sometimes what can persist is a cycle or a variety of populations. A set of strategy distributions $\mathbf{P}$ is *dynamically stable* if strategy distributions in $\mathbf{P}$ stay in $\mathbf{P}$ and the evolution for every strategy distribution $\mathbf{q}$ sufficiently close to $\mathbf{P}$ converges to $\mathbf{P}$.

We now describe an example strategy evolution system that arises from a random repeated Prisoner's Dilemma game. Let $\sigma_1 = $ AllDefect, $\sigma_2 = $ TitForTat, and $\sigma_3 = $ Pavlov be the $m = 3$ strategies playing the random repeated Prisoner's Dilemma game shown in Table 7.1 with $S = 0$, $P = 1$, $R = 3$, $T = 5$, and $\delta = 0.9$. When AllDefect faces AllDefect, both players will Defect during every stage, which results in the punishment payoff in every stage, hence,

$$V_{1,1} = P = 1.$$

When TitForTat or Pavlov faces TitForTat or Pavlov, both players will cooperate during every stage, which results in the reward payoff in every stage, hence,

$$V_{2,2} = V_{2,3} = V_{3,2} = V_{3,3} = R = 3.$$

When AllDefect faces TitForTat, both players will defect during every stage except during the first stage when TitForTat will cooperate. Hence, the payoff to AllDefect will be

$$V_{1,2} = (1 - \delta)(T + \delta P + \delta^2 P + \cdots)$$
$$= (1 - \delta)T + \delta P$$
$$= (0.1)(5) + (0.9)(1) = 1.4,$$

and the payoff to TitForTat will be

$$V_{2,1} = (1 - \delta)(S + \delta P + \delta^2 P + \cdots)$$

$$= (1 - \delta)S + \delta P$$
$$= (0.1)(0) + (0.9)(1) = 0.9.$$

When AllDefect faces Pavlov, AllDefect will defect during every stage and Pavlov will alternately cooperate and defect. Hence, the payoff to AllDefect will be

$$V_{1,3} = (1 - \delta)(T + \delta P + \delta^2 T + \delta^2 P + \cdots)$$
$$= (1 - \delta)(T + \delta P)(1 + \delta^2 + \delta^4 + \cdots)$$
$$= (1 - \delta)(T + \delta P)/(1 - \delta^2)$$
$$= (T + \delta P)/(1 + \delta)$$
$$= (5 + (0.9)(1))/(1 + 0.9) = 3.10526,$$

and the payoff to Pavlov will be

$$V_{3,1} = (1 - \delta)(S + \delta P + \delta^2 S + \delta^2 P + \cdots)$$
$$= (S + \delta P)/(1 + \delta)$$
$$= (0 + (0.9)(1))/(1 + 0.9) = 0.473684.$$

The payoffs in the two-player symmetric strategic game are summarized in the following matrix.

$$\mathbf{V} = \begin{array}{c} \\ \text{AllDefect} \\ \text{TitForTat} \\ \text{Pavlov} \end{array} \begin{array}{ccc} \text{AllDefect} & \text{TitForTat} & \text{Pavlov} \\ \left(\begin{array}{ccc} 1.000 & 1.400 & 3.105 \\ 0.900 & 3.000 & 3.000 \\ 0.474 & 3.000 & 3.000 \end{array} \right) \end{array} \qquad (7.3)$$

For our example, we will use the *replicator evolution mechanism* [114]:

$$\mathbf{F}_i(\mathbf{p}) = p_i \frac{v_i}{\bar{v}} = p_i \frac{(\mathbf{Vp})_i}{\mathbf{p}^T \mathbf{Vp}} \qquad (7.4)$$

where

$$v_i = V_{i,1}p_1 + V_{i,2}p_2 + \cdots + V_{i,m}p_m = (\mathbf{Vp})_i. \qquad (7.5)$$

is the total payoff to a player using strategy σ_i and

$$\bar{v} = p_1 v_1 + p_2 v_2 + \cdots + p_m v_m = \mathbf{p}^T \mathbf{v} = \mathbf{p}^T \mathbf{Vp} \qquad (7.6)$$

is the mean total payoff across all players. Equation 7.4 indicates that doubling either the proportion of users of a strategy or its payoff should double the proportion of users of that strategy in the next generation. This is reasonable if payoffs are proportional to the number of progeny or the spread of knowledge about the relative performance of the strategies.

If we begin with equal proportions of players using each strategy, the payoffs to and proportions of players using each strategy in the next two generations are as follows:

$$\mathbf{p}^0 = \begin{pmatrix} 0.333 \\ 0.333 \\ 0.333 \end{pmatrix}$$

$$\mathbf{v}^0 = \mathbf{Vp}^0 = \begin{pmatrix} 1.000 & 1.400 & 3.105 \\ 0.900 & 3.000 & 3.000 \\ 0.474 & 3.000 & 3.000 \end{pmatrix} \begin{pmatrix} 0.333 \\ 0.333 \\ 0.333 \end{pmatrix} = \begin{pmatrix} 1.835 \\ 2.300 \\ 2.158 \end{pmatrix}$$

$$\bar{v}^0 = (\mathbf{p}^0)^T \mathbf{v}^0 = \begin{pmatrix} 0.333 & 0.333 & 0.333 \end{pmatrix} \begin{pmatrix} 1.835 \\ 2.300 \\ 2.158 \end{pmatrix} = 2.098$$

$$\mathbf{p}^1 = \left(p_i^0 v_i^0 / \bar{v}^0 \right) = \begin{pmatrix} (0.333)(1.835/2.098) \\ (0.333)(2.300/2.098) \\ (0.333)(2.158/2.098) \end{pmatrix} = \begin{pmatrix} 0.292 \\ 0.365 \\ 0.343 \end{pmatrix}$$

$$\mathbf{v}^1 = \mathbf{V}\mathbf{p}^1 = \begin{pmatrix} 1.000 & 1.400 & 3.105 \\ 0.900 & 3.000 & 3.000 \\ 0.474 & 3.000 & 3.000 \end{pmatrix} \begin{pmatrix} 0.292 \\ 0.365 \\ 0.343 \end{pmatrix} = \begin{pmatrix} 1.868 \\ 2.388 \\ 2.263 \end{pmatrix}$$

$$\bar{v}^1 = (\mathbf{p}^1)^T \mathbf{v}^1 = \begin{pmatrix} 0.292 & 0.365 & 0.343 \end{pmatrix} \begin{pmatrix} 1.868 \\ 2.388 \\ 2.263 \end{pmatrix} = 2.194$$

$$\mathbf{p}^2 = \left(p_i^1 v_i^1 / \bar{v}^1 \right) = \begin{pmatrix} (0.292)(1.868/2.194) \\ (0.365)(2.388/2.194) \\ (0.343)(2.263/2.194) \end{pmatrix} = \begin{pmatrix} 0.248 \\ 0.398 \\ 0.354 \end{pmatrix}.$$

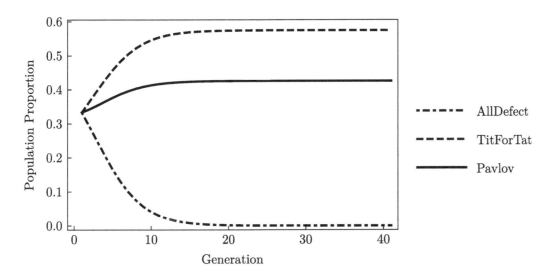

FIGURE 7.3 First Example Evolution.

Figure 7.3 plots these proportions for the first forty generations, providing one graphical view of the evolution of $(0.333, 0.333, 0.333)$. Clearly, the use of AllDefect quickly drops to 0% while use of TitForTat and Pavlov increase to and stabilize at 57% and 43%, respectively, of the population.

Different initial strategy distributions result in a different final strategy distributions. Figure 7.4 shows the evolution of the distribution of strategies from a population that is initially 40% AllDefect, 5% TitForTat, and 55% Pavlov users. Notice that the proportion of AllDefect users initially grows but the population eventually stabilizes with 0% AllDefect, 92% TitForTat, and 8% Pavlov users.

Figure 7.5 shows the evolution of the distribution of strategies from a population that is initially 4% AllDefect, 1% TitForTat, and 95% Pavlov users. The proportion of AllDefect users increases until it is the only strategy used.

For each initial strategy distribution, the evolution eventually stabilizes, although the stable strategy distributions are different. We can better understand the possibilities by

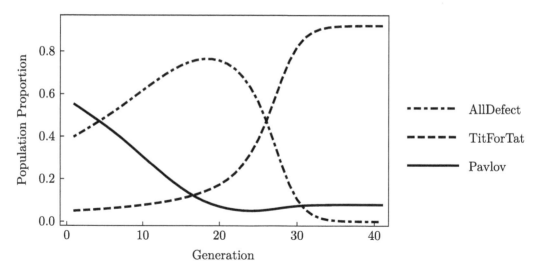

FIGURE 7.4 Second Example Evolution.

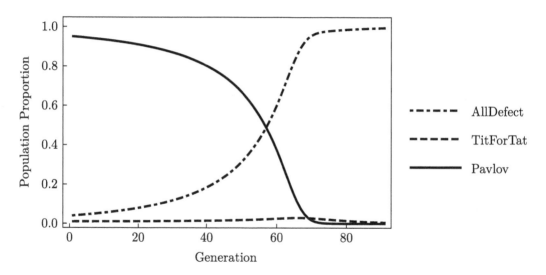

FIGURE 7.5 Third Example Evolution.

viewing all three evolutions on the plot given in Figure 7.6. The equilateral triangle shows all possible strategy distributions $\mathbf{p} = (p_1, p_2, p_3)$, where $p_1 + p_2 + p_3 = 1$, $p_1 \geq 0$, $p_2 \geq 0$, and $p_3 \geq 0$, and can be viewed as the intersection in three dimensions of the nonnegative orthant with the plane $p_1 + p_2 + p_3 = 1$. In particular, the lower left vertex corresponds to $(1, 0, 0)$ or 100% AllDefect, the lower right vertex corresponds to $(0, 1, 0)$ or 100% TitForTat, and the top vertex corresponds to $(0, 0, 1)$ or 100% Pavlov. The *barycentric* coordinates of points in the equilateral triangle change linearly. For example, the starting point $(0.33, 0.33, 0.33)$ of the first evolution appears in the center of the equilateral triangle, and the starting point $(0.04, 0.01, 0.95)$ appears near the top vertex (because $p_3 = 0.95$ is close to 1) and to the left of horizontal center (because $p_1 > p_2$). In general, a probability distribution (p_1, p_2, p_3) is mapped to $(x, y) = ((1/(\sqrt{3}))(p_2 - p_1), p_3)$ where the origin of the xy-plane is

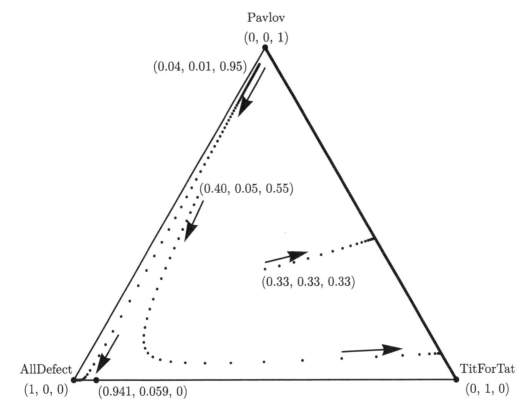

FIGURE 7.6 Example Evolutions.

at the midpoint of the bottom of the equilateral triangle. Arrows indicate the direction of evolution (i.e., $\mathbf{p}^t$ as t increases).

In addition to the three previously calculated evolutions, Figure 7.6 shows in bold the fixed strategy distributions: AllDefect alone, 94.1% of AllDefect and 5.9% of TitForTat, and any combination of TitForTat and Pavlov. The following theorem formalizes what is suggested by Figure 7.6.

Theorem 7.3.1. For the strategy evolution system consisting of the symmetric game with payoff matrix given by equation 7.3 and the replicator evolution mechanism, Figure 7.7 displays each possible evolution: (1) no movement starting from a fixed strategy distribution shown in bold, (2) convergence to AllDefect if starting from a non-fixed strategy distribution to the left of the non-border line L, (3) convergence to 0.941AllDefect + 0.059TitForTat if starting at a strategy distribution in the interior of the line segment L, or (4) convergence to $(1 - \lambda)$TitForTat + λPavlov for some $0 \leq \lambda < 0.938$ if starting elsewhere. The latter set is dynamically stable but each strategy distribution $(1 - \lambda, \lambda, 0)$ in the set is not itself dynamically stable.

> *Proof.* The fixed strategy distributions were found by substituting Equation 7.4 into the fixed strategy distribution condition $\mathbf{p} = \mathbf{F}(\mathbf{p})$ to find that a strategy distribution $\mathbf{p}$ is a fixed point if and only if $p_i(v_i - \bar{v}) - 0$ for $i = 1, 2, \ldots, m$. It immediately follows that $(1, 0, 0)$, $(0, 1, 0)$, and $(0, 0, 1)$ are fixed. We now consider strategy distributions with exactly one element equal to 0. For a strategy distribution of the form

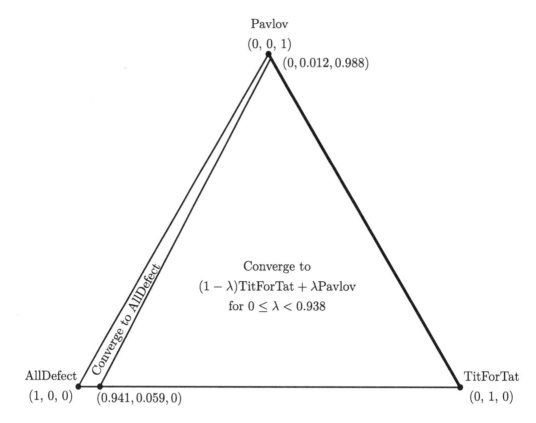

Pavlov
(0, 0, 1)
(0, 0.012, 0.988)

Converge to
$(1 - \lambda)\text{TitForTat} + \lambda\text{Pavlov}$
for $0 \leq \lambda < 0.938$

Converge to AllDefect

AllDefect
(1, 0, 0)   (0.941, 0.059, 0)

TitForTat
(0, 1, 0)

FIGURE 7.7 Example Evolution Convergence.

$\mathbf{p} = (1 - \lambda, \lambda, 0)$ with $0 < \lambda < 1$ to be fixed, the payoff vector $\mathbf{v} = \mathbf{Vp} = (1 + 0.4\lambda, 0.9 + 2.1\lambda, 0.474 + 2.526\lambda)$ must have its first two elements equal, which yields a single fixed strategy distribution $(0.941, 0.059, 0)$. For a strategy distribution of the form $\mathbf{p} = (1 - \lambda, 0, \lambda)$ with $0 < \lambda < 1$ to be fixed, the payoff vector $\mathbf{v} = \mathbf{Vp} = (1 + 2.105\lambda, 0.9 + 2.1\lambda, 0.474 + 2.526\lambda)$ must have its first and third elements equal, which yields the unique solution $(-0.25, 0, 1.25)$, which is not a strategy distribution. For any strategy distribution of the form $\mathbf{p} = (0, 1 - \lambda, \lambda)$ with $0 < \lambda < 1$, the payoff vector $\mathbf{v} = \mathbf{Vp} = (1.4 + 1.705\lambda, 3, 3)$ has its last two elements equal, hence, all such strategy distributions are fixed. Finally, for any strategy distribution $\mathbf{p}$ in which all three elements are positive, the payoff vector $\mathbf{v} = \mathbf{Vp} = (p_1 + 1.4p_2 + 3.105p_3, 0.9p_1 + 3p_2 + 3p_3, 0.474p_1 + 3p_2 + 3p_3)$ must have its three elements equal to each other and $p_1 + p_2 + p_3 = 1$, which yield the single fixed strategy distribution $(0, 0.062, 0.938)$, which is one of the fixed points found in the previous sentence.

As noted previously, Figure 7.6 suggests all evolution patterns. We can make this more precise by dividing the equilateral triangle of strategy distributions into regions where we roughly know the direction of motion going from $\mathbf{p}$ to $\mathbf{F(p)}$ as shown in Figure 7.8. If the strategy distribution $\mathbf{p}$ is on the curve labeled $v_i = \bar{v}$, then there is no movement in the p_i direction, i.e., $\mathbf{F}_i(\mathbf{p}) = p_i$. The curve labeled $v_1 = \bar{v}$ is obtained from the equation $(\mathbf{Vp})_1 = \mathbf{p}^T\mathbf{Vp}$ which is

$$(1 \quad 0 \quad 0)\begin{pmatrix} 1.000 & 1.400 & 3.105 \\ 0.900 & 3.000 & 3.000 \\ 0.474 & 3.000 & 3.000 \end{pmatrix}\begin{pmatrix} p_1 \\ p_2 \\ p_3 \end{pmatrix} = (p_1 \ p_2 \ p_3)\begin{pmatrix} 1.000 & 1.400 & 3.105 \\ 0.900 & 3.000 & 3.000 \\ 0.474 & 3.000 & 3.000 \end{pmatrix}\begin{pmatrix} p_1 \\ p_2 \\ p_3 \end{pmatrix}$$

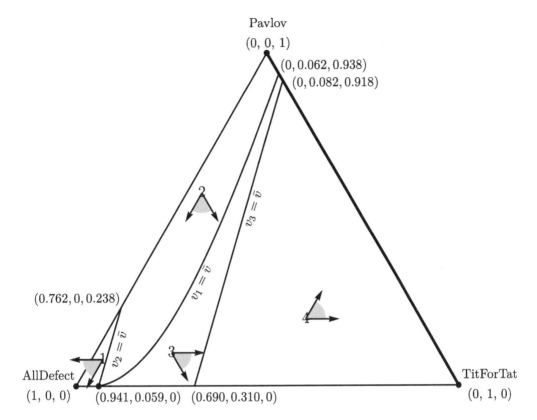

FIGURE 7.8 Example Evolution Directions.

and can be simplified to

$$p_1 + 1.4p_2 + 3.105p_3 = p_1^2 + 3p_2^2 + 3p_3^2 + 2.3p_1p_2 + 3.579p_1p_3 + 6p_2p_3.$$

Substituting in $p_3 = 1 - p_1 - p_2$ and solving for p_2, we obtain $p_2 = 0.247((1 + 3p_1 - 4p_1^2)/(4 - 3p_1))$. This results in a curve extending from $(0, 0.062, 0.938)$ to $(0.941, 0.059, 0)$. Similarly, the curve labeled $v_2 = \bar{v}$ is obtained from the equations $(\mathbf{Vp})_2 = \mathbf{p}^T\mathbf{Vp}$ and $p_1 + p_2 + p_3 = 1$, yielding $p_2 = -0.251 + 0.329p_1$, a straight line from $(0.762, 0, 0.238)$ to $(0.941, 0.059, 0)$. Finally, the curve labeled $v_3 = \bar{v}$ can be described by the equation $p_2 = 0.082 + 0.329p_1$, a straight line from $(0, 0.082, 0.918)$ to $(0.690, 0.310, 0)$.

Now that we have determined the lines where there is no movement in one of the three directions, we consider the direction of motion in the regions between these curves. If $\mathbf{p} = (0.98, 0.01, 0.01)$, then $\mathbf{Vp} = (1.025, 0.942, 0.524)$ and $\mathbf{p}^T\mathbf{Vp} = 1.019$, hence, $p_1 < \mathbf{F}_1(\mathbf{p})$ if $\mathbf{p}$ is to the left of the $v_1 = \bar{v}$ line, $p_2 > \mathbf{F}_2(\mathbf{p})$ if $\mathbf{p}$ is to the left of the $v_2 = \bar{v}$ line, and $p_3 > \mathbf{F}_3(\mathbf{p})$ if $\mathbf{p}$ is to the left of the $v_3 = \bar{v}$ line. Since the region labeled 1 is to the left of all three curves, movement must be towards AllDefect and away from TitForTat and Pavlov; this is indicated in Figure 7.8 by the range of arrows going from left (parallel to the $p_3 = 0$ line) to down and slightly left (parallel to the $p_2 = 0$ line). If $\mathbf{p} = (0.02, 0.49, 0.49)$, then $\mathbf{Vp} = (2.228, 2.958, 2.949)$ and $\mathbf{p}^T\mathbf{Vp} = 2.939$, hence, $p_1 > \mathbf{F}_1(\mathbf{p})$ if $\mathbf{p}$ is to the right of the $v_1 = \bar{v}$ line, $p_2 < \mathbf{F}_2(\mathbf{p})$ if $\mathbf{p}$ is to the right of the $v_2 = \bar{v}$ line, and $p_3 < \mathbf{F}_3(\mathbf{p})$ if $\mathbf{p}$ is to the right of the $v_3 = \bar{v}$ line. Since the region labeled 4 is to the right of all three lines, movement must be away from AllDefect and towards TitForTat and Pavlov; this is indicated in Figure 7.8 by the range of arrows going from right (parallel to the $p_3 = 0$ line) to up and slightly right (parallel to the $p_2 = 0$ line). Since region 3 is to the left of the $v_3 = \bar{v}$ curve and to the right of the other two curves, movement must be away from AllDefect and Pavlov and towards TitForTat. Since region 2 is to the right of the $v_2 = \bar{v}$ curve and to the left of the other two curves, movement must be towards AllDefect and TitForTat and away from Pavlov.

Knowing the meaning and derivation of the labeling in Figure 7.8, we can determine what evolutions are possible. A strategy distribution in region 4 must move to the right and somewhat upward, hence, its evolution cannot cross the $v_3 = \bar{v}$ line (which would require movement in an impossible direction), cannot cycle (which would require movement in some step in an impossible direction), cannot converge to a point inside region 4 (otherwise the strategy distribution to which it converges would be fixed by the continuity of the replicator evolution mechanism), and so must converge to a point on the right side boundary below $(0, 0.082, 0.918)$, a fixed strategy distribution that is a mixture of only TitForTat and Pavlov strategies. Similarly, the evolution of a strategy distribution in region 3 must eventually either converge to a point on the right boundary between $(0, 0.062, 0.938)$ exclusive and $(0, 0.082, 0.918)$ inclusive or cross into region 4; in either case, the evolution will converge to some mixture of TitForTat and Pavlov having more than 0.062 TitForTat. The evolution of a strategy distribution in region 1 must converge to AllDefect. Finally, the evolution of a strategy distribution in region 2 must either cross into region 1 and converge to AllDefect, converge to the fixed strategy distribution $(0.941, 0.059, 0)$, or cross into region 4 and converge to some mixture of TitForTat and Pavlov. The line L connecting $(0, 0.012, 0.988)$ with $(0.941, 0.059, 0)$ divides region 2 into these three possible evolution behaviors because for a strategy distribution $\mathbf{p}$ on (respectively, to the left of or to the right of) the line L, the strategy distribution $\mathbf{F}(\mathbf{p})$ is also on (respectively, to the left of or to the right of) L, and so the evolution converges to $(0.941, 0.059, 0)$ (respectively, AllDefect or a mixture of only TitForTat and Pavlov).

We observe that while the set $\{(1 - \lambda, \lambda, 0) : 0 \leq \lambda < 0.938\}$ is dynamically stable, each strategy distribution $(1 - \lambda, \lambda, 0)$ in the set is not itself dynamically stable because evolutions of other strategy distributions in the set do not converge to the focal strategy distribution.

While this completes the proof of the theorem as stated, we now provide an alternate approach because it is often difficult to carry out a complete global analysis as we have done above. It is sufficient to use local arguments to determine whether a fixed strategy distribution $\mathbf{p}$ is dynamically stable. The idea is to linearize the evolution mechanism near $\mathbf{p}$ and check whether that linearized mechanism is stable by examining its eigenvalues. Specifically,

$$\mathbf{F}(\mathbf{q}) \approx \mathbf{J}(\mathbf{p})\mathbf{q}$$

for $\mathbf{q}$ near $\mathbf{p}$ where

$$\mathbf{J}(\mathbf{p}) = \begin{pmatrix} \dfrac{\partial \mathbf{F}_1}{\partial q_1} & \cdots & \dfrac{\partial \mathbf{F}_1}{\partial q_m} \\ \vdots & \ddots & \vdots \\ \dfrac{\partial \mathbf{F}_m}{\partial q_1} & \cdots & \dfrac{\partial \mathbf{F}_m}{\partial q_m} \end{pmatrix} \Bigg|_{\mathbf{q}=\mathbf{p}}.$$

An eigenvalue of $\mathbf{J}(\mathbf{p})$ tells us approximately what to multiply the distance from $\mathbf{p}$ to $\mathbf{q}$ by in the direction of the corresponding eigenvector to obtain the distance from $\mathbf{p}$ to $\mathbf{F}(\mathbf{q})$ in the direction of the corresponding eigenvector (with rotational movement happening when there is a pair of complex conjugate eigenvalues). Thus, if the absolute values of the eigenvalues are all less than 1, the strategy distribution $\mathbf{p}$ is dynamically stable. If any eigenvalue absolute value is greater than 1, the strategy distribution $\mathbf{p}$ is not dynamically stable. It is easiest to use computer software to calculate the Jacobian and its eigenvalues.

For our example,

$$\mathbf{J}(1,0,0) = \begin{pmatrix} 0 & -0.9 & -0.474 \\ 0 & 0.9 & 0 \\ 0 & 0 & 0.474 \end{pmatrix}$$

which has eigenvalues with absolute values 0.9, 0.474, and 0. Since they are all less than 1, $(1,0,0)$ is dynamically stable, as we have already seen. Also

$$\mathbf{J}(0.941, 0.059, 0) = \begin{pmatrix} 0.064 & -1.028 & -0.567 \\ -0.064 & 1.028 & -0.042 \\ 0 & 0 & 0.608 \end{pmatrix}$$

which has eigenvalues with absolute values 1.09, 0.608, and 0. Since there is at least one greater than 1, $(0.941, 0.059, 0)$ is not dynamically stable, as we have already seen. □

We now interpret the results of our example in the context of our original Natural Evolution scenario. The mathematical model is a good representation of animals that (1) engage in repeated Prisoner's Dilemma interactions with each other over the course of their lifetime, (2) adopt one of the three strategies AllDefect, TitForTat, or Pavlov genetically, and (3) have reproduction rates that are proportional to the total payoffs obtained by the animals. Except in highly exceptional cases, either AllDefect (if there are few TitForTat players in the population) or a mixture of TitForTat and Pavlov (in most circumstances) will emerge. These qualitative results will not change if there are small changes in the values of the Prisoner's Dilemma payoffs $S < P < R < T$ or the probability of continuation δ because the payoff matrix in the two-player symmetric strategic game in the more general case is (as derived earlier).

$$\mathbf{V} = \begin{array}{c} \\ \text{AllDefect} \\ \text{TitForTat} \\ \text{Pavlov} \end{array} \begin{array}{c} \text{AllDefect} \\ \begin{pmatrix} P \\ (1-\delta)S + \delta P \\ (S + \delta P)/(1+\delta) \end{pmatrix} \end{array} \begin{array}{c} \text{TitForTat} \\ (1-\delta)T + \delta P \\ R \\ R \end{array} \begin{array}{c} \text{Pavlov} \\ (T + \delta P)/(1+\delta) \\ R \\ R \end{array} \begin{array}{c} \\ \\ \end{pmatrix} \end{array}$$

and the qualitative results only depend on the rank order of these payoffs. This tells us that the qualitative conclusions are somewhat robust.

The replicator evolution mechanism has some important properties. Strategies cannot spontaneously generate, strategies with high payoffs will increase their proportion in the population, and strategies with low payoffs will decrease their proportion in the population. The evolution mechanism $\mathbf{F}$ is *mutation-free* if $\mathbf{F}_i(\mathbf{p}) = 0$ whenever $p_i = 0$ and is *payoff monotone* if $\mathbf{F}_i(\mathbf{p})/p_i \geq \mathbf{F}_j(\mathbf{p})/p_j$ whenever $p_i > 0$, $p_j > 0$, and $v(\sigma_i, \mathbf{p}) \geq v(\sigma_j, \mathbf{p})$. If some in the population could anticipate how others in the population will change and then choose their strategy as a best response, then new strategies could appear in the population and currently low payoff strategies could become high payoff with the anticipated strategy changes elsewhere. Near the end of this section, we will explore how allowing mutations changes the evolution mechanism. For now, let us consider more generally the behavior of mutation-free and payoff-monotone evolution mechanisms.

As we have already noted, a strategy evolution system is not a game. Nonetheless, it includes a symmetric two-player strategic game as part of its definition. Given a strategy distribution $\mathbf{p} = (p_1, p_2, \ldots, p_m)$, we have been interpreting p_i as the fraction of players in a large population of players that use strategy σ_i; however, one could also interpret $\mathbf{p}$ as the mixed strategy $\sigma = p_1\sigma_1 + p_2\sigma_2 + \cdots + p_m\sigma_m$ used by a single player in the two-player symmetric strategic game. Since the payoff to a player using the pure strategy σ_i against the other player using the mixed strategy $p_1\sigma_1 + p_2\sigma_2 + \cdots + p_m\sigma_m$ is $v(\sigma_i, \mathbf{p}) = p_1 V_{i,1} + p_2 V_{i,2} + \cdots + p_m V_{i,m} = (\mathbf{V}\mathbf{p})_i$, the payoff to a player using the mixed strategy $q_1\sigma_1 + q_2\sigma_2 + \cdots + q_m\sigma_m$ against the other player using the mixed strategy $p_1\sigma_1 + p_2\sigma_2 + \cdots + p_m\sigma_m$ is

$$v(\mathbf{q}, \mathbf{p}) = q_1 v(\sigma_1, \mathbf{p}) + q_2 v(\sigma_2, \mathbf{p}) + \cdots + q_m v(\sigma_m, \mathbf{p}) = \mathbf{q}^T \mathbf{V} \mathbf{p}.$$

This motivates finding a relationship between symmetric Nash equilibria of the game and evolutions for the evolution system.

Suppose $(\sigma_1, \sigma_2, \ldots, \sigma_m, \mathbf{V}, \mathbf{F})$ is an evolution system. A strategy distribution $\mathbf{p}$ is *Nash* if (σ, σ), where $\sigma = p_1\sigma_1 + p_2\sigma_2 + \cdots + p_m\sigma_m$, is a Nash equilibrium for the symmetric two-player strategic game with payoffs given by the matrix V. A *subgame* of a strategic game is obtained by eliminating some of the strategies while maintaining the same payoffs for the strategy profiles that remain.

Theorem 7.3.2. Suppose $(\sigma_1, \ldots, \sigma_m, \mathbf{V}, \mathbf{F})$ is a mutation-free and payoff-monotone evolution system. The strategy distribution $\mathbf{p}$ is fixed if and only if $\mathbf{p}$ is Nash on the subgame that includes only the strategies σ_i for which $p_i > 0$. Furthermore, if $\mathbf{p}$ is dynamically stable, then $\mathbf{p}$ is Nash.

Proof. Let $\sigma = p_1\sigma_1 + p_2\sigma_2 + \cdots + p_m\sigma_m$. The strategy pair (σ, σ) is a Nash equilibrium for the symmetric two-player strategic game with payoffs given by the matrix $\mathbf{V}$ if and only if σ is a best response to σ if and only if payoffs $\mathbf{p}^T \mathbf{V} \mathbf{p} \geq \mathbf{q}^T \mathbf{V} \mathbf{p}$ for any mixed strategy $q_1\sigma_1 + q_2\sigma_2 + \cdots + q_m\sigma_m$ if and only if

$$(\mathbf{V}\mathbf{p})_i = \mathbf{p}^T \mathbf{V} \mathbf{p} \qquad\qquad \text{if } p_i > 0 \qquad\qquad (7.7)$$

$$(\mathbf{V}\mathbf{p})_i \leq \mathbf{p}^T \mathbf{V} \mathbf{p} \qquad\qquad \text{if } p_i = 0 \qquad\qquad (7.8)$$

for all $i = 1, 2, \ldots, m$. Furthermore, if $S = \{i : p_i > 0\}$, then (σ, σ) is a Nash equilibrium for the subgame on the strategies in $\{\sigma_i : i \in S\}$ if and only if Condition 7.7 holds.

Suppose $\mathbf{p}$ is a Nash strategy distribution. Then Conditions 7.7 and 7.8 hold. By Condition 7.7, it follows that $v(\sigma_i, \mathbf{p}) = v(\sigma_j, \mathbf{p})$ whenever $p_i > 0$ and $p_j > 0$. Since $\mathbf{F}$ is evolution monotone, $\mathbf{F}_i(\mathbf{p}, \mathbf{V})/p_i = \mathbf{F}_j(\mathbf{p}, \mathbf{V})/p_j$ whenever $p_i > 0$ and $p_j > 0$. Naming the common ratio λ, we obtain $\mathbf{F}_i(\mathbf{p}, \mathbf{V}) = \lambda p_i$ whenever $p_i > 0$. Since $\mathbf{F}$ is mutation free, $\mathbf{F}_i(\mathbf{p}, \mathbf{V}) = 0 = \lambda p_i$ whenever $p_i = 0$. Since $\mathbf{F}(\mathbf{p}, \mathbf{V})$ and $\mathbf{p}$ are probability distributions, $1 = \mathbf{F}_1(\mathbf{p}, \mathbf{V}) + \mathbf{F}_2(\mathbf{p}, \mathbf{V}) + \cdots + \mathbf{F}_m(\mathbf{p}, \mathbf{V}) = \lambda p_1 + \lambda p_2 + \cdots + \lambda p_m = \lambda(p_1 + p_2 + \cdots + p_m) = \lambda$. Since $\lambda = 1$, it follows that $\mathbf{F}_i(\mathbf{p}, \mathbf{V}) = p_i$ for all $i = 1, 2, \ldots, m$, that is, $\mathbf{p}$ is fixed.

Suppose $\mathbf{p}$ is a fixed strategy distribution. Then $\mathbf{F}_i(\mathbf{p}, \mathbf{V}) = p_i$ for all $i = 1, 2, \ldots, m$. Hence, $\mathbf{F}_i(\mathbf{p}, \mathbf{V})/p_i = 1$ whenever $p_i > 0$. Since $\mathbf{F}$ is payoff monotone, $v(\sigma_i, \mathbf{p}) = v(\sigma_j, \mathbf{p})$ whenever $p_i > 0$ and $p_j > 0$. Naming the common value λ, we obtain $v(\sigma_i, \mathbf{p}) = \lambda$ whenever $p_i > 0$, hence, $p_i v(\sigma_i, \mathbf{p}) = p_i \lambda$ for all $i = 1, 2, \ldots, m$. Thus, $\mathbf{p}^T \mathbf{V} \mathbf{p} = p_1 v_1(\mathbf{p}) + p_2 v_2(\mathbf{p}) + \cdots + p_m v_m(\mathbf{p}) = p_1 \lambda + p_2 \lambda + \cdots + p_m \lambda = \lambda$. This yields Condition 7.7, which we have already seen implies that (σ, σ) is a Nash equilibrium for the subgame on the strategies in $\{\sigma_i : p_i > 0\}$.

Finally, we prove the contrapositive of the last statement. Suppose $\mathbf{p}$ is not Nash. Then there exists a strategy σ_j satisfying $v(\sigma_j, \mathbf{p}) > v(\mathbf{p}, \mathbf{p})$, which is equivalent to $(\mathbf{V}\mathbf{p})_j > \mathbf{p}^T \mathbf{V} \mathbf{p}$. This implies that $(\mathbf{V}\mathbf{q})_j > \mathbf{q}^T \mathbf{V} \mathbf{q}$ for $\mathbf{q}$ sufficiently close to $\mathbf{p}$. Hence, $\mathbf{F}_j(\mathbf{q}) > q_j$ for $\mathbf{q}$ sufficiently close to $\mathbf{p}$. In particular, for $\mathbf{q}$ sufficiently close to $\mathbf{p}$ and

$q_j > p_j$, $\mathbf{F}(\mathbf{q})$ is further away from $\mathbf{p}$ in the jth coordinate than is $\mathbf{q}$. Thus, $\mathbf{p}$ is not dynamically stable. □

Since Nash strategy distributions are fixed, we can check whether a fixed strategy distribution is Nash by using Condition 7.8. Returning to our example, we can verify that $(1, 0, 0)$ and $(0, 1, 0)$ are Nash (in the matrix $\mathbf{V}$ in Equation 7.3, row 1 has the maximum value in column 1 and row 2 has the maximum value in column 2) but $(0, 0, 1)$ is not Nash (row 3 is not the maximum value in column 3). Similarly, $\mathbf{v} = \mathbf{V}(0.941, 0.059, 0)^T = (1.024, 1.024, 0.622)^T$ shows that $(0.941, 0.059, 0)$ is Nash. Finally, $\mathbf{v} = \mathbf{V}(0, 1 - \lambda, \lambda)^T = (1.4 + 1.705\lambda, 3, 3)^T$ and $1.4 + 1.705\lambda \le 3$ if and only if $\lambda \le 0.938$ shows that $(0, 1 - \lambda, \lambda)$ is Nash if and only if $0 \le \lambda \le 0.938$. Observe that as stated in the theorem, the dynamically stable strategy distribution $(1, 0, 0)$ is Nash. The fact that $(0.941, 0.059, 0)$ is Nash but not dynamically stable shows that the converse of the last statement in the theorem does not hold.

In Chapter 3, we defined an evolutionarily stable strategy to be one that forms a symmetric Nash equilibrium and has an additional condition that intuitively ensures that no other strategy can displace the stable one. In the current context, a strategy distribution $\mathbf{p}$ is *evolutionarily stable* if (σ, σ), where $\sigma = p_1\sigma_1 + p_2\sigma_2 + \cdots + p_m\sigma_m$, is a Nash equilibrium for the symmetric two-player strategic game with payoffs given by the matrix $\mathbf{V}$, and $v(\mathbf{q}, \mathbf{q}) < v(\mathbf{p}, \mathbf{q})$ whenever $v(\mathbf{q}, \mathbf{p}) = v(\mathbf{p}, \mathbf{p})$. For our example, the only evolutionarily stable strategy distribution is the dynamically stable strategy distribution $(1, 0, 0)$. Unfortunately, the two stability concepts are not equivalent. There are replicator evolution systems in which there are evolutionarily stable strategy distributions that are not dynamically stable, and there are other replicator evolution systems in which there are dynamically stable strategy distributions that are not evolutionarily stable. While replicator evolution systems typically have dynamically stable strategy distributions (sometimes as a set), symmetric two-player strategic games often have no evolutionarily stable strategy distributions. Since these two stability concepts, at first blush, seem to be describing the same intuitive concept, these actual differences in results emphasize the importance in taking great care in formalizing our intuitions.

Returning to our example, our results change if a small amount of noise is added to the system. If we let $S = 0$, $P = 1$, $R = 3$, $T = 5$, and $\delta = 0.9$ as before but make the probability of a perception error $\xi = 0.01$ instead of 0, then the payoffs in the two-player symmetric strategic game are summarized in the following matrix.

$$
\mathbf{V} = \begin{array}{c} \\ \text{AllDefect} \\ \text{TitForTat} \\ \text{Pavlov} \end{array}
\begin{array}{c} \text{AllDefect} \quad \text{TitForTat} \quad \text{Pavlov} \end{array}
\left(\begin{array}{ccc} 1.000 & 1.436 & 3.106 \\ 0.891 & 2.913 & 2.853 \\ 0.473 & 2.865 & 2.959 \end{array} \right) \tag{7.9}
$$

These numbers are not much different from the numbers in the payoff matrix when there is no noise (Equation 7.3), however, the fact that the four payoffs in the bottom right corner are no longer equal results in profound changes in the possible replicator evolutions. Figures 7.9 and 7.10 are two representative replicator evolutions. Note the oscillations before the populations stabilize. In Figure 7.9, the initial strategy distribution is 16.9% AllDefect, 12.9% TitForTat, and 70.2% Pavlov, and the evolution converges to 100% TitForTat. In Figure 7.10, the initial strategy distribution is 16.8% AllDefect, 13.0% TitForTat, and 70.2% Pavlov — slightly less AllDefect and slightly more TitForTat, yet the evolution converges to 100% AllDefect! It turns out that essentially all initial strategy distribution evolutions converge to 100% TitForTat or 100% AllDefect, but the separating curve spirals into the completely mixed fixed strategy distribution $(0.164, 0.133, 0.703)$ as indicated in Figure 7.12.

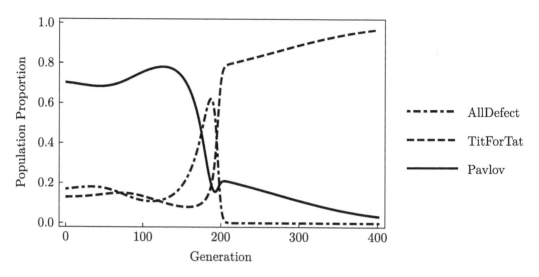

FIGURE 7.9 First Example Evolution with Noise.

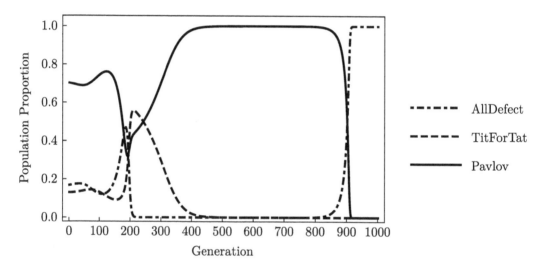

FIGURE 7.10 Second Example Evolution with Noise.

We now provide an outline of a more rigorous argument justifying the claims of the previous paragraph. The calculations described were performed using the software Mathematica. The fixed strategy distributions were found by solving the equations and inequalities $p_1 + p_2 + p_3 = 1$, $p_i(v_i - \bar{v}) = 0$, and $p_i \geq 0$ for $i = 1, 2, 3$. This resulted in $(1, 0, 0)$, $(0, 1, 0)$, $(0, 0, 1)$, $(0, 0.689, 0.311)$, $(0.931, 0.069, 0)$, and $(0.164, 0.133, 0.703)$. All except $(0, 0, 1)$ are Nash, and only $(1, 0, 0)$ and $(0, 1, 0)$ are evolutionarily and dynamically stable. The equations $v_i = \bar{v}$ for $i = 1, 2, 3$ lead to some messy quadratic equations that can be used to obtain parametric curves as shown in Figure 7.11. By checking the values of v_1, v_2, v_3, and $\bar{v}$ at a couple of points, we can determine in each region whether (1) $v_i > \bar{v}$ and so $\mathbf{F}_i(\mathbf{p}) > p_i$, or (2) $v_i < \bar{v}$ and so $\mathbf{F}_i(\mathbf{p}) < p_i$; this is indicated in Figure 7.11 by the range of arrows icons. It is then clear that if an evolution enters the lower part of region 1, then it will converge to AllDefect; if an evolution enters region 5, then it will converge to TitForTat; the only evolutions that converge to Pavlov are those that initially have no AllDefect players;

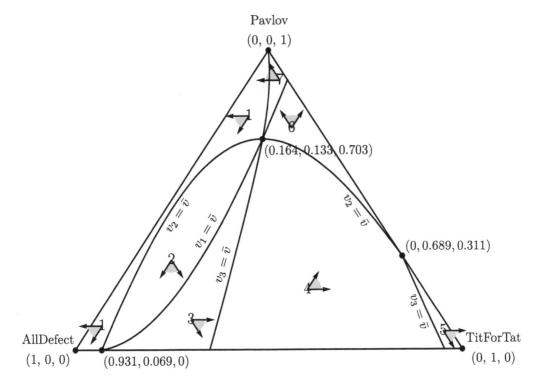

FIGURE 7.11 Example with Noise Evolution Directions.

and evolutions otherwise circulate counter-clockwise about the fixed strategy distribution $(0.164, 0.133, 0.703)$.

Figure 7.12 was obtained by examining the evolution for up to 2000 stages of each strategy distribution $\mathbf{p}$ in which each component was a multiple of 0.01. The point $\mathbf{p}$ was colored dark gray if the tth stage strategy distribution ever had at least 99.9999% AllDefect and colored light gray if the tth stage strategy distribution ever had at least 99.9999% TitForTat.

We now consider what happens if mutations can occur. We make the simple assumption that a player using strategy σ_i with probability $1 - \mu$ replicates at a rate proportional to its total expected payoff against everyone in the population (this is simply the replicator evolution mechanism) and with probability μ replicates into any of the m strategies with equal probability. This yields the *replicator with mutations evolution mechanism*:

$$\mathbf{F}_i(\mathbf{p}) = (1 - \mu) p_i \frac{v_i}{\bar{v}} + \mu \frac{1}{m}$$

where as before

$$v_i = V_{i,1} p_1 + V_{i,2} p_2 + \cdots + V_{i,m} p_m = (\mathbf{V}\mathbf{p})_i$$

is the total payoff to a player using strategy σ_i and

$$\bar{v} = p_1 v_1 + p_2 v_2 + \cdots + p_m v_m = \mathbf{p}\mathbf{V}\mathbf{p}$$

is the mean total payoff across all players. If we let $S = 0$, $P = 1$, $R = 3$, $T = 5$, $\delta = 0.9$, and $\xi = 0.01$ as before but make the probability of a mutation $\mu = 0.01$ instead of 0, then the payoffs in the two-player symmetric strategic game are as in Equation 7.9, but we must

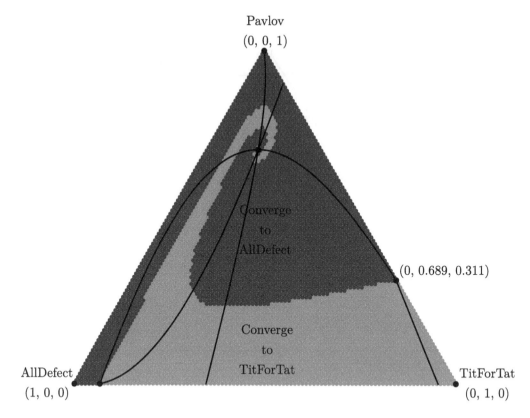

FIGURE 7.12 Example with Noise Evolution Convergence.

use the replicator with mutations evolution mechanism described above. Figure 7.13 shows four evolutions.

The fixed strategy distribution equation $\mathbf{F}(\mathbf{p}) = \mathbf{p}$ for the replicator with mutations evolution mechanism with $m = 3$ is equivalent to the equations $(1 - \mu)p_i(\mathbf{Vp})_i + \mu(1/3)(\mathbf{p}^T\mathbf{Vp}) = p_i(\mathbf{p}^T\mathbf{Vp})$ for $i = 1, 2, 3$. Solving the system consisting of these equations and $p_1 + p_2 + p3 = 1$ yields the unique solution $\mathbf{p} = (0.076, 0.181, 0.743)$. Notice that this strategy distribution is not Nash. The introduction of mutations has obviated the correspondence between the fixed and Nash strategy distributions we proved earlier for mutation-free and payoff-monotone evolution mechanisms.

The eigenvalues of the Jacobian of the replicator with mutations evolution mechanism evaluated at $\mathbf{p}$ have absolute values 0.986, 0.986, and 0.003. Since they are all less than 1, $\mathbf{p}$ is dynamically stable.

In addition to $\mathbf{p}$ being the unique fixed strategy distribution and all evolutions starting sufficiently close to $\mathbf{p}$ converging to $\mathbf{p}$, Figure 7.13 suggests that every strategy distribution not on the triangular boundary will spiral counter-clockwise towards the fixed point. We can provide additional theoretical support for this global stability conjecture. The positive introduction of every strategy because of the μ/m term in the evolution mechanism implies that strategy distributions sufficiently close to the boundary of the triangle in Figure 7.13 will be repelled. With evolutions being pushed away from the boundaries and being locally attracted to $\mathbf{p}$, it is reasonable to presume that our global stability conjecture is correct. Unfortunately, we can also imagine that evolutions of strategy distributions near the triangle border never get close to $\mathbf{p}$, perhaps because there is a dynamically stable closed loop

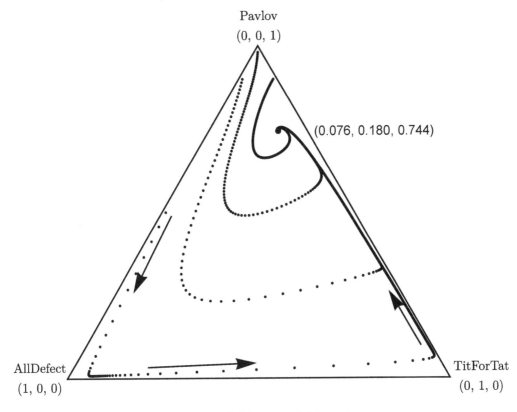

FIGURE 7.13 Example Evolutions with Noise and Mutations.

enclosing **p**. More advanced mathematical techniques would be required to fully confirm our global stability conjecture.

We now summarize what we have learned about our repeated Prisoner's Dilemma example using replicator and replicator with mutations evolution mechanisms. With no noise or mutations, most strategy distributions evolve to a combination of TitForTat and Pavlov, and almost all of the remaining strategy distributions evolve to AllDefect. With the introduction of some noise, there is an increasing oscillation among the three strategies leading eventually to either AllDefect or TitForTat. For initial strategy distributions near the completely mixed fixed one, the strategy to which the evolution converges is highly sensitive. With the introduction of mutations (with or without noise), all interior initial strategy distribution evolutions converge to the completely mixed fixed one. The very different behaviors of these three evolution systems despite the constancy of repeated Prisoner's Dilemma strategies used shows the necessity of determining which assumptions best represent the scenario under consideration.

One valuable step in the modeling process is to carefully review the assumptions inherent in the model. This can uncover assumptions that have been made implicitly. The model we developed for the Natural Evolution scenario minimally uses the following assumptions:

1. The population of players is infinite. If the population of players were finite, then very small proportions of the population using a particular strategy are likely to translate into zero players with that strategy.

2. Each player interacts with every other player. This may not be realistic, since some

players may have opportunities to interact with only some of the players from the other population. This could affect the players' ability to pass on information learned from their interactions with the whole population.

3. New members of the population begin their interactions with others immediately, but a time-delayed entry into interactions could be more realistic.

4. The interaction among players is pairwise. Animals band together in herds, nations form alliances, and people join organizations. Such types of interactions are not considered.

5. The interaction between two players is a random repeated Prisoner's Dilemma. We have observed that our qualitative conclusions would not change with sufficiently small changes in the quantitative values of the parameters. Nonetheless, there could be interactions among players that could be modeled by different strategic games. For example, we could explore independently treating the propagation of genes for behavior in random repeated Prisoner's Dilemma from the genes for behavior in random repeated Battle of the Sexes.

Of course, the hope is that a simple model may provide useful insights even if the full complexity of a scenario is not included, and once a simple model is understood, it could form the basis for more complex and realistic models. This idea was illustrated by our addition of noise and mutations to the original model. Readers interested in more sophisticated models of this type can consult [90], [100], [49], and [59].

EXERCISES

7.1 Describe advantages, disadvantages, and possible motivations for each of the strategies listed in the text for the repeated Prisoner's Dilemma game.

7.2 *OneTitForTwoConsecutiveTats* is the following strategy for the repeated Prisoner's Dilemma game: Cooperate in the first stage, and in subsequent stages, defect whenever the other player defected in the two previous stages. Table 7.6 shows what Random-Cooperate(0.4) might do in a finite repeated Prisoner's Dilemma game with seven stages. Complete the table with the other player's actions and both players' payoffs. Assume that $S = 0$, $P = 1$, $R = 3$, and $T = 5$.

TABLE 7.6   Exercise 7.2 Actions and Payoffs

Stage	1	2	3	4	5	6	7	Mean
RandomCooperate(0.4)'s Action	D	C	D	D	D	C	C	
OneTitForTwoConsecutiveTats' Action								
RandomCooperate(0.4)'s Payoff								
OneTitForTwoConsecutiveTats' Payoff								

7.3 Construct a table of actions and payoffs (similar to Table 7.6) for Pavlov versus OneTitForTwoConsecutiveTats in a finite repeated Prisoner's Dilemma game with seven stages. Assume that $S = 0$, $P = 1$, $R = 3$, and $T = 5$.

7.4 Construct a table of actions and payoffs (similar to Table 7.6) for AllDefect versus OneTitForTwoConsecutiveTats in a finite repeated Prisoner's Dilemma game with seven stages. Assume that $S = 0$, $P = 1$, $R = 3$, and $T = 5$.

7.5 Find a best response to OneTitForTwoConsecutiveTats in a finite repeated Prisoner's Dilemma game with seven stages. Assume that $S = 0$, $P = 1$, $R = 3$, and $T = 5$.

7.6 *AllCooperate* is the following strategy for the repeated Prisoner's Dilemma game: Cooperate in every stage. Prove directly (without using Theorem 7.1.1) that in the finite repeated Prisoner's Dilemma game with m-periods that (AllCooperate, AllCooperate) is not a Nash equilibrium.

7.7 Prove directly (without using Theorem 7.1.1) that in the finite repeated Prisoner's Dilemma game with m-periods that (TitForTat, TitForTat) is not a Nash equilibrium.

7.8 Suppose for a pair of strategies in a finite repeated game with m stages that $\mathbf{A}$ is the state transition matrix, $\mathbf{x}$ is the initial state probability distribution vector, and $\mathbf{\Gamma}$ is the state probability distribution to payoffs matrix. Show that if the matrix $\mathbf{I} - \mathbf{A}$ has an inverse, then the expected payoffs per stage is $\mathbf{v} = (1/m)(\mathbf{\Gamma})(\mathbf{I} - \mathbf{A})^{-1}(\mathbf{I} - \mathbf{A}^m)\mathbf{x}$.

7.9 Suppose Rose uses RandomCooperate(ρ) and Colin uses Pavlov in a random repeated Prisoner's Dilemma game with continuation probability δ.

    a. Construct the transition diagram (similar to Figure 7.2).

    b. State the transition matrix and initial vector (similar to Equations 7.1).

    c. Find the state probability distribution vectors in stages $k = 2, 3, \ldots$.

    d. Find the expected payoffs per stage.

    e. Find the expected payoffs per stage if $S = 0$, $P = 1$, $R = 3$, $T = 5$, $\rho = 0.6$, and $\delta = \frac{5}{6}$.

7.10 Prove that in the random repeated Prisoner's Dilemma game with continuation probability $0 \le \delta < 1$, (AllDefect, AllDefect) is a Nash equilibrium with expected payoff per stage vector (P, P).

7.11 Prove that in the random repeated Prisoner's Dilemma game, (TitForTat, TitForTat) is a Nash equilibrium with expected payoffs per stage of (R, R) if $T + S < 2R$ and the continuation probability is sufficiently large.

7.12 Let *MostlyCooperative* be the strategy that Cooperates except when the player is suckered (i.e., when the player cooperates and the opponent defects) in the most recent stage. Prove that in the random repeated Prisoner's Dilemma game, (MostlyCooperative, MostlyCooperative) is a Nash equilibrium if $P + T < 2R$ and the continuation probability is sufficiently large.

7.13 Choose a 2-by-2 strategic game such as Stag Hunt, Battle of the Sexes, or Chicken.

    a. Find all the Nash equilibria and corresponding payoffs of the strategic game.

    b. Find the prudential strategies and security levels of the strategic game.

    c. Discuss possible equilibria of the random repeated game based on this strategic game.

    d. There are Nash equilibria in the random repeated Prisoner's Dilemma game that are better for both players than the unique Nash equilibrium in the single stage Prisoner's Dilemma game. Is there a similar result for your game? What is your interpretation with respect to the original game?

7.14 Prove the finite geometric series formula

$$1 + x + x^2 + \cdots + x^m = (1 - x^{m+1})/(1 - x)$$

if $x \neq 1$. Hint: Expand $(1 - x)(1 + x + x^2 + \cdots + x^m)$.

7.15 Prove the infinite geometric series formula

$$1 + x + x^2 + \cdots + x^m + \cdots = 1/(1 - x)$$

if $|x| < 1$. Hint: Use the result of the previous exercise.

7.16 Suppose $\mathbf{A}$ is a state transition matrix and $0 \leq \delta < 1$. Prove that the matrix $\mathbf{I} - \delta\mathbf{A}$ has an inverse.

7.17 Suppose for some pair of strategies, the initial state probability distribution vector is $\mathbf{x}$ and the state probability distribution vectors satisfy $\mathbf{x}^k = \mathbf{y}$ for $k = 2, 3, \ldots$. Suppose also that $\delta = 1 - (1/m)$ for some positive integer m. Show that the expected payoffs per stage in the finite repeated game with m stages and the random repeated game with continuation probability δ are the same.

7.18 Suppose Rose uses AllDefect and Colin uses TitForTat in a noisy repeated game with a positive probability ξ of a perception error (and no probability of an implementation error). Create the state transition matrix and initial state probability distribution vector. Find the expected payoffs per stage.

7.19 Suppose Rose uses AllDefect and Colin uses TitforTat in a noisy repeated game with a positive probability η of an implementation error (and no probability of a perception error). Create the state transition matrix and initial state probability distribution vector. Find the expected payoffs per stage as δ approaches 1 and as η approaches 0.

7.20 Suppose Rose uses AllCooperate and Colin uses TitforTat in a noisy repeated game with a positive probability ξ of a perception error (and no probability of an implementation error). Create the state transition matrix and initial state probability distribution vector. Find the expected payoffs per stage.

7.21 Suppose Rose uses AllCooperate and Colin uses TitforTat in a noisy repeated game with a positive probability η of an implementation error (and no probability of a perception error). Create the state transition matrix and initial state probability distribution vector. Find the expected payoffs per stage as δ approaches 1 and as η approaches 0.

7.22 Suppose both Rose and Colin use Pavlov in a noisy repeated game with a positive probability η of an implementation error (and no probability of a perception error). Create the state transition matrix and initial state probability distribution vector. Find the expected payoffs per stage as δ approaches 1 and as η approaches 0.

*7.23 Provide a sketch of a proof of the Another Folk Theorem 7.2.1.

7.24 Verify that the replicator evolution mechanism is mutation free and payoff monotone.

7.25 For the example with noise considered in the text, defined by Equation 7.9, verify that $(1, 0, 0)$, $(0, 1, 0)$, $(0, 0, 1)$, $(0, 0.689, 0.311)$, $(0.931, 0.069, 0)$, and $(0.164, 0.133, 0.703)$ are fixed. Also verify that all fixed strategy distributions except $(0, 0, 1)$ are Nash, only $(1, 0, 0)$ and $(0, 1, 0)$ are dynamically stable (using the Jacobian of the replicator evolution mechanism), and only $(1, 0, 0)$ and $(0, 1, 0)$ are evolutionarily stable.

7.26 *Rock Paper Scissors* Two people simultaneously shout one of the words "rock," "paper," or "scissors." If both persons shout the same word, they tie. If they shout different words, then rock breaks and so wins against scissors, scissors cuts and so wins against paper, and paper covers and so wins against rock. We can arbitrarily assign a utility of 0 for a loss and 2 for a win. A tie then has utility a where $0 < a < 2$.

    a. What does the value of a tell us about each player's stance towards risk?

    b. Model this scenario as a strategic game. Show that there is no pure-strategy Nash equilibrium.

    c. Create an evolution system by adding the replicator evolution mechanism to the strategic game. Find all of the fixed strategy distributions.

    d. Find all of the Nash strategy distributions.

    e. Find all of the evolutionarily stable distributions.

    f. Find all of the dynamically stable distributions.

    g. What happens when $a = 0$?

    h. From these examples, what conclusions can you draw about the existence of evolutionarily stable and dynamically stable strategy distributions and the relationship between the two stability concepts?

7.27 Change the Rock Paper Scissors scenario described in the previous exercise by setting the utility for a win to be 8, the utility for a tie to be 5, and the utility for a loss to be 4. Continue to use the replicator evolution mechanism with the following.

    a. Show that the strategy distribution $\mathbf{p} = (1/3, 1/3, 1/3)$ is evolutionarily stable.

    b. Show that the strategy distribution $\mathbf{p} = (1/3, 1/3, 1/3)$ is dynamically stable.

    c. Since a positive linear transformation of a vNM utility function yields an equivalent vNM utility function with respect to player choices, for what value of a is the vNM utility function described in this exercise equivalent to the vNM utility function described in the previous exercise? For this choice of a, determine whether $\mathbf{p} = (1/3, 1/3, 1/3)$ is evolutionarily stable and/or dynamically stable in the original Rock Paper Scissors evolution system.

    d. From these examples, what conclusions can you draw about the existence of evolutionarily stable and dynamically stable strategy distributions in evolution systems with strategic games having equivalent vNM utilities?

7.28 Gintis [42] uses game theory to model the interactions of three types of male side-blotched lizards. He notes that in pairwise competitions for mating, orange-throats beat out blue-throats who beat out yellow-stripes. Yellow-stripes in turn win over orange-throats. Gintis modeled this using the Rock Paper Scissors game. Explain why Rock Paper Scissors is an appropriate game to model this scenario. Based on the analysis done for Rock Paper Scissors in the previous two exercises, what should biologists observe about these three types of male side-blotched lizards.

7.29 Consider the two three-player symmetric strategic games with the payoff matrices

$$
\mathbf{U} = \begin{array}{c} \\ A \\ B \\ C \end{array} \begin{array}{c} A \quad B \quad C \\ \begin{pmatrix} 6 & 8 & 5 \\ 5 & 6 & 8 \\ 11 & 2 & 6 \end{pmatrix} \end{array} \quad \text{and} \quad \mathbf{V} = \begin{array}{c} \\ A \\ B \\ C \end{array} \begin{array}{c} A \quad B \quad C \\ \begin{pmatrix} 4 & 6 & 3 \\ 3 & 4 & 6 \\ 9 & 0 & 4 \end{pmatrix} \end{array}.
$$

a. Describe how the two payoff matrices relate to each other.

b. Show that no strategy distribution in either strategic game is evolutionarily stable.

c. Show that the strategy distribution $\mathbf{p} = (1/3, 1/3, 1/3)$ is dynamically stable for the replicator evolution system based on U and is not dynamically stable for the replicator evolution system based on V.

d. From these examples, what conclusions can you draw about the existence of evolutionarily stable and dynamically stable strategy distributions and the relationship between the two stability concepts?

7.30 Consider the replicator evolution system with payoffs

$$
\mathbf{V} = \begin{array}{c} \\ A \\ B \\ C \end{array} \begin{array}{c} A \qquad B \qquad C \\ \begin{pmatrix} 10.00 & 11.50 & 9.00 \\ 9.00 & 10.00 & 11.50 \\ 15.85 & 4.65 & 10.00 \end{pmatrix} \end{array}.
$$

Show empirically that there is a dynamically stable set that appears to be on a simple closed loop [121].

*7.31 Write a computer program to generate and plot evolutions given an evolution system that uses the replicator with mutations evolution mechanism. Verify the correctness of one or more of the figures in Section 7.3 that display evolutions.

*7.32 Choose a different two-player game and set of strategies and analyze the behavior of the resulting evolution system using the replicator evolution mechanism. Consider what happens when a small amount of noise or mutations are added to the system.

*7.33 Describe a real-world scenario in which two or more agents repeatedly interact with each other in similar ways. Model as a repeated game and/or an evolution system. Analyze with appropriate mathematical tools. Interpret your results.

CHAPTER 8

Multilateral Agreements

We have seen how allowing two players to enter into binding agreements can lead to more preferred outcomes for both. We now consider scenarios which permit binding agreements for more than two players. We define a coalition game in which the payoffs are known for all possible coalitions of players that agree to collaborate. We discuss how players negotiate to determine the coalitions that are formed and the distribution of the payoffs among players in the coalitions. For effective negotiations, players must form arguments justifying their proposed coalitions and allocations. Finally, we examine how coalition games and allocation methods can be used to model voting systems and voting power.

In strategic, sequential, extensive, and Bayesian games, players choose among well-defined possible actions that collectively determine the outcome of the game. While they may communicate, it is assumed that players cannot make binding agreements with each other about what actions they will choose. In bargaining and coalition games, players make binding agreements, and these agreements specify the outcome of the game. The actions players take to reach the outcome specified by the agreement is not a central focus of these games. Instead we focus on the value of different outcomes to the players, and how they use their individual negotiating power to achieve an agreement and an outcome of value to them.

Definition 8.0.1. A *coalition game* (N, w) consists of the following:

1. A set $N = \{1, 2, \ldots, n\}$ of at least two players along with all nonempty subsets $S \subseteq N$, called *coalitions*. A *coalition structure* $\mathcal{S}$ is a partition of N into coalitions.

2. For each coalition S, a determined *worth* $w(S) \in \mathbb{R}$.

3. Any coalition S can come to an agreement to divide $w(S)$ amongst themselves as payoffs. The game ends when there is a coalition structure of agreements, or a pre-specified time limit has passed. In the latter case, each player i that has not entered into a nontrivial coalition receives a payoff of $w(\{i\})$.

4. Utilities u_i are given by the payoff received by player i .

The agreed-upon payoffs taken together are called an allocation.

Definition 8.0.2. An *allocation* for the coalition game (N, w) is a payoff vector $x = (x_1, x_2, \ldots, x_n)$ satisfying $\sum_{i \in S} x_i = w(S)$ for each coalition S in some partition $\mathcal{S}$ of N.

Throughout this chapter, we will assume that each coalition game is *superadditive*: $w(S) + w(T) \leq w(S \cup T)$ for all disjoint coalitions S and T. An interpretation of this condition is that one way the players in $S \cup T$ can collaborate is by having the players in S collaborate to achieve a joint payoff of $w(S)$ and the players in T to achieve a joint payoff of $w(T)$; any other mode of collaboration might achieve an even larger joint payoff.

Unlike with bargaining games, in which players can agree to any payoff pair in an arbitrary set of possible payoff pairs, we have made the simplifying assumption that the available payoff vectors must sum to the coalition's worth. This would be a reasonable assumption if the worth of a coalition is an amount of money the coalition receives by collaborating, and the players are self-interested and risk neutral. The more general case (called NTU games) is treated in more advanced texts such as [83].

8.1 SHAPLEY ARGUMENT

EPA. The Environmental Protection Agency (EPA) has mandated improvements in the sewage treatment facilities in the cities of Avon, Barport, Claron, and Delmont. Each city could work separately, but savings would be obtained if two or more cities are willing to collaborate. Determine which cities should collaborate and a division of the obtained savings that can be considered best and fair to all cities [41].

We can model the EPA scenario as a coalition game: the players are the four cities, which we will abbreviate with their first letters, and the worth of a coalition is the amount of money saved when the cities in the coalition collaborate. To fully specify the model, we must have values for the savings achieved in each coalition. We could determine these savings values by researching budgets of sewage projects for small towns, but in Table 8.1 we arbitrarily assign values (in millions of dollars) without too much concern for the fidelity of the model. We also remove set notation from the coalitions. Note that only coalitions

TABLE 8.1   EPA Coalition Game

Coalition S	ABCD	ABC	ABD	ACD	AB	any other
Savings $w(S)$	140	108	96	84	24	0

containing Avon can save positive amounts. Therefore, when Avon negotiates, it may be justified in asking for a bigger share.

There are plenty of allocations from which to choose, including the allocation $(35, 35, 35, 35)$ achieved with the coalition structure $\{ABCD\}$, the allocation $(38, 36, 34, 0)$ achieved with the coalition structure $\{ABC, D\}$, and the allocation $(20, 4, -1, 1)$ achieved with the coalition structure $\{AB, CD\}$. Of course, some of these allocations are better than others for a variety of reasons. For example, since payoffs correspond to utilities, player C should not agree to split $w(CD) = 0$ by receiving -1 while player D is receiving 1 because player C can always choose to collaborate with no one and receive $w(C) = 0$. In general, a player should receive a payoff no less than what the player could obtain without collaborating with any other player. Such an allocation is said to be player rational.

Definition 8.1.1. An allocation x for the coalition game (N, w) is *player rational* if $x_i \geq w(\{i\})$ for all $i \in N$.

While the allocation $(38, 36, 34, 0)$ is player rational, the allocation $(40, 40, 35, 25)$ provides each player with a higher payoff. Suppose players A, B, and C have agreed to $(38, 36, 34, 0)$ but would be willing to collaborate with player D if the allocation is $(40, 40, 35, 25)$. However player D demands to receive at least 40 to collaborate with the others. Such intransigence in negotiating often occurs. Players A, B, and C, have a strong argument to exclude D because adding D at 40 does not add value for each of them. However, settling for $(38, 36, 34, 0)$ when $(40, 40, 35, 25)$ is available would not be best. To avoid these dilemmas we now make the assumption that only efficient allocations will be considered.

Definition 8.1.2. An allocation is *efficient* if it is impossible to increase the payoff of one player without decreasing the payoff to another player.

For the EPA coalition game, the most money will be saved if the coalition ABCD comes to an agreement. In general, for a superadditive game, it is always best for all players to collaborate, as is shown by the following theorem.

Theorem 8.1.1. Suppose the coalition game (N, w) is superadditive. The allocation x is efficient if and only if $\sum_{i \in N} x_i = w(N)$.

Proof. Suppose x is an allocation such that $\sum_{i \in N} x_i = w(N)$ but is not efficient. Then there is an allocation y satisfying $x_i \leq y_i$ for all $i \in N$ and $x_j < y_j$ for some $j \in N$. Hence, $w(N) = \sum_{i \in N} x_i < \sum_{i \in N} y_i$. Since y is an allocation, there is a partition $\{S_1, S_2, \ldots, S_m\}$ of N satisfying $\sum_{i \in S_k} y_i = w(S_k)$ for $k = 1, 2, \ldots, m$. Since the game is superadditive, $\sum_{i \in N} y_i = w(S_1) + w(S_2) + \cdots + w(S_m) \leq w(N)$, contradicting an earlier inequality.

Conversely, suppose x is an allocation such that $\sum_{i \in N} x_i \neq w(N)$. Then there is a partition $\{S_1, S_2, \ldots, S_m\}$ of N satisfying $\sum_{i \in S_k} x_i = w(S_k)$ for $k = 1, 2, \ldots, m$. Since the game is superadditive, $\sum_{i \in N} x_i = w(S_1) + w(S_2) + \cdots + w(S_m) \leq w(N)$. Combined with our initial supposition, $\sum_{i \in N} x_i < w(N)$. Let $\epsilon = w(N) - \sum_{i \in N} x_i > 0$, and $y_i = x_i + \epsilon/n$ for each $i \in N$. Since $\sum_{i \in N} y_i = w(N)$, the vector y is an allocation. Since $y_i > x_i$ for each $i \in N$, the allocation x is not efficient. □

For the EPA coalition game, the efficient and player-rational allocations comprise the set

$$\{x \in \mathbb{R}^4 : x_A + x_B + x_C + x_D = 140, x_A \geq 0, x_B \geq 0, x_C \geq 0, x_D \geq 0\}.$$

As shown in Figure 8.1, the efficient and player-rational allocations can be viewed as a tetrahedron in $\mathbb{R}^3$. Each vertex corresponds to one player receiving all of the benefits of collaboration (e.g., the top vertex has player A receiving 140), and each face corresponds to one player receiving none of the benefits of collaboration (e.g., the bottom face has player A receiving 0).

Shapley [97] gives the following method for creating allocations based on an idea known as marginal contribution.

Definition 8.1.3. The *Shapley allocation* φ gives each player that player's marginal contribution to all coalitions, averaged over all player orders. Given an ordering of the players, a player i's *marginal contribution* to a coalition is the difference between (1) the worth of the coalition containing i and all players before i in the ordering, and (2) the worth of the coalition containing all players before i in the ordering.

Table 8.2 outlines the calculations for the EPA coalition game. A player order can be considered the order in which the players decide that they want to cooperate. For example, the player order BADC represents player B choosing to cooperate first, player A choosing to cooperate second, player D choosing to cooperate third, and player C choosing to cooperate last. A player's marginal contribution is the additional worth that player generates by

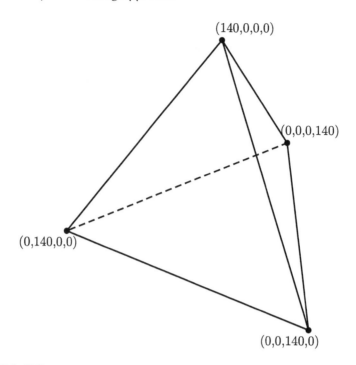

FIGURE 8.1 EPA Efficient and Player-Rational Allocations.

choosing to cooperate. For the player order BADC, player B alone has a worth of 0, and so generates 0 additional worth. The pair BA has a worth of 24, and so player A generates $24 - 0 = 24$ additional worth. The triple BAD has a worth of 96, and so player D generates $96 - 24 = 72$ additional worth. Finally, BADC has a worth of 140, and so player C generates $140 - 96 = 44$ additional worth. These marginal contributions are recorded in Table 8.2 in the row for the BADC order. Table 8.2 includes all $4! = 24$ ways to order the four players. The marginal contributions for each player have been summed and then averaged.

A marginal contribution formalizes the notion of how much a player adds by joining a coalition. It would not seem fair to specify a particular ordering of the players, but by averaging over all possible orders, each player is being treated in the same manner. The differences in player payoffs occur because of the real differences in how much each player contributes.

Figure 8.2 provides a geometric view of the Shapley allocation.

The diagram shows the marginal contribution vectors (i.e., the rows of Table 8.2) as gray points whose sizes are proportional to the number of identical marginal contribution vectors, and the Shapley allocation is the black point that is the center of mass of the gray points.

Player i's marginal contribution to coalition S containing s players occurs in $(s-1)!(n-s)!$ of the player orderings because the players in $S \setminus \{i\}$ can come in any order before player i and the players in $N \setminus S$ can come in any order after player i. This yields the following formula for the Shapley payoff to player i.

$$\varphi_i(N, w) = \sum_{s=1}^{n} \frac{(s-1)!(n-s)!}{n!} \sum_{\substack{|S|=s \\ i \in S}} (w(S) - w(S \setminus \{i\}))$$

where $|S|$ is the number of players in the coalition S.

TABLE 8.2   Shapley Allocation for EPA

Order	\multicolumn{4}{c}{Marginal Contribution}			
	A	B	C	D
ABCD	0	$24 - 0 = 24$	$108 - 24 = 84$	$140 - 108 = 32$
ABDC	0	$24 - 0 = 24$	$140 - 96 = 44$	$96 - 24 = 72$
ACBD	0	$108 - 0 = 108$	$0 - 0 = 0$	$140 - 108 = 32$
ACDB	0	$140 - 84 = 56$	$0 - 0 = 0$	$84 - 0 = 84$
ADBC	0	$96 - 0 = 96$	$140 - 96 = 44$	$0 - 0 = 0$
ADCB	0	$140 - 84 = 56$	$84 - 0 = 84$	$0 - 0 = 0$
BACD	$24 - 0 = 24$	0	$108 - 24 = 84$	$140 - 108 = 32$
BADC	$24 - 0 = 24$	0	$140 - 96 = 44$	$96 - 24 = 72$
BCAD	$108 - 0 = 108$	0	$0 - 0 = 0$	$140 - 108 = 32$
BCDA	$140 - 0 = 140$	0	$0 - 0 = 0$	$0 - 0 = 0$
BDAC	$96 - 0 = 96$	0	$140 - 96 = 44$	$0 - 0 = 0$
BDCA	$140 - 0 = 140$	0	$0 - 0 = 0$	$0 - 0 = 0$
CABD	$0 - 0 = 0$	$108 - 0 = 108$	0	$140 - 108 = 32$
CADB	$0 - 0 = 0$	$140 - 84 = 56$	0	$84 - 0 = 84$
CBAD	$108 - 0 = 108$	$0 - 0 = 0$	0	$140 - 108 = 32$
CBDA	$140 - 0 = 140$	$0 - 0 = 0$	0	$0 - 0 = 0$
CDAB	$84 - 0 = 84$	$140 - 84 = 56$	0	$0 - 0 = 0$
CDBA	$140 - 0 = 140$	$0 - 0 = 0$	0	$0 - 0 = 0$
DABC	$0 - 0 = 0$	$96 - 0 = 96$	$140 - 96 = 44$	0
DACB	$0 - 0 = 0$	$140 - 84 = 56$	$84 - 0 = 84$	0
DBAC	$96 - 0 = 96$	$0 - 0 = 0$	$140 - 96 = 44$	0
DBCA	$140 - 0 = 140$	$0 - 0 = 0$	$0 - 0 = 0$	0
DCAB	$84 - 0 = 84$	$140 - 84 = 56$	$0 - 0 = 0$	0
DCBA	$140 - 0 = 140$	$0 - 0 = 0$	$0 - 0 = 0$	0
φ_i	$1464/24 = 61$	$792/24 = 33$	$600/24 = 25$	$504/24 = 21$

Our description of the Shapley allocation as the average marginal contribution has an air of fairness. Our claim that the Shapley allocation is efficient and player rational provides some justification for this; however, as shown in Figure 8.1 there are other efficient and player-rational allocations. A better justification requires properties that completely characterize the Shapley allocation. For example, if two players are interchangeable based on the worths of coalitions, then they should each be allocated the same amount. This idea is formalized by the next definition.

Definition 8.1.4. An allocation x for the coalition game (N, w) is *unbiased* if whenever for some $i, j \in N$ and $w(S \cup \{i\}) = w(S \cup \{j\})$ for all coalitions $S \subseteq N \backslash \{i, j\}$, it follows that $x_i = x_j$.

TABLE 8.3   ABCD Unanimity Game

Coalition S	ABCD	any other
Savings $w(S)$	100	0

In the ABCD Unanimity Game (defined by Table 8.3), each pair of players is interchangeable. If the allocation is unbiased, then each player should be allocated the same amount. If the allocation is also efficient, then the payoffs need to sum to 100. Thus, the unique efficient and unbiased allocation is $(25, 25, 25, 25)$. Each player is equally valuable to the coalition and receives the same amount.

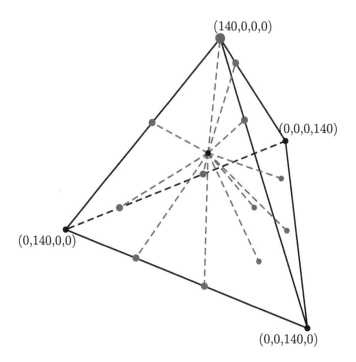

FIGURE 8.2 EPA Marginal Contribution Vectors and Shapley Allocation.

On the other hand, if a player never contributes to nor detracts from the worth of a coalition, then that player should be allocated zero. This idea is formalized by the next definition.

Definition 8.1.5. Let $w(\emptyset) = 0$ by convention. An allocation x for the coalition game (N, w) is *subsidy free* if whenever for some $i \in N$ and $w(S \cup \{i\}) = w(S)$ for all $S \subseteq N\backslash\{i\}$, it follows that $x_i = 0$.

TABLE 8.4 ABC Unanimity Game

Coalition S	ABCD	ABC	any other
Savings $w(S)$	150	150	0

In the ABC Unanimity Game (defined by Table 8.4), player D never contributes to nor detracts from the worth of a coalition. If the allocation is subsidy free, then player D should be allocated 0. Each pair of players A, B, and C is interchangeable. If the allocation is unbiased, then players A, B, and C should be allocated the same amount. Finally, if the allocation is also efficient, then the payoffs need to sum to 150. Thus, the unique efficient, unbiased, and subsidy-free allocation is $(50, 50, 50, 0)$.

The previous two coalition games can be generalized.

Definition 8.1.6. Given a set of players N, any coalition T, and any positive real number c, the *unanimity* game is (N, u_T) where for any coalition $S \subset N$, $u_T(S) = c$ if $T \subseteq S$ and $u_T(S) = 0$ otherwise.

In the ABC Unanimity game, $T = ABC$ and $c = 150$.

A unanimity game is superadditive and has a unique efficient, unbiased, and subsidy-free allocation for the coalition N, since each player in $N\backslash T$ neither contributes to nor detracts from the worth of any coalition, and each pair of players in T is interchangeable. Specifically, this allocation is $x = (x_1, x_2, \ldots, x_n)$ where $x_i = c/|T|$ if $i \in T$ and $x_i = 0$ if $i \notin T$.

The properties we have discussed for individual allocations are extended to properties for allocation methods with the following definition.

Definition 8.1.7. An allocation method μ for coalition games is *efficient, player rational, unbiased,* and *subsidy free* if μ always yields such allocations, respectively.

If players plan to play multiple coalition games, it may be useful to combine them into one coalition game. For example, four cities may plan to cooperate on sewage treatment improvements, road repair, library book acquisition, and law enforcement. The cities, for accounting purposes, may wish to keep these games separate. But they may also wish to combine them for the purpose of arriving at a fair allocation of the savings. The cities might expect that the savings allocation obtained for the separate games should sum to the savings allocation for the combined game. We must consider, then, how an allocation method behaves on a combined coalition game.

Definition 8.1.8. An allocation method μ for coalition games is *additive* if $\mu(N, v + w) = \mu(N, v) + \mu(N, w)$ for all coalition games (N, v) and (N, w).

TABLE 8.5   Combined Coalition Game

Coalition S	ABCD	ABC	any other
Savings $w(S)$	250	150	0

In the Combined Coalition Game (defined by Table 8.5), each pair of players A, B, and C is interchangeable. If the allocation is unbiased, then players A, B, and C should be allocated the same amount. If the allocation is also efficient, then the payoffs need to sum to 250. The subsidy-free property is not directly applicable because $w(ABCD) > w(ABC)$. Nonetheless, the Combined Coalition Game is the sum of the ABCD Unanimity Game and ABC Unanimity Game. Thus, if the allocation method μ is additive and yields efficient, unbiased, and subsidy-free allocations, then μ of the Combined Coalition Game should be the sum of the allocations for the ABCD Unanimity Game and the ABC Unanimity Game: $(25, 25, 25, 25) + (50, 50, 50, 0) = (75, 75, 75, 25)$.

An efficient, unbiased, subsidy-free, and additive allocation method determines the allocation for the EPA coalition game uniquely. In Table 8.6, (N, w_5) is the EPA Game, and we

TABLE 8.6   EPA Coalition Game with an Efficient, Unbiased, Subsidy-Free, and Additive Allocation Method μ

S	ABCD	ABC	ABD	ACD	AB	i	A	B	C	D
$w_1(S)$	24	24	24	0	24	$\mu(N, w_1)$	12	12	0	0
$w_2(S)$	84	0	0	84	0	$\mu(N, w_2)$	28	0	28	28
$w_3(S)$	72	0	72	0	0	$\mu(N, w_3)$	24	24	0	24
$w_4(S)$	84	84	0	0	0	$\mu(N, w_4)$	28	28	28	0
$w_5(S)$	140	108	96	84	24	$\mu(N, w_5)$	61	33	25	21
$w_6(S)$	124	0	0	0	0	$\mu(N, w_6)$	31	31	31	31

have created five simpler unanimity games (N, w_i), each with a unique allocation $\mu(N, w_i)$ that satisfies the first three of the desired properties. For example, in game (N, w_1), players C and D must be allocated zero in order for the allocation to be subsidy free; players A and B must be allocated the same amount in order for the allocation to be unbiased; and finally the sum of the payoffs must be 24 so that the allocation is efficient.

Now observe that

$$w_1 + w_2 + w_3 + w_4 = w_5 + w_6.$$

If μ is additive,

$$\mu(N, w_1) + \mu(N, w_2) + \mu(N, w_3) + \mu(N, w_4) = \mu(N, w_5) + \mu(N, w_6),$$

and since the allocations for the other games have already been uniquely determined, we can obtain $\mu(w_5)$. Finally, observe that the allocation obtained is the Shapley allocation.

Given any coalition game, we can write it as the sum and difference of unanimity games by starting with the smallest coalitions and working our way up in the size of the coalitions. For example, with the EPA game, the smallest coalition with a positive worth is AB, and that worth is 24; hence, w_1 was defined to be the unanimity game (N, u_{AB}) with $c = 24$. There are three coalitions having three players with positive worth. Since the worth of the coalition ACD is 84, w_2 was defined to be the unanimity game (N, u_{ACD}) with $c = 84$. Since the worth of the coalition ABD is 96 and $u_{AB}(ABD) = 24$, w_3 was defined to be the unanimity game (N, u_{ABD}) with $c = 96 - 24 = 72$. Since the worth of the coalition ABC is 108 and $u_{AB}(ABC) = 24$, w_4 was defined to be the unanimity game (N, u_{ABC}) with $c = 108 - 24 = 84$. Finally, there is one coalition having four players. Since the worth of $ABCD$ is 140 in the EPA game and the sum of the worth of $ABCD$ in the already constructed unanimity games is $24 + 84 + 72 + 84 = 264$, w_6 was defined to be the unanimity game (N, u_{ABCD}) with $c = 264 - 140 = 124$. The following theorem shows that we always obtain the Shapley allocation using this process.

Theorem 8.1.2 (Shapley Characterization). The Shapley allocation method is the only allocation method that is efficient, unbiased, subsidy-free, and additive on superadditive coalition games. Furthermore, the Shapley allocation method is player rational on superadditive coalition games.

> *Proof.* ⇒: We first show that the Shapley allocation method satisfies the five properties. For each player order, the sum of the marginal contributions equals $w(N)$, and so the average of these marginal contributions must equal $w(N)$; hence, the method yields efficient allocations. If players i and j are interchangeable, then the marginal contribution of player i in some player order is the same as the marginal contribution of player j in the order in which players i and j are interchanged; hence, the method yields the same payoffs for interchangeable players and so the allocation is unbiased. If player i never contributes to nor detracts from the worth of a coalition, then player i's marginal contribution in every player order is zero; hence, the method yields a payoff of zero for player i and so the allocation is subsidy free. For any particular player order, the marginal contribution of a player in the game $v + w$ is the sum of the marginal contribution of that player in the game v and the marginal contribution of that player in the game w; hence, the method is additive. Finally, for any player i by superadditivity, $w(S) \geq w(S \setminus \{i\}) + w(\{i\})$, and so $w(S) - w(S \setminus \{i\}) \geq w(\{i\})$, for all coalitions S that contain i; player i's average marginal contribution is at least $w(\{i\})$ showing that the method is player rational.
>
> ⇐: Now suppose an allocation method μ satisfies the four properties. Given any coalition T and worth c, the coalition game in which $u^{c,T}(S) = c$ if $T \subseteq S$ and $u^{c,T}(S) = 0$ otherwise is superadditive and has a unique efficient, unbiased, and subsidy-free allocation because each player in $N \setminus T$ neither contributes to nor detracts from the worth of any coalition, and each pair of players in T is interchangeable. Given an arbitrary coalition game w, we can write $w = \sum_{T \subseteq N} u^{c(T),T}$ where $c(T) = \sum_{R \subseteq T} (-1)^{|T| - |R|} w(R)$ by

working upward from the smallest to largest coalitions. This equality can also be verified algebraically: $\sum_{T \subseteq N} u^{c(T),T}(S) = \sum_{T \subseteq S} c(T) = \sum_{T \subseteq S} \sum_{R \subseteq T} (-1)^{|T|-|R|} w(R) = \sum_{R \subseteq S} w(R) \sum_{T:R \subseteq T \subseteq S} (-1)^{|T|-|R|} = w(S)$ because the inner sum is 1 if $R = S$ and is 0 if $R \neq S$ by the binomial formula. By additivity, $\mu(w) = \sum_{T \subseteq N} \mu(u^{c(T),T})$. □

This characterization theorem may be enough to convince players to use the Shapley allocation method. If Avon, Barport, Claron, and Delmont do so, it is interesting to imagine what the allocation would be if one or more of the coalition worths change, say because of a cost overrun. The only difference between the EPA and EPA Cost Overrun (defined in

TABLE 8.7 EPA Cost Overrun Coalition Game

Coalition S	ABCD	ABC	ABD	ACD	AB	any other
Savings $w(S)$	120	108	96	84	24	0

Table 8.7) is that the worth of the coalition of all players is 20 less in the latter game. By the additivity property, the Shapley allocation for the EPA Cost Overrun game simply subtracts 5 from each player's payoff in the EPA game: $(61 - 5, 33 - 5, 25 - 5, 21 - 5) = (56, 28, 20, 16)$. Observe that the combined payoffs to players A, B, and C is $56 + 28 + 20 = 104$, which is less than what the three players could have saved without player D: $w(ABC) = 108$. Thus we see that the Shapley allocation violates the following strengthening of the player-rational property.

Definition 8.1.9. An allocation x for the coalition game (N, w) is *coalition rational* if $\sum_{i \in S} x_i \geq w(S)$ for all coalitions $S \subseteq N$. An allocation method μ is *coalition rational* if $\mu(w)$ is coalition rational whenever the coalition game w has coalition-rational allocations.

Since $(60, 30, 20, 10)$ is coalition rational for the EPA Cost Overrun coalition game, but the Shapley allocation does not give a coalition-rational allocation, the Shapley allocation method is not coalition rational.

The only case in which a coalition-rational method would not give a coalition-rational allocation is when no such allocation exists. For example, consider the three-player game in which $w(S) = 1$ if S has two or three players and $w(S) = 0$ otherwise. An allocation (a, b, c) must satisfy $a + b + c \leq 1$, and if (a, b, c) is coalition rational, then $a + b \geq 1$, $a + c \geq 1$, and $b + c \geq 1$. Combining these, we obtain $2 \geq 2(a+b+c) = (a+b)+(a+c)+(b+c) \geq 1+1+1 = 3$, which is a contradiction to $2 < 3$.

8.2 NUCLEOLUS ARGUMENT

The fact that the Shapley allocation method is not coalition rational is a serious objection to the perceived fairness of the method. A second objection to the method is its reliance on the additive property which, while seemingly reasonable, relies on games that feel somewhat artificial. Perhaps it would be more reasonable to compare the original game allocation with allocations obtained in games more naturally related to the original game. The nucleolus is a response to these two objections.

Returning to the EPA coalition game given in Table 8.1, recall that the Shapley allocation has an economic interpretation stemming from the marginal contributions of players. The nucleolus allocation has a political interpretation: maximize the excess benefits given to the least well-off coalitions. Suppose that the Shapley allocation $(61, 33, 25, 21)$ was proposed by a player. Coalition ABC could obtain 108 on its own but actually receives $61 + 33 + 25 = 119$ in the proposed allocation. The excess of the proposed coalition is simply the difference in the total amount received by players A, B, and C in that allocation versus what the coalition ABC could have obtained on its own.

Definition 8.2.1. Suppose (N, w) is a coalition game. The *excess* of coalition S at the allocation x is

$$e(S, x) = \sum_{i \in S} x_i - w(S).$$

Table 8.8 shows the calculation of coalition excesses with the Shapley allocation $(61, 33, 25, 21)$. Notice that the last four excesses duplicate the allocations to the individual players.

TABLE 8.8   Excesses for the Shapley Allocation of the EPA Coalition Game

| Coalition | Allocation to | | | | $-$ | Worth | $=$ | Excess |
	A	B	C	D				
ABC	61 $+$	33 $+$	25		$-$	108	$=$	11
ABD	61 $+$	33		$+$ 21	$-$	96	$=$	19
ACD	61		$+$ 25	$+$ 21	$-$	84	$=$	23
BCD		33 $+$	25	$+$ 21	$-$	0	$=$	79
AB	61 $+$	33			$-$	24	$=$	70
AC	61		$+$ 25		$-$	0	$=$	86
AD	61			$+$ 21	$-$	0	$=$	82
BC		33 $+$	25		$-$	0	$=$	58
BD		33		$+$ 21	$-$	0	$=$	54
CD			25	$+$ 21	$-$	0	$=$	46
A	61				$-$	0	$=$	61
B		33			$-$	0	$=$	33
C			25		$-$	0	$=$	25
D				21	$-$	0	$=$	21

All coalitions have a positive excess, but coalition ABC has the smallest excess. One possible criterion for fairness is to maximize the minimum excess, that is, players should try to negotiate allocations that make the smallest excess as big as possible. This approach is a central tenet of the approach taken by Rawls [87] to social justice: the least well-off group in society should be made as well-off as possible.

Table 8.9 shows the coalition excesses with four different proposed allocations (the allocations can be recovered by looking at the last four rows of excesses).

TABLE 8.9   Excesses for Different EPA Allocations

S	$w(S)$	$e(S, q)$	$e(S, r)$	$e(S, s)$	$e(S, t)$
ABC	108	11	16	16	16
ABD	96	19	18	22	22
ACD	84	23	21	23	28
BCD	0	79	77	71	66
AB	24	70	74	78	78
AC	0	86	89	91	96
AD	0	82	79	85	90
BC	0	58	61	55	50
BD	0	54	51	49	44
CD	0	46	42	38	38
A	0	61	63	69	74
B	0	33	35	33	28
C	0	25	26	22	22
D	0	21	16	16	16

Starting with the Shapley allocation $q = (61, 33, 25, 21)$, we can improve the well-being of coalition ABC by taking from D and distributing to A, B, and C. Notice that if we take 5 from D, the excess of ABC will increase from 11 to 16 and the excess of D will decrease from 21 to 16, so balancing these two excesses. We distribute the 5 from D somewhat uniformly to A, B, and C to obtain the allocation $q = (61, 33, 25, 21)$. It is not possible to increase the smallest excess any further because to increase ABC's excess above 16 would require a transfer from D to the players in ABC, to increase D's excess above 16 would require a transfer to D from the players in ABC, and it is impossible to both transfer from and to D simultaneously.

While it is impossible to make the smallest two excesses greater than 16, it is possible to increase ABD's excess from 18 to 22 while simultaneously decreasing C's excess from 26 to 22 by taking 4 from C and distributing it between A and B (as noted in the previous paragraph we no longer want to change D's payoff). But this would result in ACD's excess becoming smaller than 21. After some experimentation, the allocation $s = (69, 33, 22, 16)$ is found to be an improvement over r because the first and second smallest excesses remain at 16 but the third smallest excess increases from 18 to 22. Now it is impossible to increase both ABD's excess and C's excess any further because that would require that C receive both more and less than 22.

The allocation $t = (74, 28, 22, 16)$ is even better because the smallest four excesses are the same for s and t; however, the fifth smallest excess for s and t are 23 and 28, respectively. Now it is impossible to increase both ACD's excess and B's excess any further because that would require that B receive both more and less than 28.

At this point, we have argued that changing the payoff to players D, C, or B would result in a worse allocation. Thus, the efficient allocation $t = (74, 28, 22, 16)$ successively maximizes the smallest excesses: the least happy coalition is as happy as possible; given that, the second least happy coalition is as happy as possible; given that, the third least happy coalition is as happy as possible; and so forth. This allocation method was first described by Schmeidler [94] and the resulting allocation is known as the prenucleolus or nucleolus.

Definition 8.2.2. The *prenucleolus* ν^0 [resp., *nucleolus* ν] is the efficient [resp., and player-rational] allocation that successively maximizes the smallest excesses.

On superadditive games, the prenucleolus and nucleolus yield the same allocations, and so we will use the shorter word except when necessary to distinguish between the two. The nucleolus allocation is defined in terms of a goal, unlike the Shapley allocation which was defined by an algorithm. To find the nucleolus for the EPA game, we made a sequence of informed guesses and provided a reasoned argument to verify our claim. If we had miraculously started with $(74, 28, 22, 16)$, rather than the Shapley allocation, we would have been done after one paragraph of verification. Any sequence of informed guesses should eventually end in the same place. Of course, this assumes that the goal is uniquely achievable. This was proved by Schmeidler [94] but will not be proven here. A formal algorithm for computing the nucleolus involves solving a sequence of linear programs.

The nucleolus method, like the Shapley method, is efficient, unbiased, and subsidy free. Unlike the Shapley method, which is not coalition rational, we have the following.

Theorem 8.2.1. The nucleolus method is coalition rational.

The second objection to the Shapley method was that the games used in conjunction with the additive property seemed somewhat artificial and unrelated to the original game. We will now introduce games that arise naturally from the original game and consider the resulting implications.

Suppose that the players from the EPA game have agreed to use the Shapley allocation method resulting in the allocation $(61, 33, 25, 21)$. Claron and Delmont are happy to receive their $25 million and $21 million in savings, respectively, but Avon and Barport do not agree on how to share the remaining \$94 million. Reasonably, they might agree to use the Shapley allocation method on a reduced game $(\{A, B\}, w^*)$ involving only Avon and Barport saving \$94 million together.

In the reduced game $w^*(A)$ and $w^*(B)$ need not be zero because either Avon or Barport may separately cooperate with the other players in the original game. For example, Avon could have formed a coalition with Claron and Delmont, leading to a savings for Avon of \$38 = \$84 − \$25 − \$21 million. Thus Avon may reasonably argue that $w^*(A) \geq 38$. Barport cannot obtain any savings other than zero without the help of Avon, and Avon cannot do any better than \$38 million without Barport. The following definition captures this argument in a general context.

Definition 8.2.3. The *reduced coalition game* on a coalition S with respect to an allocation x is played among the players in S, the worth of S is the sum of the amounts allocated in the original game to the players in S, and the worth of each subcoalition R is the maximum obtainable by R joining with zero or more players outside of S and giving those outside players what they were allocated in the original game.

The reduced game involving Avon and Barport then must have $w^*(AB) = 94$, $w^*(A) = 38$, and $w^*(B) = 0$. Table 8.10 computes the Shapley allocation for this reduced game. The unanimity game associated with w_1 has Avon receiving the \$38 million it could receive in the original game without Barport. The unanimity game associated with w_2 gives AB \$56 million so that, additively, we have AB receiving \$94 million.

TABLE 8.10   Shapley Allocation for the Reduced **EPA** Coalition Game

S	AB	A	B	Player	A	B
$w_1(S)$	38	38	0	$\varphi(w_1)$	38	0
$w_2(S)$	56	0	0	$\varphi(w_2)$	28	28
$w^*(S)$	94	38	0	$\varphi(w^*)$	66	28

The Shapley allocation of the reduced game (\$66 million for Avon and \$28 for Barport) is not the same as their Shapley allocations in the original game (\$61 and \$33 millions). Thus, the Shapley method is not consistent on reduced games, an idea formalized in the following definition.

Definition 8.2.4. An allocation method is *consistent* if the allocation for any reduced game is the same as the allocation for the original game.

Now consider the EPA coalition game with the nucleolus allocation $(74, 28, 22, 16)$. The reduced coalition game $(\{A, B\}, v)$ with respect to the nucleolus allocation has $v(AB) = 74 + 28 = 102$, $v(A) = 84 − 22 − 16 = 46$, and $v(B) = 0$. By equalizing the excesses for the two singleton coalitions (both equal 28), the nucleolus allocation for the reduced game is $x = (74, 28)$.

The reduced coalition game $(\{B, C, D\}, v)$ with respect to the nucleolus allocation satisfies $v(BCD) = 28 + 22 + 16 = 66$, $v(BC) = 108 − 74 = 34$, $v(BD) = 96 − 74 = 22$, $v(CD) = 84 − 74 = 10$, and $v(B) = v(C) = v(D) = 0$. The nucleolus allocation for this reduced game is $x = (28, 22, 16)$.

In both cases, the nucleolus allocation on the reduced game matched the allocation in the original game. To verify that the nucleolus of the EPA coalition game is consistent would require a similar check for every other coalition. Fortunately, Sobolev [103] provided a characterization theorem for the prenucleolus in terms of fairness properties, which was strengthened by Orshan [77].

Theorem 8.2.2 (Prenucleolus Characterization). The prenucleolus method is the only allocation method that is efficient, unbiased, scale invariant, and consistent.

Recall that scale invariant was defined for bargaining games, and its definition can be extended to coalition games.

Definition 8.2.5. An allocation method is *scale invariant* if (1) multiplying all gains by the same number M results in the method multiplying all payoffs by M, and (2) if changing all worths containing a player i by the same amount B results in the method changing player i's payoff by B.

Interestingly, while the nucleolus of the EPA coalition game was found to be $(74, 28, 22, 16)$, the nucleolus of the EPA Cost Overrun coalition game is $(84, 18, 12, 6)$. Thus, a \$20 million cost overrun results in a \$10 million increase in Avon's savings! This highly non-intuitive result shows that the nucleolus method violates the following property.

Definition 8.2.6. An allocation method is *coalition monotone* if increases (decreases) in a single coalition's worth do not result in a decrease (increase) in the payoff for any player who is in the coalition.

The Shapley method can be shown to be coalition monotone. Thus, the Shapley method is efficient, player rational, unbiased, subsidy free, scale invariant, additive, and coalition monotone. The nucleolus method is efficient, player rational, unbiased, subsidy free, scale invariant, consistent, and coalition rational. Unfortunately, the Shapley method is neither consistent nor coalition rational, and the nucleolus method is neither additive nor coalition monotone. Figure 8.3 summarizes the properties satisfied by each method.

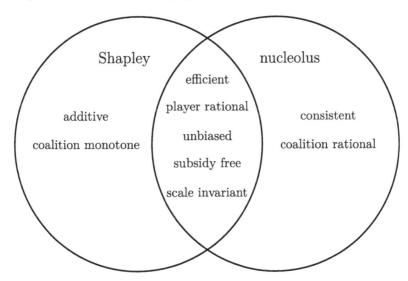

FIGURE 8.3 Properties Satisfied by the Shapley and Nucleolus Methods.

Is there a method that satisfies all of these fairness properties? We have already seen that the answer is no. It is sometimes surprising how few properties are needed to obtain mutual incompatibility.

Theorem 8.2.3 (Coalition Game Impossibility). There is no allocation method for coalition games with four or more players that is efficient, rational, and coalition monotone.

This impossibility was shown for five or more players by Young [125]. Housman and Clark [52] proved the impossibility for four or more players and showed that there are many efficient, rational, and monotone allocation methods (the nucleolus method being one) when restricted to three-player games.

The mathematical theorems show us that we must think carefully about what we mean by fairness, and they also help to delineate possible fairness properties. We have argued that fair allocation methods should be efficient, coalition rational, and coalition monotone. The Coalition Game Impossibility Theorem tells us that there is no allocation method that satisfies all three properties. Therefore, negotiators need to decide which properties are most important for the situation in which the method will be used. If players make and break agreements freely, then violations of the coalition-rational property would be disastrous for collaborative agreements. However, if cooperation will be enforced by government or moral decree, then violations of the coalition-rational property (when they occur) may not be too objectionable. If the gains change often or if players can unilaterally sabotage projects, then violations of the coalition monotone property could be psychologically detrimental to the players. However, if the gains are clear and impossible to manipulate, then the coalition monotone property may be less relevant.

8.3   BARGAINING ARGUMENT

In the previous sections, players proposed allocations that were supported by various fairness criteria. In this section, players attempt to achieve allocations that can be defended against any objections.

Corporate Home. Negosh Advertising is interested in relocating its headquarters and has identified properties on the boundaries of Avon, Barport, and Claron as possible locations. Based on their relative locations, Negosh Advertising would pay $840,000 in property taxes if it located convenient to all three, $800,000 if near to just Avon and Barport, $600,000 if just near to Avon and Claron, and $400,000 if near to just Barport and Claron. This information is provided to representatives of Avon, Barport, and Claron, who have been asked to present the company with one site proposal for the headquarters.

We model the Corporate Home scenario as a coalition game: the players are the three cities, which we will abbreviate with their first letters, and the worth of a coalition is the amount of money (in tens of thousands of dollars) the cities in the coalition can earn, as shown in Table 8.11.

TABLE 8.11   Corporate Home Coalition Game

Coalition S	ABC	AB	AC	BC	A	B	C
Earnings $w(S)$	84	80	60	40	0	0	0

Suppose the negotiation proceeded in the following manner. Table 8.12 summarizes this conversation. (Town names stand for their respective negotiators.)

Barport: We have worked together well many times previously, Claron. How about the two of us agree to work together again with me receiving $300,000 and you receiving $100,000 of the $400,000 that is being offered?

Claron: I have enjoyed working with you in the past, Barport, but I need to earn more money. I could earn $200,000 by working with Avon and her receiving a generous $400,000. How about giving me a greater share of the $400,000?

Barport: Claron, be reasonable! I also could work with Avon instead of you and still earn $300,000 while allowing Avon to earn $500,000, but I would prefer to work with you.

Claron: Barport, if you aren't going to budge, I need to find a more lucrative opportunity. Avon, I would be willing to work with you if I could earn $200,000. Sound good?

Avon: Actually, I prefer Barport's suggestion. Barport, let's work together with me earning $500,000 and you earning $300,000.

Barport: I wasn't really making a proposal. I was merely trying to convince Claron of the reasonableness of my initial proposal. I would really like for Claron to be part of this proposal. I now propose that the three of us work together and allocate the earnings as follows: $380,000 for Avon, $280,000 for me, and $180,000 for Claron.

Avon: I believe that the following is a better allocation: $480,000 for me, $280,000 for Barport, and $80,000 for Claron.

Claron: Avon, I would be willing to work with you as long as I would earn $110,000, which would allow you to earn $490,000, and Barport could even earn $240,000 if we work together.

Avon: Claron, that is an enticing proposal. Barport, are you sure you are not interested in working with me and earning $300,000 while I earn $500,000? We could even allow Claron to join us and earn $40,000.

TABLE 8.12 Corporate Home

Allocation	Coalition Structure	Comment
$(0, 30, 10)$	$\{A, BC\}$	Proposed by B
$(40, 0, 20)$	$\{AC, B\}$	Objection by C against B
$(50, 30, 0)$	$\{AB, C\}$	Counter-objection by B to C's objection
$(40, 0, 20)$	$\{AC, B\}$	Proposed by C
$(50, 30, 0)$	$\{AB, C\}$	Proposed by A
$(38, 28, 18)$	$\{ABC\}$	Proposed by B; Shapley allocation
$(48, 28, 8)$	$\{ABC\}$	Proposed by A; nucleolus
$(49, 0, 11)$	$\{AC, B\}$	Proposed by C
$(49, 24, 11)$	$\{ABC\}$	Proposed by C
$(50, 30, 0)$	$\{AB, C\}$	Proposed by A
$(50, 30, 4)$	$\{ABC\}$	Proposed by A

This negotiation could go on for a long time, because for any proposal, there will always be a pair of players who can ensure themselves greater earnings. That is, for any allocation x there is another allocation y for which two players receive more. Players use this fact in their negotiations in two ways. Most of the time, one of the players who could benefit actually proposed y; however, at the beginning of the negotiation recorded above, C used the possibility of $y = (40, 0, 20)$ as an argument to extract more from B than in the proposal $x = (0, 30, 10)$, and then B used the possibility of $z = (50, 30, 0)$ as a counter-argument.

While it would be nice to find an allocation against which there would be no argument, that cannot happen in the Corporate Home coalition game. Indeed, suppose $x = (a, b, c)$ is an allocation. Then $a + b + c \leq w(ABC)$. Now if $a + b < w(AB)$, then A and B could ensure themselves $a + (w(AB) - a - b)/2 > a$ and $b + (w(AB) - a - b)/2 > b$, respectively; hence, a necessary condition for A and B to not be able to ensure themselves greater earnings is $a + b \geq w(AB)$. By similar arguments, for the other pairs of players to not be able to ensure themselves greater earnings, $a + c \geq w(AC)$ and $b + c \geq w(BC)$. Summing the last three inequalities together, we obtain $2(a + b + c) \geq w(AB) + w(AC) + w(BC)$. By the first inequality $2w(ABC) \geq w(AB) + w(AC) + w(BC)$; however, plugging in the worths for our game, we obtain $2(84) \not\geq 80 + 60 + 40$. This contradiction shows that some pair must be able to ensure themselves greater earnings.

Therefore, we will suggest a weaker form of stability: allocations for which any argument can be countered. A group of players may object to a proposed allocation because they can do better with another allocation. However, another group of players can counter by showing a third allocation in which they do at least as well or better than in either of the previously proposed allocations, even though their counter may hurt the players that initially objected. An objection without a counter suggests the original proposal will be replaced. Otherwise, the cycle of objections and counter-objections prevents any changes from the originally proposed allocation. If every objection has a counter-objection, then, the initially proposed allocation has a weak form of stability: it is not that a player could not obtain a higher payoff with a different arrangement, but the danger that such a move may result in the objecting player ending up worse off keeps players from changing from the original allocation. The following definition formalizes this idea.

Definition 8.3.1. Let x be a player-rational allocation in a game (N, w) with the coalition structure $\mathcal{S}$. Let j and k be two players in a coalition $S \in \mathcal{S}$. An *objection* of j against k at x is a pair (Y, y) satisfying the following:

1. Y is a coalition, $j \in Y$, and $k \notin Y$.

2. y is a vector of payoffs for players in Y satisfying $\sum_{i \in Y} y_i = w(Y)$.

3. $y_i > x_i$ for all players $i \in Y$.

A *counter-objection* to this objection is a pair (Z, z) satisfying the following:

1. Z is a coalition, $k \in Z$, and $j \notin Z$.

2. z is a vector of payoffs for players in Z satisfying $\sum_{i \in Z} z_i = w(Z)$.

3. $z_i \geq y_i$ for all players $i \in Z \cap Y$.

4. $z_i \geq x_i$ for all players $i \in Z \setminus Y$.

A *bargaining allocation* is a player-rational allocation for which every objection has a counter-objection.

That is, the objection (Y, y) of j to k forms a coalition Y that excludes player k and makes every player in Y, including j, better off. The counter-objection (Z, z) forms a third coalition that excludes j and makes every player in Z, including k, at least as well off as in the original allocation or the objection. Bargaining allocations have no objections that can't be countered.

Finding bargaining allocations is a fairly complex computational task. For a three-player game in which two players collaborate, there is a simple algebraic characterization for bargaining allocations.

Theorem 8.3.1. A player-rational allocation $x = (x_i, x_j, x_k)$ with the coalition structure $\{jk, i\}$ is a bargaining allocation if and only if at least one expression from each of the following lines holds:

$$x_j + x_i \geq w(ji) \qquad x_k = w(k) \qquad w(ji) - x_j \leq w(ki) - x_k \qquad (8.1)$$

$$x_k + x_i \geq w(ki) \qquad x_j = w(j) \qquad w(ki) - x_k \leq w(ji) - x_j \qquad (8.2)$$

Proof. Suppose x is a player-rational allocation with the coalition structure $\{jk, i\}$. Because $x_l \geq w(l)$ for each $l \in N$, if (Y, y) is an objection, then either (i) $Y = \{j, i\}$ and the objection is j against k, or (ii) $Y = \{k, i\}$ and the objection is k against j. We will show that every objection by j against k has a counter-objection if and only if at least one of Conditions 8.1 holds.

$\Leftarrow$: Suppose at least one of Conditions 8.1 holds. If $x_j + x_i \geq w(ji)$, then $y_j > x_j$ and $y_i > x_i$ implies $y_j + y_i > w(ji)$; hence, there is no objection of j against k. If $x_k = w(k)$, then $(k, w(k))$ is a counter-objection to any objection of j against k. If $w(ji) - x_j \leq w(ki) - x_k$ and (Y, y) is an objection of j against k, then $Z = \{k, i\}$ contains k but not j, and $z_k = w(ki) - y_i$ and $z_i = y_i$ satisfy $z_k + z_i = w(ki)$, $z_i \geq y_i$, and

$$
\begin{aligned}
z_k &= w(ki) - y_i & \text{(definition of } z_k) \\
&\geq w(ji) - x_j + x_k - y_i & \text{(assumption } w(ji) - x_j \leq w(ki) - x_k) \\
&> w(ji) - y_j - y_i + x_k & \text{(objection condition } y_j > x_j) \\
&= x_k & \text{(objection condition } y_j + y_i = w(ji)),
\end{aligned}
$$

which implies that (Z, z) is a counter-objection. Thus, every objection by j against k has a counter-objection.

$\Rightarrow$: Conversely, suppose none of Conditions 8.1 hold. Since $x_j + x_i < w(ji)$ and $w(ji) - x_j > w(ki) - x_k$, there is a positive $\epsilon < \min\{w(ji) - x_j - x_i, w(ji) - x_j - w(ki) + x_k\}$. Let $y_j = x_j + \epsilon$ and $y_i = w(ji) - x_j - \epsilon$. Then $y_j + y_i = w(ji)$, $y_j > x_j$, and $y_i = w(ji) - x_j - \epsilon > w(ji) - x_j - (w(ji) - x_j - x_i) = x_i$; hence, (ji, y) is an objection by j against k. Suppose (ki, z) is a counter-objection. Then $w(ki) = z_k + z_i \geq x_k + y_i = x_k + w(ji) - x_j - \epsilon > x_k + w(ji) - x_j - (w(ji) - x_j - w(ki) + x_k) = w(ki)$, a contradiction that shows that there is no counter-objection.

Therefore, every objection by j against k has a counter-objection if and only if at least one of Conditions 8.1 holds. By a similar argument, every objection by k against j has a counter-objection if and only if at least one of Conditions 8.2 holds. □

Table 8.13 lists all bargaining allocations for the Corporate Home coalition game. We

TABLE 8.13 Corporate Home Bargaining Allocations

Allocation	Coalition Structure
$(0, 0, 0)$	$\{A, B, C\}$
$(0, 30, 10)$	$\{A, BC\}$
$(50, 0, 10)$	$\{AC, B\}$
$(50, 30, 0)$	$\{AB, C\}$
$(48, 28, 8)$	$\{ABC\}$

now present a lengthy verification of this fact. For the coalition structure $\{A, B, C\}$, the only player-rational allocation is $x = (0, 0, 0)$ at which any objection by player j against player k has the counter-objection (Z, z) where $Z = \{k\}$ and $z_k = 0$. The bargaining allocations for the next three coalition structures in Table 8.13 are found using Theorem 8.3.1.

For coalition structure $\{AB, C\}$, the vector x is a player-rational allocation if and only if $x_A + x_B = 80$, $x_C = 0$, $x_A \geq 0$, and $x_B \geq 0$. By Theorem 8.3.1, x is a bargaining allocation if and only if at least one expression from each of the following lines holds:

$$
\begin{array}{lll}
x_A + x_C \geq w(AC) & x_B = w(B) & w(AC) - x_A \leq w(BC) - x_B \\
x_B + x_C \geq w(BC) & x_A = w(A) & w(BC) - x_B \leq w(AC) - x_A
\end{array}
$$

Using each coalition's worth from Table 8.11 and the player-rational allocation conditions, we obtain that x is a bargaining allocation if and only if at least one expression from each of the following lines holds:

$$x_A \geq 60 \qquad\qquad x_A = 80 \qquad\qquad 50 \leq x_A$$
$$x_A \leq 40 \qquad\qquad x_A = 0 \qquad\qquad x_A \leq 50$$

The only one of the nine possible pairs of conditions that yields a nonempty solution is $50 \leq x_A$ and $x_A \leq 50$, which is equivalent to $x_A = 50$. Thus, x is a bargaining allocation for the coalition structure $\{AB, C\}$ if and only if $x = (50, 30, 0)$. The verifications that $(50, 0, 10)$ is the unique bargaining allocation for the coalition structure $\{AC, B\}$ and that $(50, 30, 0)$ is the unique bargaining allocation for the coalition structure $\{AB, C\}$ are similar.

Now we verify that $(48, 28, 8)$ is a bargaining allocation for the coalition structure $\{ABC\}$ using a different argument. Since $x_A + x_B + x_C = 48 + 28 + 8 = 84 = w(ABC)$, x is an allocation. Since $x_A = 48 \geq 0 = w(A)$, $x_B = 28 \geq 0 = w(B)$, and $x_C = 8 \geq 0 = w(C)$, the allocation x is player rational. Since there is no coalition-rational allocation, any of the three players could have an objection against any of the other players. Since each $x_i > w(i)$, an objection (Y, y) by j against k must satisfy $Y = ji$ where i is the third player, and Table 8.14 shows that for all possibilities, there is a counter-objection.

TABLE 8.14 Showing $(48, 28, 8)$ Is a Bargaining Allocation

Players	Objection	Counter-objection
A against B	$y_A > 48$ and $y_A + y_C = 60$	$z_B = 28$ and $z_C = 12$
	implies $y_C = 60 - y_A < 12$	implies $x_B \leq z_B$ and $y_C \leq z_C$
B against A	$y_B > 28$ and $y_B + y_C = 40$	$z_A = 48$ and $z_C = 12$
	implies $y_C = 40 - y_B < 12$	implies $x_A \leq z_A$ and $y_C \leq z_C$
A against C	$y_A > 48$ and $y_A + y_B = 80$	$z_C = 8$ and $z_B = 32$
	implies $y_B = 80 - y_A < 32$	implies $x_C \leq z_A$ and $y_B \leq z_B$
C against A	$y_C > 8$ and $y_C + y_B = 40$	$z_A = 48$ and $z_B = 32$
	implies $y_B = 40 - y_C < 32$	implies $x_A \leq z_A$ and $y_B \leq z_B$
B against C	$y_B > 28$ and $y_B + y_A = 80$	$z_C = 8$ and $z_A = 52$
	implies $y_A = 80 - y_B < 52$	implies $x_C \leq z_C$ and $y_A \leq z_A$
C against B	$y_C > 8$ and $y_C + y_A = 60$	$z_B = 28$ and $z_A = 52$
	implies $y_A = 60 - y_C < 52$	implies $x_B \leq z_B$ and $y_A \leq z_A$

Finally, we verify that if $x \neq (48, 28, 8)$ is a player-rational allocation for the coalition structure $\{ABC\}$, then x is not a bargaining allocation. Since x is a player-rational allocation for the coalition structure $\{ABC\}$, it follows that $x_A + x_B + x_C = w(ABC) = 84 = 48 + 28 + 8$, and since $x \neq (48, 28, 8)$, it follows that at least one of the differences $x_A - 48$, $x_B - 28$, or $x_C - 8$ is positive and at least one is negative. For each case, Table 8.15 provides an objection for which there is no counter-objection. For example, in the first case $x_A - 48 \leq x_C - 8 \leq x_B - 28$ which implies that $x_A - 48 < 0$ and $x_B - 28 > 0$; hence, $y_A = x_A + x_B - 28 > x_A$, $y_C = x_C + 4 > x_C$, and $y_A + y_C = (x_A + x_B - 28) + (x_C + 4) = (x_A + x_B + x_C) + (-28 + 4) = 84 - 24 = 60 = w(AC)$ implies (AC, y) is an objection of A against B, and if $z_B \geq x_B$ and $z_C \geq y_C$, then $z_B + z_C \geq (x_C + 4) + x_B = (x_B + x_C) + 4 = (84 - x_A) + 4 = 40 + (48 - x_A) > 40 = w(BC)$ implies that there is no counter-objection.

The Corporate Home game, therefore, has exactly one bargaining allocation for each coalition structure.

We end this section with two theorems which connect this work with that of the earlier sections of this chapter.

TABLE 8.15   Showing $x \neq (48, 28, 8)$ Is Not a Bargaining Allocation

Minimum	Maximum	Objection of Min Against Max
$x_A - 48$	$x_B - 28$	$y_A = x_A + x_B - 28$ and $y_C = x_C + 4$
$x_A - 48$	$x_C - 8$	$y_A = x_A + x_C - 8$ and $y_B = x_B + 4$
$x_B - 28$	$x_A - 48$	$y_B = x_B + x_A - 48$ and $y_C = x_C + 4$
$x_B - 28$	$x_C - 8$	$y_B = x_B + x_C - 8$ and $y_A = x_A + 4$
$x_C - 8$	$x_A - 48$	$y_C = x_C + x_A - 48$ and $y_B = x_B + 4$
$x_C - 8$	$x_B - 28$	$y_C = x_C + x_B - 28$ and $y_A = x_A + 4$

Theorem 8.3.2. Coalition-rational allocations are bargaining allocations.

Theorem 8.3.3. The nucleolus is a bargaining allocation.

8.4   VOTING POWER

One important application of coalition game theory is to voting systems instituted to make group decisions, in which voting power is not necessarily proportional to the number of votes a player has. For example, a stockholder owning 51% of the company has 100% of the voting power on measures that pass with a simple majority. Solution concepts such as the Shapley method provide a way to measure voting power. Court rulings in New York force county boards to have each member represent the same population or be assigned voting weights so that the voting power of board members is proportional to the populations represented. Here we present a fictitious scenario similar to the one faced by one county in New York [65].

> **Tonkins County.** This relatively small county has five towns and their rural surroundings. Alkane, Broome, Chemung, Dutches, and Essex (including each of their rural surroundings) have populations of 50, 40, 20, 20, and 10 thousand, respectively. The County Board has had a member elected from each of the towns. Because different members represent different numbers of people, the County Board has given weights to its members proportional to the number of people represented. Thus, the member from Alkane has 5 votes, the member from Broome has 4 votes, the member from Chemung has 2 votes, the member from Dutches has 2 votes, the member from Essex has 1 vote, and bills require a simple majority of at least 8 of the 14 weighted votes for passage. Voters in the different communities, however, may be concerned that their voices are not equally represented on the board.

The Tonkins County scenario can be modeled as a coalition game with the five towns acting as the players, and the worth of a coalition equal to 1 if the coalition can pass a bill and 0 otherwise. For example, $w(AB) = 1$ because $5 + 4 \geq 8$, and $w(BCE) = 0$ because $4+2+1 < 8$. We limit our discussion to situations that satisfy all of the following properties.

1. Each coalition is winning or losing.

2. The coalition of all players is winning.

3. Supersets of winning coalitions are winning.

4. Two disjoint coalitions cannot both be winning.

These four conditions are formalized in the following definition.

Definition 8.4.1. A *voting game* is a coalition game (N, w) for which:

1. $w(S) = 1$ or $w(S) = 0$ for each coalition S.

2. $w(N) = 1$.

3. $w(T) = 1$ whenever $S \subset T$ and $w(S) = 1$.

4. $w(N \setminus S) = 0$ whenever $w(S) = 1$.

The coalitions S satisfying $w(S) = 1$ are called *winning*, and the coalitions S satisfying $w(S) = 0$ are called *losing*.

The winning coalitions for the Tonkins County game are listed in the first column of Table 8.16. By condition (3), a voting game can be characterized by the *minimal winning coalitions*: those winning coalitions for which every proper subset is losing. For the Tonkins County game, the set of minimal winning coalitions is $\{AB, ACD, ACE, ADE, BCD\}$. Conditions 1, 2, and 3 in the following theorem correspond with conditions 2, 3, and 4, respectively, in the voting game definition.

Theorem 8.4.1. The set of minimal winning coalitions $\mathcal{M}$ of a voting game (N, w) satisfies

1. $\mathcal{M} \neq \emptyset$,

2. S is not a subset of T whenever S and T are different elements of $\mathcal{M}$, and

3. $S \cap T \neq \emptyset$ whenever $S, T \in \mathcal{M}$.

Conversely, any collection $\mathcal{M}$ of subsets of players N that satisfies the above conditions defines a voting game (N, w) where $w(S) = 1$ if S is a superset of some element of $\mathcal{M}$ and $w(S) = 0$ otherwise.

Proof. $\Rightarrow$: Suppose $\mathcal{M}$ is the set of minimal winning coalitions of a voting game (N, w). We verify that the three conditions in the theorem are satisfied. Since $w(N) = 1$, $\mathcal{M}$ must contain at least one coalition, verifying theorem condition (1). If S and T are different elements of $\mathcal{M}$ and $S \subset T$, then both S and T are winning but T is not minimal, a contradiction that verifies theorem condition (2). If $S, T \in \mathcal{M}$ and $S \cap T = \emptyset$, then both S and T are winning, and so by voting game condition (3), $N \setminus S$ (which is a superset of T) is also winning, a contradiction to voting game condition (4) that verifies theorem condition (3).

$\Leftarrow$: Conversely, suppose that $\mathcal{M}$ is a collection of subsets of players N that satisfies the theorem conditions. Define $w(S) = 1$ if S is a superset of some element of $\mathcal{M}$ and $w(S) = 0$ otherwise. We verify that the coalition game (N, w) so defined satisfies the voting game conditions. Directly from our definition, voting game condition (1) is satisfied. Since $\mathcal{M} \neq \emptyset$, $w(S) = 1$ for some coalition S which implies by the definition of w that $w(N) = 1$, verifying voting game condition (2). If $S \subset T$ and $w(S) = 1$, then there must be a coalition $R \in \mathcal{M}$ such that $R \subset S$ which implies $R \subset T$, and so $w(T) = 1$ by the definition of w, verifying voting game condition (3). If $w(S) = 1$, then by theorem condition (3) and the fact that S and $N \setminus S$ are disjoint, $w(N \setminus S) = 0$, verifying voting game definition condition (4). $\qquad \square$

Many voting games can be represented in the manner that the Tonkins County scenario was originally described, $[8; 5, 4, 2, 2, 1]$, eight votes to pass, with members having 5, 4, 2, 2, and 1 votes, respectively.

Definition 8.4.2. A *weighted voting system*, denoted by $[q; v_1, v_2, \ldots, v_n]$, consists of a positive quota q and nonnegative voting weights $v_1, v_2, \ldots, v_n$ for the n voters. The weighted voting system $[q; v_1, v_2, \ldots, v_n]$ is called a *representation* of the n-player voting game for which coalition S is winning if and only if $\sum_{i \in S} v_i \geq q$.

An allocation of power in a voting game or weighted voting system can be thought of as an allocation of payoffs in the corresponding coalition game. One method we will use is the *Shapley-Shubik power index* [98], which is the specialization of the Shapley method to voting games.

Recall that the Shapley allocation method gives a player their average marginal contribution over all orders of the players. In the Tonkins County coalition game in which coalitions that can pass a bill have worth 1, one player order would be $BDAEC$, in which the marginal contributions are

- $w(B) - w(\emptyset) = 0 - 0 = 0$ for B,

- $w(BD) - w(B) = 0 - 0 = 0$ for D,

- $w(BDA) - w(BD) = 1 - 0 = 1$ for A,

- $w(BDAE) - w(BDA) = 1 - 1 = 0$ for E, and

- $w(BDAEC) - w(BDAE) = 1 - 1 = 0$ for C.

So in the player order $BDAEC$, player A has a marginal contribution of 1 and all other players have a marginal contribution of 0. In general, since each coalition is worth either 0 or 1 and supersets of winning coalitions are winning, in each player order there will be exactly one player that has a nonzero marginal contribution, which must be 1. We describe this player as *pivotal*.

This suggests the following procedure for calculating the Shapley-Shubik power index: List the $n!$ player orders, determine which player is pivotal in each player order, count the number of times a player is pivotal, and divide by $n!$. Since there are five players in the Tonkins County voting game, there are $5! = 120$ player orders to consider.

The procedure can be shortened by examining only the winning coalitions. For example, ABE is a winning coalition in the Tonkins County game and BE is losing; hence, A is pivotal in every order in which players B and E come before A and players C and D come after A: $BEACD$, $BEADC$, $EBACD$, and $EBADC$. This is captured in Table 8.16: We list the winning coalition ABE as a row, and underline A to indicate that A is pivotal anytime B and E come before A. This happens 4 times, recorded in A's pivot column. Likewise, coalition AE is losing while ABE is winning, so B is underlined in this row and its pivot count is also 4. However E is never pivotal when A and B come before E, since AB is already a winning coalition. Thus, E is not underlined and has no pivot count in this row.

In general, if a winning coalition S contains a player i for which $S \setminus \{i\}$ is losing, i will be pivotal in all orders in which players in $S \setminus \{i\}$ come before i, and the players in $N \setminus S$ come after i. Thus, if $s = |S|$ is the number of players in S, then there will be $(s-1)!(n-s)!$ orders for which i is pivotal.

Table 8.16 shows all relevant calculations for the Tonkins County game Shapley-Shubik power index. The columns are summed to find the number of times each player is pivotal, and these numbers are then divided by $5! = 120$ to obtain the Shapley-Shubik power indices given in the second to last line.

The last two rows of Table 8.16 allow us to easily compare the voting power of each board member to the relative population represented (e.g., the Alkane board member represents

TABLE 8.16   Tonkins County Winning Coalitions and Shapley-Shubik Power

Winning Coalition	Pivots				
	A	B	C	D	E
ABCDE					
ABCD					
<u>A</u>BC E	6				
<u>A</u>B DE	6				
<u>A</u> CDE	6				
<u>BC</u>DE		6	6	6	
A<u>B</u>C	4	4			
A<u>B</u> D	4	4			
A<u>B</u> E	4	4			
A <u>CD</u>	4		4	4	
A <u>C</u> E	4		4		4
A <u>DE</u>	4			4	4
B<u>CD</u>		4	4	4	
<u>AB</u>	6	6			
Total	48	28	18	18	8
Shapley-Shubik Power	0.400	0.233	0.150	0.150	0.067
Population Proportion	0.357	0.286	0.143	0.143	0.071

$50/140 \approx 0.357$ of the total population). Although similar, the voting powers and relative populations do not match: the Alkane representative has an absolute difference of $0.400 - 0.357 = 0.043$ in power or a relative difference of $0.043/0.357 \approx 12\%$ more power than the population represented. Similarly, the Broome representative has an absolute difference in power of $0.233 - 0.286 = -0.053$, or $0.053/0.286 \approx 18\%$ less power than the population represented.

When the voting power of a representative aligns with the represented population, we can make the argument that the individual voting power of voters across the county is roughly equal. If town i has population p_i and the people in the town select their board member by simple majority with each person having one vote, then the Shapley-Shubik index for person i is clearly $1/p_i$. This provides a justification for our search for a voting game whose players have voting power as nearly as possible proportional to the populations of the represented towns. By doing so, we are attempting to give each person in Tonkins County roughly equal voting power in matters addressed by the county board.

This brings us back to the question of whether or not the differences in voting power versus population given by the Shapley-Shubik power index and this weighted voting system are acceptable for Tonkins County. While the answer depends on cultural norms, it also depends on whether better alignment is possible. We consider this possibility now.

Observe that since voting games are characterized by their collections of minimal winning coalitions, there can only be a finite number of such games with five players. Table 8.17 is a list of all such games (up to renaming the players) along with corresponding weighted voting system representations when they exist and the corresponding Shapley-Shubik power indices. (This information is available electronically at [51].)

If we identify A with 1, B with 2, C with 3, D with 4, and E with 5, we see that game 55 in Table 8.17 has the same set of minimal winning coalitions as the original Tonkins County voting system $[8; 5, 4, 2, 2, 1]$, although the weighted voting system representation provided in the table is different. A voting game with a weighted voting system representation will have an infinite number of such representations. The ones given in the table minimize the quota with nonnegative integer weights.

TABLE 8.17 Five-Player Voting Games

ID	Minimal Winning Coalitions	Weighted	Shapley-Shubik Power
1	1	$[1; 1, 0, 0, 0, 0]$	$(1.000, 0.000, 0.000, 0.000, 0.000)$
2	12	$[2; 1, 1, 0, 0, 0]$	$(0.500, 0.500, 0.000, 0.000, 0.000)$
3	123	$[3; 1, 1, 1, 0, 0]$	$(0.333, 0.333, 0.333, 0.000, 0.000)$
4	12, 13	$[3; 2, 1, 1, 0, 0]$	$(0.667, 0.167, 0.167, 0.000, 0.000)$
5	12, 13, 23	$[2; 1, 1, 1, 0, 0]$	$(0.333, 0.333, 0.333, 0.000, 0.000)$
6	1234	$[4; 1, 1, 1, 1, 0]$	$(0.250, 0.250, 0.250, 0.250, 0.000)$
7	12, 134	$[5; 3, 2, 1, 1, 0]$	$(0.583, 0.250, 0.083, 0.083, 0.000)$
8	123, 124	$[5; 2, 2, 1, 1, 0]$	$(0.417, 0.417, 0.083, 0.083, 0.000)$
9	12, 13, 14	$[4; 3, 1, 1, 1, 0]$	$(0.750, 0.083, 0.083, 0.083, 0.000)$
10	12, 13, 234	$[5; 3, 2, 2, 1, 0]$	$(0.417, 0.250, 0.250, 0.083, 0.000)$
11	12, 134, 234	$[4; 2, 2, 1, 1, 0]$	$(0.333, 0.333, 0.167, 0.167, 0.000)$
12	123, 124, 134	$[4; 2, 1, 1, 1, 0]$	$(0.500, 0.167, 0.167, 0.167, 0.000)$
13	12, 13, 14, 234	$[3; 2, 1, 1, 1, 0]$	$(0.500, 0.167, 0.167, 0.167, 0.000)$
14	123, 124, 134, 234	$[3; 1, 1, 1, 1, 0]$	$(0.250, 0.250, 0.250, 0.250, 0.000)$
15	12345	$[5; 1, 1, 1, 1, 1]$	$(0.200, 0.200, 0.200, 0.200, 0.200)$
16	12, 1345	$[7; 4, 3, 1, 1, 1]$	$(0.550, 0.300, 0.050, 0.050, 0.050)$
17	123, 145		$(0.467, 0.133, 0.133, 0.133, 0.133)$
18	123, 1245	$[8; 3, 3, 2, 1, 1]$	$(0.383, 0.383, 0.133, 0.050, 0.050)$
19	1234, 1235	$[7; 2, 2, 2, 1, 1]$	$(0.300, 0.300, 0.300, 0.050, 0.050)$
20	12, 13, 145	$[7; 5, 2, 2, 1, 1]$	$(0.700, 0.117, 0.117, 0.033, 0.033)$
21	12, 13, 2345	$[8; 5, 3, 3, 1, 1]$	$(0.467, 0.217, 0.217, 0.050, 0.050)$
22	12, 134, 135	$[8; 5, 3, 2, 1, 1]$	$(0.617, 0.200, 0.117, 0.033, 0.033)$
23	12, 134, 235		$(0.333, 0.333, 0.167, 0.083, 0.083)$
24	12, 134, 2345	$[9; 5, 4, 2, 2, 1]$	$(0.383, 0.300, 0.133, 0.133, 0.050)$
25	12, 1345, 2345	$[6; 3, 3, 1, 1, 1]$	$(0.350, 0.350, 0.100, 0.100, 0.100)$
26	123, 124, 125	$[7; 3, 3, 1, 1, 1]$	$(0.450, 0.450, 0.033, 0.033, 0.033)$
27	123, 124, 135		$(0.500, 0.167, 0.167, 0.083, 0.083)$
28	123, 124, 345		$(0.217, 0.217, 0.217, 0.217, 0.133)$
29	123, 124, 1345	$[9; 4, 3, 2, 2, 1]$	$(0.467, 0.217, 0.133, 0.133, 0.050)$
30	123, 145, 2345		$(0.267, 0.183, 0.183, 0.183, 0.183)$
31	123, 1245, 1345	$[7; 3, 2, 2, 1, 1]$	$(0.433, 0.183, 0.183, 0.100, 0.100)$
32	1234, 1235, 1245	$[6; 2, 2, 1, 1, 1]$	$(0.350, 0.350, 0.100, 0.100, 0.100)$
33	12, 13, 14, 15	$[5; 4, 1, 1, 1, 1]$	$(0.800, 0.050, 0.050, 0.050, 0.050)$
34	12, 13, 14, 2345	$[7; 5, 2, 2, 2, 1]$	$(0.550, 0.133, 0.133, 0.133, 0.050)$
35	12, 13, 145, 234	$[8; 5, 3, 3, 2, 1]$	$(0.450, 0.200, 0.200, 0.117, 0.033)$
36	12, 13, 234, 235	$[7; 4, 3, 3, 1, 1]$	$(0.367, 0.283, 0.283, 0.033, 0.033)$
37	12, 13, 145, 2345	$[6; 4, 2, 2, 1, 1]$	$(0.500, 0.167, 0.167, 0.083, 0.083)$
38	12, 134, 135, 145	$[6; 4, 2, 1, 1, 1]$	$(0.650, 0.150, 0.067, 0.067, 0.067)$
39	12, 134, 135, 234	$[9; 5, 4, 3, 2, 1]$	$(0.367, 0.283, 0.200, 0.117, 0.033)$
40	12, 134, 135, 245		$(0.367, 0.283, 0.117, 0.117, 0.117)$
41	12, 134, 135, 2345	$[7; 4, 3, 2, 1, 1]$	$(0.417, 0.250, 0.167, 0.083, 0.083)$
42	123, 124, 125, 134	$[9; 5, 3, 2, 2, 1]$	$(0.533, 0.200, 0.117, 0.117, 0.033)$
43	123, 124, 125, 345		$(0.250, 0.250, 0.167, 0.167, 0.167)$
44	123, 124, 134, 235		$(0.250, 0.250, 0.250, 0.167, 0.083)$
45	123, 124, 135, 145		$(0.533, 0.117, 0.117, 0.117, 0.117)$
46	123, 124, 135, 245		$(0.250, 0.250, 0.167, 0.167, 0.167)$
47	123, 124, 125, 1345	$[6; 3, 2, 1, 1, 1]$	$(0.500, 0.250, 0.083, 0.083, 0.083)$

TABLE 8.17 Five-Player Voting Games (continued)

ID	Minimal Winning Coalitions	Weighted	Shapley-Shubik Power
48	123, 124, 134, 2345	$[7; 3, 2, 2, 2, 1]$	$(0.300, 0.217, 0.217, 0.217, 0.050)$
49	123, 124, 135, 2345		$(0.300, 0.217, 0.217, 0.133, 0.133)$
50	123, 124, 1345, 2345	$[8; 3, 3, 2, 2, 1]$	$(0.267, 0.267, 0.183, 0.183, 0.100)$
51	123, 1245, 1345, 2345	$[6; 2, 2, 2, 1, 1]$	$(0.233, 0.233, 0.233, 0.150, 0.150)$
52	1234, 1235, 1245, 1345	$[5; 2, 1, 1, 1, 1]$	$(0.400, 0.150, 0.150, 0.150, 0.150)$
53	12, 13, 14, 15, 2345	$[4; 3, 1, 1, 1, 1]$	$(0.600, 0.100, 0.100, 0.100, 0.100)$
54	12, 13, 145, 234, 235	$[5; 3, 2, 2, 1, 1]$	$(0.400, 0.233, 0.233, 0.067, 0.067)$
55	12, 134, 135, 145, 234	$[7; 4, 3, 2, 2, 1]$	$(0.400, 0.233, 0.150, 0.150, 0.067)$
56	12, 134, 135, 234, 235	$[6; 3, 3, 2, 1, 1]$	$(0.317, 0.317, 0.233, 0.067, 0.067)$
57	12, 134, 135, 234, 245		$(0.317, 0.317, 0.150, 0.150, 0.067)$
58	12, 134, 135, 145, 2345	$[5; 3, 2, 1, 1, 1]$	$(0.450, 0.200, 0.117, 0.117, 0.117)$
59	123, 124, 125, 134, 135	$[7; 4, 2, 2, 1, 1]$	$(0.567, 0.150, 0.150, 0.067, 0.067)$
60	123, 124, 125, 134, 234	$[7; 3, 3, 2, 2, 1]$	$(0.283, 0.283, 0.200, 0.200, 0.033)$
61	123, 124, 125, 134, 235		$(0.283, 0.283, 0.200, 0.117, 0.117)$
62	123, 124, 125, 134, 345		$(0.283, 0.200, 0.200, 0.200, 0.117)$
63	123, 124, 135, 145, 234		$(0.283, 0.200, 0.200, 0.200, 0.117)$
64	123, 124, 135, 245, 345		$(0.200, 0.200, 0.200, 0.200, 0.200)$
65	123, 124, 125, 134, 2345	$[8; 4, 3, 2, 2, 1]$	$(0.333, 0.250, 0.167, 0.167, 0.083)$
66	123, 124, 135, 145, 2345		$(0.333, 0.167, 0.167, 0.167, 0.167)$
67	123, 124, 125, 1345, 2345	$[5; 2, 2, 1, 1, 1]$	$(0.300, 0.300, 0.133, 0.133, 0.133)$
68	1234, 1235, 1245, 1345, 2345	$[4; 1, 1, 1, 1, 1]$	$(0.200, 0.200, 0.200, 0.200, 0.200)$
69	12, 134, 135, 145, 234, 235	$[9; 5, 4, 3, 2, 2]$	$(0.350, 0.267, 0.183, 0.100, 0.100)$
70	123, 124, 125, 134, 135, 145	$[5; 3, 1, 1, 1, 1]$	$(0.600, 0.100, 0.100, 0.100, 0.100)$
71	123, 124, 125, 134, 135, 234	$[8; 4, 3, 3, 2, 1]$	$(0.317, 0.233, 0.233, 0.150, 0.067)$
72	123, 124, 125, 134, 135, 245		$(0.317, 0.233, 0.150, 0.150, 0.150)$
73	123, 124, 125, 134, 234, 345		$(0.233, 0.233, 0.233, 0.233, 0.067)$
74	123, 124, 125, 134, 235, 345		$(0.233, 0.233, 0.233, 0.150, 0.150)$
75	124, 125, 134, 135, 234, 235		$(0.233, 0.233, 0.233, 0.150, 0.150)$
76	123, 124, 125, 134, 135, 2345	$[6; 3, 2, 2, 1, 1]$	$(0.367, 0.200, 0.200, 0.117, 0.117)$
77	12, 134, 135, 145, 234, 235, 245	$[4; 2, 2, 1, 1, 1]$	$(0.300, 0.300, 0.133, 0.133, 0.133)$
78	123, 124, 125, 134, 135, 145, 234	$[6; 3, 2, 2, 2, 1]$	$(0.350, 0.183, 0.183, 0.183, 0.100)$
79	123, 124, 125, 134, 135, 234, 235	$[5; 2, 2, 2, 1, 1]$	$(0.267, 0.267, 0.267, 0.100, 0.100)$
80	123, 124, 125, 134, 135, 234, 245		$(0.267, 0.267, 0.183, 0.183, 0.100)$
81	123, 124, 125, 134, 135, 245, 345		$(0.267, 0.183, 0.183, 0.183, 0.183)$
82	123, 124, 125, 134, 135, 145, 2345	$[4; 2, 1, 1, 1, 1]$	$(0.400, 0.150, 0.150, 0.150, 0.150)$
83	123, 124, 125, 134, 135, 145, 234, 235	$[8; 4, 3, 3, 2, 2]$	$(0.300, 0.217, 0.217, 0.133, 0.133)$
84	123, 124, 125, 134, 135, 234, 245, 345		$(0.217, 0.217, 0.217, 0.217, 0.133)$
85	123, 124, 125, 134, 135, 145, 234, 235, 245	$[7; 3, 3, 2, 2, 2]$	$(0.250, 0.250, 0.167, 0.167, 0.167)$
86	123, 124, 125, 134, 135, 145, 234, 235, 245, 345	$[3; 1, 1, 1, 1, 1]$	$(0.200, 0.200, 0.200, 0.200, 0.200)$

Since all five-player voting games have been identified, we can simply compare their voting powers with the relative populations to find the one with the best alignment; however, there are many reasonable ways that we might measure alignment. When we initially examined the powers and relative populations in Table 8.16, we noted both the absolute and relative differences in power versus population for Alkane and Broome. Not only must we decide on whether to measure absolute or relative differences in power versus population, we must then choose a method to combine them for all players: the maximum, mean, and root mean squared are all possible ways to model this.

Game 24 in Table 8.17 with representation $[9; 5, 4, 2, 2, 1]$ minimizes the maximum absolute difference over all players with the value

$$\max\{|0.383 - 0.357|, |0.300 - 0.286|, |0.133 - 0.143|, |0.133 - 0.143|, |0.050 - 0.071|\} \approx 0.026$$

and minimizes the mean absolute difference over all players with the value

$$\tfrac{1}{5}(|0.383 - 0.357| + |0.300 - 0.286| + |0.133 - 0.143| + |0.133 - 0.143| + |0.050 - 0.071|) \approx 0.016.$$

As a point of comparison, the original weighted voting game $[8; 5, 4, 2, 2, 1]$ is the sixth smallest maximum absolute difference with the value 0.052, twice the corresponding game 24 value, and the fourth smallest mean absolute difference with the value 0.023, more than 40% higher than the corresponding game 24 value.

Another pleasant property of game 24 is that the weights are proportional to the populations; however, a $9/14 = 64\%$ majority of weighted votes is required to pass a measure. This objection can be remedied with a different weighted representation: $[702; 698, 697, 2, 2, 1]$ yields exactly the same voting game, and only a $702/1400 = 50.1\%$ majority of weighted votes is required to pass a measure. On the other hand, the Tonkins County Board might find the large disparity of weights unacceptable even though there would be no difference in what coalitions need to form to pass a measure. This concern highlights issues in transferring a perfectly reasonable mathematical model to a real-world context.

Examining relative differences, we see that game 57 minimizes the maximum relative difference with the value

$$\max\{\tfrac{|0.317-0.357|}{0.357}, \tfrac{|0.317-0.286|}{0.286}, \tfrac{|0.150-0.143|}{0.143}, \tfrac{|0.150-0.143|}{0.143}, \tfrac{|0.067-0.071|}{0.071}\} \approx 0.113$$

and minimizes the mean relative difference with the value

$$\tfrac{1}{5}\left(\tfrac{|0.317-0.357|}{0.357} + \tfrac{|0.317-0.286|}{0.286} + \tfrac{|0.150-0.143|}{0.143} + \tfrac{|0.150-0.143|}{0.143} + \tfrac{|0.067-0.071|}{0.071}\right) \approx 0.078.$$

(So that readers can more easily perform their own calculations and comparisons, a spreadsheet with the information contained in Table 8.17 is available from the book's web page.) As a point of comparison, the original weighted voting game $[8; 5, 4, 2, 2, 1]$ is the third smallest maximum relative difference with the value 0.183, 62% higher than the corresponding game 57 value, and the second smallest mean relative difference with the value 0.094, 21% higher than the corresponding game 57 value.

Unfortunately, for game 57, there is no way to create a weighted voting system with the winning coalitions given by the game. Citizens might find it odd that although the members from Chemung and Dutches represent the same number of people, "yes" votes by the members from Alkane, Chemung, and Essex can pass a measure while "yes" votes by the members from Alkane, Dutches, and Essex do not. Similarly, they may find it odd that "yes" votes by the members from Broome, Chemung, and Essex do not pass a measure while "yes" votes by the members from Broome, Dutches, and Essex do. Again, this highlights the modeling problem of interpreting a mathematical solution in the real world.

If a weighted voting system is required, then game 55 with weighted representations $[8; 5, 4, 2, 2, 1]$ or $[7; 4, 3, 2, 2, 1]$ minimizes the mean relative difference in power versus population and game 65 with weighted representation $[8; 4, 3, 2, 2, 1]$ minimizes the maximum relative difference.

In summary, we have identified four different voting games whose Shapley-Shubik power indices most closely align with the relative populations represented, depending upon what definition of alignment is chosen and whether voting games without weighted representations can be considered. These results are summarized in the first three numerical rows of Table 8.18 where the numerical columns correspond to the four comparison criteria (maximum absolute, mean absolute, maximum relative, and mean relative difference), row $\varphi(G55) - p$ compares the Shapley-Shubik power indices for game 55 with the relative populations, row $\varphi(\text{best}) - p$ compares the Shapley-Shubik power indices for the games that most closely match the relative populations (game number in parentheses), and row $\varphi(\text{bwgt}) - p$ compares the Shapley-Shubik power indices for the games with weighted voting representations that most closely match the relative populations (game number in parentheses). This variety of mathematical results highlights once again the importance of the assumptions made at the front end of the modeling process.

TABLE 8.18   Tonkins County Voting Game Comparisons

Comparison	Max Abs	Mean Abs	Max Rel	Mean Rel
$\varphi(G55) - p$	0.052	0.023	0.183	0.094
$\varphi(\text{best}) - p$	(24) 0.026	(24) 0.016	(57) 0.113	(57) 0.078
$\varphi(\text{bwgt}) - p$	(24) 0.026	(24) 0.016	(65) 0.167	(55) 0.094
$\beta(G55) - p$	0.055	0.022	0.192	0.100
$\beta(\text{best}) - p$	(29) 0.023	(29) 0.017	(57) 0.138	(57) 0.089
$\beta(\text{bwgt}) - p$	(29) 0.023	(29) 0.017	(65) 0.167	(55) 0.100

While we are examining assumptions, we might ask if the Shapley allocation method is the best way to measure voting power or whether the nucleolus is a better measure. In New York State, county legislative bodies that choose to use a weighted voting system must find one that aligns voting power with relative populations represented, but voting power must be measured by the Banzhaf power index [7]. Both are proportional to the number of times a player can change a winning coalition into a losing coalition; however, the Shapley-Shubik power index considers each ordering of the players while the Banzhaf power index considers each winning coalition.

The relative Banzhaf power index β is calculated using the following procedure: List the winning coalitions, for each winning coalition determine which players can swing the coalition to losing by dropping out, count the number of times η_i that player i is a swinger, and divide by the total number of swings $\beta_i = \eta_i / (\eta_1 + \cdots + \eta_n)$. For the Tonkins County voting game, the winning coalitions with swingers underlined are given in Table 8.19. Notice that the underlined swingers correspond precisely with the underlined pivotal players in the first column of Table 8.16. Players A, B, C, D, and E have 10, 6, 4, 4, and 2 swings,

TABLE 8.19   Tonkins County Winning Coalitions and Swingers

$ABCDE$	$ABCD$	$ABCE$	$ABDE$	$ACDE$
$BCDE$	ABC	ABD	ABE	ACD
ACE	ADE	BCD	AB	

respectively, and so have relative Banzhaf power indices

$$\beta = \left(\tfrac{10}{26}, \tfrac{6}{26}, \tfrac{4}{26}, \tfrac{4}{26}, \tfrac{2}{26} \right) \approx (0.385, 0.231, 0.154, 0.154, 0.077).$$

The last three rows of Table 8.18 summarize an analysis of the Tonkins County scenario using relative Banzhaf power indices in same manner as the Shapley-Shubik power indices were used previously. (The spreadsheet [51] contains all voting games with no more than five players and the corresponding Shapley-Shubik and Banzhaf power indices.) We can see that for game 55 there is not much difference between the two power indices when it comes to power versus population differences. With respect to both power indices, game 57 minimizes the maximum and mean relative differences between power indices and relative populations, game 65 minimizes the maximum relative differences among weighted voting games, and game 55 minimizes the mean relative differences among weighted voting games. The absolute maximum and mean differences are minimized by different games dependent upon the power index used: game 24 with the Shapley-Shubik power index and game 29 with the relative Banzhaf power index.

Our goal has been to design a voting system for the Tonkins County Board so that the voting powers of its members are closely proportional to the number of people they represent. We have modeled each aspect of our goal in two ways: voting system by a set of minimal winning coalitions or a weighted voting system, voting power by the Shapley-Shubik or relative Banzhaf power indices, single member closeness by absolute or relative differences between a power index and the relative population, and overall closeness by maximum or mean of the single member measures of closeness. The second, third, fifth, and sixth numerical rows of Table 8.18 summarize our mathematical results: the best voting system for each of the sixteen model choices. This illustrates how results are sensitive to the modeling assumptions that are made.

There are no real cost or flexibility trade-offs to be made by choosing any combination of the different assumptions considered here; hence, fidelity to the real-world scenario is of utmost importance. In particular, one might wonder whether the citizens of Tonkins County can accept a voting system based on minimal winning coalitions rather that a weighted voting system. Similarly, one might wonder if the citizens of Tonkins County understand the differences between absolute and relative differences, or between maximum and mean differences.

The defining difference between the two power indices is what is considered equally probable: player orders for Shapley-Shubik and coalitions for Banzhaf. Straffin [107] has provided a different way of distinguishing between the two power indices, which we describe below.

Suppose that player i votes "yes" on a bill with probability p_i. Then the probability that player i could change the outcome by changing their individual vote equals

$$\sum_{\substack{i \text{ swings} \\ \text{for } S}} \left(\prod_{j \in S \setminus \{i\}} p_j \right) \left(\prod_{j \in N \setminus S} (1 - p_j) \right). \tag{8.3}$$

Shapley-Shubik and Banzhaf each calculate the probability in Equation 8.3 but make different assumptions about how the p_i are chosen: Shapley-Shubik chooses the same value for each player and Banzhaf chooses each value independently of the other values.

Theorem 8.4.2 (Shapley-Shubik Characterization). If p is chosen from the uniform distribution on the unit interval and we set $p_i = p$ for all players i, then the probability that player i could change the outcome by changing their vote equals their Shapley-Shubik power index φ_i.

Here, the fact that all p_i's are equal reflects the fact that the Shapley-Shubik power index treats all players as having homogenous viewpoints (i.e., there are shared cultural norms that all voters share). In the next theorem, the fact that the p_i's are chosen independently reflects the fact that the Banzhaf power index assumes players have independent viewpoints (i.e., voters' opinions are developed in isolation from each other).

Theorem 8.4.3 (Banzhaf Characterization). If each p_i is chosen independently from the uniform distribution on the unit interval, then the probability that player i could change the outcome by changing their vote equals their absolute Banzhaf power index $\beta_i' = \eta_i/2^{n-1}$.

Observe that this theorem characterizes the absolute Banzhaf power index $\beta_i' = \eta_i/2^{n-1}$ rather than the relative Banzhaf power index $\beta_i = \eta_i/(\eta_1 + \cdots + \eta_n)$. For a fixed number of voters, we have been interested in the relative power of voters; hence, the absolute and relative Banzhaf power indices have been indistinguishable.

We close this section with two more scenarios that are often investigated using a power index analysis.

United Nations Security Council. The United Nations Security Council consists of five permanent members (United States, Russian Federation, France, China, and United Kingdom) and ten temporary members elected by the General Assembly. A motion on a substantive issue passes with nine "yes" votes and no "no" votes by the permanent members. What is the advantage that permanent members have over the temporary members?

We can model this scenario as a voting game in which the winning coalitions are those with all five permanent members and four or more temporary members. A permanent member (P) is a swinger for all of these coalitions, while a temporary member (T) is a swinger for winning coalitions having exactly three of the other nine temporary members. Thus,

$$
\begin{aligned}
\eta_P &= \sum_{j=4}^{10} \binom{10}{j} &&= &&848 \\
\eta_T &= \binom{9}{3} &&= &&84 \\
\eta_{Total} &= 5\eta_P + 10\eta_P &&= &&5080 \\
\beta_P &= \eta_P/\eta_{Total} &&\approx &&16.69\% \\
\beta_T &= \eta_T/\eta_{Total} &&\approx &&1.65\% \\
\beta_P' &= \eta_P/2^{14-1} &&\approx &&2.59\% \\
\beta_T' &= \eta_T/2^{14-1} &&\approx &&0.26\%.
\end{aligned}
$$

The Shapley-Shubik calculations are the same as the Banzhaf calculations except for the addition of multiplicative factors to count the number of pivots associated with each swing:

$$
\begin{aligned}
\varphi_P &= \tfrac{1}{15!}\sum_{j=4}^{10}(4+j)!(10-j)!\binom{10}{j} &&\approx &&19.63\% \\
\varphi_T &= \tfrac{1}{15!}8!6!\binom{9}{3} &&\approx &&0.19\%.
\end{aligned}
$$

Alternatively, these power indices can be calculated using programs implemented by Leech and Leech [62] after recognizing that the United Nations Security Council voting game has the weighted voting system representation [39; 7, 7, 7, 7, 7, 1, 1, 1, 1, 1, 1, 1, 1, 1, 1].

As might be expected, a permanent member has more power than a temporary member; however, it is interesting to observe the difference between the two power indices. The Banzhaf indices show a permanent member to be $16.69/1.65 \approx 10$ times as powerful as a temporary member, while the Shapley-Shubik indices show a permanent member to be $19.63/0.19 \approx 100$ times as powerful as a temporary member. It would be interesting to ask diplomats or political scientists which ratio seems to be more accurate. It could also be interesting to expand our notion of a power index to take into account the often used abstention from voting used by permanent members.

Legislative Gridlock. In recent decades, many observed that there was "gridlock" in the legislative process because of an increasing partisan divide. The President was from one party, while the other party held simple majorities in the House of Representatives and the Senate, and Senate rules made it nearly impossible for any bill to be considered without at least a 60% majority, which they did not possess. How much voting power did the President have under these circumstances?

It is possible to treat the United States legislative process as a voting game with the President, Vice President, 100 Senators, 435 Representatives, and perhaps others as players; however, it is important for this scenario to take into account voting blocs and the political dispositions of those who actually filled the roles during this historical moment. With an enormous amount of simplification, this scenario could be modeled by a five-player game consisting of the President (P), Senate Republicans (SR), Senate Democrats (SD), House Republicans (HR), and House Democrats (HD), and the minimal winning coalitions $\{P, textSR, textSD, textHR\}$ and $\{P, textSR, textSD, textHD\}$. This is equivalent to game 19 with $\varphi_P = 0.30$, $\beta'_P = 0.3125$, and $\beta_P \approx 0.3333$. According to this model, the President had significantly more voting power in the legislative process with the parties in place than would be expected. A more sophisticated model would again treat Senators and Representatives as individuals but would place each in a policy space (e.g., where each places on a fiscal liberal-to-conservative spectrum, military interventionist-to-isolationist spectrum, etc.) and the probability that a particular ordering of players or coalition of players would form would be based upon players' policy space placement in relationship with each other. However, this would be *a posteriori* voting power: the influence a voter has given the inclinations of the voters. Our focus in this section has been on *a priori* voting power: the influence a voter has knowing only the mechanism that translates individual votes into bill passage.

Keijzer, Klos, and Zhang [58] provide a short overview of computational aspects of power indices, especially with respect to the Tonkins County problem. Lucas [64] and Straffin [109] are good sources for suggesting possible applications. Felsenthal and Machover [36] and [37] consider the theoretical underpinnings for measuring voting power. Edelman and Chen and [32] suggest a data-based (and so *a posteriori*) power index that they use to measure the power of Justices in the U.S. Supreme Court.

EXERCISES

8.1 **TVA**. Ransmeier [86] and Parker [81] have analyzed a decision of the Tennessee Valley Authority to allocate the cost of building a dam to its various uses. Here we describe a simplified scenario in which the dam is built to provide electricity, flood control, and recreation facilities. By planning the dam for all three purposes, there will be a savings of $120 million over what it would have cost to provide for each of the three purposes separately. If the TVA had left out recreation, the savings would have been only $100 million. If the TVA had left out flood control, the savings would have been only $80 million. If the TVA had left out electricity, the savings would have been only $60 million.

a. Model this situation as a coalition game.

b. Compute the Shapley allocation using the average marginal contribution approach illustrated in Table 8.2.

c. Use the computational approach suggested by the proof of the Shapley Characterization Theorem to produce an allocation.

8.2 **Joint Police Force.** The towns of Abeje, Blipster, and Crawford have decided to save money by having a joint police force. The current annual cost (in thousands of dollars) for the three individual cities and the projected annual cost for the three cities together and each pair together are given in Table 8.20.

TABLE 8.20   Joint Police Force Costs

Coalition	ABC	AB	AC	BC	A	B	C
Cost	1980	1110	1340	1570	500	700	900

a. Model this situation as a coalition game in which the worth of a coalition is the savings obtained through collaboration.

b. Find the Shapley allocation for the coalition game and the corresponding allocation of costs.

c. Model this situation as a coalition game in which the worth of a coalition is the negative of the cost.

d. Find the Shapley allocation for the second coalition game and the corresponding allocation of costs.

8.3 In game theory circles, a scenario attributed to [27] is known as Maschler's Aunt. Here an aunt can complete with any one of her four nephews for a fixed amount. The four nephews together (without the aunt) can also complete the same job. Here we assume the payment for a completed job, regardless of how many people help, is $900.

a. Suppose (N, w) is the coalition game where $N = \{1, 2, 3, 4, 5\}$ and $w(S) = 1$ if S is a superset of $\{1, 2\}$, $\{1, 3\}$, $\{1, 4\}$, $\{1, 5\}$, $\{2, 3, 4, 5\}$, and $w(S) = 0$ otherwise. Describe the assumptions being made for $(N, w)M$ to be a model for the Maschler's Aunt scenario.

b. Find the Shapley allocation.

8.4 Analyze the TVA scenario using the nucleolus method.

a. Find the nucleolus using the guess-and-verify method illustrated by Table 8.9 and the corresponding text.

b. Find the reduced game on some pair of players and verify that the nucleolus of the reduced game corresponds to the nucleolus on the original game.

c. Discuss whether you believe that the Shapley or nucleolus allocation is fairer.

8.5 Analyze the Joint Police Force scenario using the nucleolus method.

a. Find the nucleolus for the first coalition game, where the worth of a coalition is the savings due to collaboration, and give the corresponding allocation of costs.

b. Find the nucleolus for the second coalition game, where the worth of a coalition is the negative of its cost, and give the corresponding allocation of costs.

c. Explain how the allocation of costs for both of these games illustrates the scale invariance property.

8.6 Find the nucleolus for the coalition game modeling the Maschler's Aunt scenario.

8.7 Verify that the nucleolus of the EPA Cost Overrun coalition game is $(84, 18, 12, 6)$.

8.8 Consider a group of voters voting for a political office as a coalition game, where coalitions of voters are identified special interest groups. Explain why the candidate in office might choose the nucleolus method for allocating gains such as tax breaks or economic benefits to these coalitions.

8.9 Compare and contrast the Shapley and the nucleolus methods. Explain why each might be preferred by certain groups or in certain situations.

8.10 Prove that the prenucleolus and nucleolus yield the same allocation on superadditive games, that is, if the coalition game is superadditive, then the prenucleolus is player rational. Show that the prenucleolus is not player rational for the three-player coalition game $w(123) = 5$, $w(12) = w(13) = 6$, $w(1) = 3$, and $w(23) = w(2) = w(3) = 1$. Find the nucleolus for this game.

8.11 Prove that the prenucleolus method is coalition rational.

8.12 Prove that the Shapley allocation method is coalition monotone.

8.13 Prove that the prenucleolus allocation method is scale invariant.

*8.14 We observed that while the nucleolus is not coalition monotone on the set of all superadditive coalition games, it is coalition monotone on the set of all three-player coalition games. Thus, if the situation in which we find ourselves involves only three-player games and fairness dictates both coalition rational and coalition monotone, then the nucleolus method would be a good choice. Similarly, the Shapley allocation method is coalition rational on convex games: (N, w) is *convex* if $w(S) - w(S \setminus \{i\}) \leq w(T) - w(T \setminus \{i\})$ whenever $i \in S \subset T$. Thus, if the situation in which we find ourselves involves only convex games and fairness dictates both coalition rational and coalition monotone, then the Shapley method would be a good choice. Prove these results or explore other restrictions on the set of coalition games under which there are allocation methods that satisfy all fairness properties of interest.

*8.15 For the EPA coalition game, the (pre)nucleolus is the centroid of the set of efficient and group rational allocations. This is not true in general. One reason is that the set of efficient and group rational allocations is empty for many coalition games. The set

$$\{(x_1, x_2, \ldots, x_n) : \sum_{i \in N} x_i = w(N), \sum_{i \in S} x_i \geq w(S) + \epsilon \text{ for all coalitions } S \subsetneq N\},$$

which generalizes the group rationality constraints by adding an ϵ to each coalition worth, is always nonempty when ϵ is chosen to be sufficiently small, negative when there are no efficient and group rational allocations. The *least core* is the above set for the largest possible value of ϵ for which the set is nonempty. The prenucleolus, while not necessarily the centroid, is a certain type of geometric center of the least core as described in [68]. Provide a description in your own diagrams and words.

*8.16 We can write the problem of finding the maximum of the minimum excess in the nucleolus method with the following linear program.

$$
\begin{aligned}
\max \quad & \epsilon \\
\text{s.t.} \quad & \epsilon & \leq & \sum_{i \in S} x_i - w(S), && \text{for each } S \subsetneq N \\
& w(N) & = & \sum_{i \in N} x_i \\
& w(\{i\}) & \leq & \; x_i, && \text{for each } i \in N
\end{aligned}
$$

The first set of constraints state that all of the excesses are at least ϵ, which is then maximized, subject to the other constraints for efficiency and player rationality. For the prenucleolus, the player-rational constraints would be eliminated. Subsequent linear programs would replace any excess inequality from the previous linear program that can no longer change with the corresponding equality. Create these linear programs and apply them to TVA and Joint Police Force.

8.17 Verify the remaining rows of Table 8.13. That is, for the Corporate Home coalition game given in Table 8.11, verify that $(50, 0, 10)$ is the unique bargaining allocation with coalition structure $\{AC, B\}$ and $(0, 30, 10)$ is the unique bargaining allocation with coalition structure $\{A, BC\}$.

8.18 Show that Barport's proposal of $(38, 28, 18)$ in line six of the negotiations in the Corporate Home scenario is the Shapley value of the coalition game given in Table 8.11.

8.19 Show that Avon's proposal of $(48, 28, 8)$ in line seven of the negotiations in the Corporate Home scenario is the nucleolus of the coalition game given in Table 8.11.

8.20 Analyze the Maschler's Aunt scenario by finding the bargaining allocations for the corresponding coalition game.

  a. Find the bargaining allocations for the coalition structure $\{1, 2345\}$ in which the nephews collaborate without Maschler's Aunt.

  b. Find the bargaining allocations for the coalition structure $\{N\}$ in which everyone collaborates.

  c. Find the bargaining allocations for any coalition structure, such as $\{12, 345\}$, in which Maschler's Aunt collaborates with one nephew.

8.21 Prove that coalition-rational allocations are bargaining allocations.

*8.22 Prove that the nucleolus is a bargaining allocation for superadditive coalition games.

*8.23 Describe a real-world scenario in which three or more agents plan or are involved in a collaborative endeavor. Model the scenario as a coalition game. Analyze the coalition game by finding the Shapley, nucleolus, and/or bargaining allocations. Interpret your results.

8.24 Consider the weighted voting game representations in Table 8.17.

  a. Verify that the weighted voting game representation for game 24 yields the corresponding minimal winning coalitions.

  b. Verify that game 17 does not have a weighted voting game representation. It is helpful to consider both winning and losing coalitions.

c. Verify the Shapley-Shubik power indices for game 24.

d. Find the Banzhaf power indices for game 24.

8.25 Analyze the Maschler's Aunt scenario by modeling it as a weighted voting system.

a. Find a weighted voting system that represents the Maschler's Aunt coalition game.

b. Find the Shapley-Shubik power indices.

c. Find the absolute and relative Banzhaf power indices.

8.26 Table 8.21 shows the allocations to the aunt and each of the nephews in the winning coalition for each of the previously calculated Maschler's Aunt solutions. Nephews is for the coalition structure in which the nephews form a coalition without the aunt, All is for the coalition structure in which everyone collaborates, and Pair is for the coalition structures in which the aunt pairs with one nephew. For All and Pair, there are line segments of bargaining allocations with end points given by Lo and Hi. Interpret each of the solutions in the context of the scenario. Under what circumstances would each be a good prediction for what would actually happen?

TABLE 8.21 Solutions for Maschler's Aunt

Solution	Aunt	Nephew
Shapley	0.600	0.100
Nucleolus	0.429	0.143
Bargaining Nephews	0.000	0.250
Bargaining All Lo	0.429	0.143
Bargaining All Hi	0.692	0.077
Bargaining Pair Lo	0.500	0.500
Bargaining Pair Hi	0.750	0.250
Shapley-Shubik	0.600	0.100
Relative Banzhaf	0.636	0.091
Absolute Banzhaf	0.875	0.125

8.27 We have suggested two ways to choose an allocation method: decide whether it makes more sense to consider marginal contributions, excesses, or objections, and decide which properties best model fairness. We have suggested two ways to choose a power index: decide whether we think it is more natural to think of player orders or coalitions as being equally likely of formation, or decide whether players' propensities to agree to a position are homogeneous or independent. A third way to choose an allocation method or power index is to examine which best matches our intuition for specific examples. We ask you to do this for coalition games in which one player can veto any action but needs some of the other players to carry out an action.

Suppose n and r are integers satisfying $2 \leq r \leq n - 1$. Let (N, w) be the coalition game in which $N = \{1, 2, \ldots, n\}$, $w(S) = 1$ if $1 \in S$ and $|S| \geq r$, and $w(S) = 0$ otherwise. Player 1 is said to have *veto power* because he is needed for a coalition to have a positive worth, but player 1 is not a dictator because he cannot obtain a positive worth on his own but instead needs at least $r - 1$ of the other players.

a. The worth of a coalition in the Veto Power game can be interpreted either as a prize the coalition can obtain or as whether the coalition wins in a voting system. Before doing any calculations, describe what player 1's payoff or voting power should be. Does it depend on whether the game is modeling an allocation problem or a voting system? In particular, should player 1 receive a payoff of less than 1 and is his power less than 1? Should player 1's payoff or power change as r increases? If so, in what manner should it change?

b. Find the Shapley allocation, or equivalently, the Shapley-Shubik power indices for the general game in terms of n and r.

c. Find the absolute and relative Banzhaf power indices for the general game in terms of n and r.

d. Find the nucleolus for the general game in terms of n and r.

e. For $n = 20$ and $2 \leq r \leq 19$, plot the Shapley, relative Banzhaf, and nucleolus payoffs to player 1.

f. For $r = n - 1$ and $3 \leq n \leq 15$, plot the Shapley, relative Banzhaf, and nucleolus payoffs to player 1.

g. For $n = 20$ and $2 \leq r \leq 19$, plot the Shapley payoff to player 1, absolute Banzhaf to player 1, and absolute Banzhaf to any player $i \neq 1$.

h. Show that the only coalition-rational allocation is $x = (1, 0, \ldots, 0)$.

i. Show that for any coalition structure $\mathcal{S}$ in which there is a coalition $S \in (S)$ such that $w(S) = 1$, the unique bargaining allocation is $x = (1, 0, \ldots, 0)$.

j. Consider the scenario in which any coalition containing player 1 and at least $r - 1$ of the other $n - 1$ players can split \$1,000. Interpret the above solutions in the context of this scenario.

k. Consider the scenario in which the players are members of an organization which can pass a motion if and only if player 1 and at least $r - 1$ of the other $n - 1$ players vote for the motion. Interpret the above solutions in the context of this scenario.

8.28 As introduced, the Shapley allocation was interpreted as a fair way to share the fruits of collaboration. Hence, in the context of voting games, the Shapley-Shubik index could be interpreted as voters' fair shares of the "prize" of political power. Felsenthal and Machover ([36], [37]) take a different point of view and claim the Banzhaf power index better encapsulates the ideas they call "power as influence," whereas the Shapley-Shubik power index measures "power as a prize." Research their argument and discuss which of the three power indices (Shapley-Shubik, relative Banzhaf, or absolute Banzhaf) you believe to be the best way to measure voting power, supporting your argument with mathematics.

*8.29 Verify that Table 8.17 contains all five-player voting games up to renaming of the players.

8.30 Prove that a player in a weighted voting game is never pivotal if and only if the player is not in any minimal winning coalition. We call such players *null players*.

8.31 Suppose player i is a null player in the weighted voting game (N, w), and let $(N \setminus \{i\}, w_i)$ be the game restricted to the players without i, that is, $w_i(S) = w(S)$ for all coalitions S of $N \setminus \{i\}$. Prove that the Shapley-Shubik power index for each player $j \neq i$ is the same in both games.

8.32 Suppose (N, w) is the coalition game arising from the weighted voting system $[3; 1, 1, 1, 1, 1, 0]$.

a. Determine whether any of the players in the game are null players.

b. Verify that $(1/5, 1/5, 1/5, 1/5, 1/5, 0)$ is the Shapley allocation, Banzhaf allocation, and nucleolus.

c. Verify that $(1/7, 1/7, 1/7, 1/7, 1/7, 2/7)$ is a bargaining allocation for the coalition structure $\{N\}$.

d. Discuss the reasonableness of the allocations obtained in the previous two parts.

*8.33 **U.S. Legislative System.** There are two primary routes for a new federal law to be created in the United States: (1) a bill could be passed by simple majorities in the House of Representatives and Senate and then approved by the President, or (2) the President vetoes the bill but the veto is overridden by two-thirds majorities in the House of Representatives and Senate. There are 435 Representatives in the House of Representatives and 100 Senators in the Senate with the Vice President able to break tie votes in the Senate. Assume that the original and potential veto override votes by Representatives and Senators are the same, and the Vice President when voting to break ties in the Senate acts at the behest of the President (so at least 50 Senators is a simple majority when the President supports the bill and at least 51 Senators is a simple majority when the President is against the bill). Furthermore, ignore the possibility of "pocket vetoes" whereby a President prevents passage of a bill with no possibility of a veto override if Congress adjourns within ten days after the bill is sent to the President, the possibility that a simple majority of the Supreme Court can effectively veto a bill by declaring it unconstitutional, and the possibility that Congress and the states amend the Constitution to nullify a Supreme Court ruling. This yields a voting game with 536 players: 1 President, 100 Senators, and 435 Representatives.

a. Find the absolute and relative Banzhaf power indices for the *a priori* voting powers of the President, a Senator, and a Representative.

b. Find the Shapley-Shubik power indices for the *a priori* voting powers of the President, a Senator, and a Representative.

c. Suppose the House of Representatives and Senate cannot override a veto. Find the absolute and relative Banzhaf power indices for the *a priori* voting powers of the President, a Senator, and a Representative in this revised game.

d. Suppose the House of Representatives and Senate cannot override a veto. Find the Shapley-Shubik power indices for the *a priori* voting powers of the President, a Senator, and a Representative in this revised game.

e. Discuss how to interpret the different powers ascribed to the President by the different power indices and whether the President can veto legislation.

*8.34 **Rock County.** During the 1970s, the Wisconsin Rock County Board of County Supervisors consisted of 40 members. The cities of Janesville and Beloit included 14 and 11 districts, respectively. Bloc voting by board members from the two cities had not emerged, but had occasionally been urged upon the board members from Beloit. Straffin, a mathematician in Beloit, analyzed the voting power of board members in an article for *Mathematics Magazine* [108].

a. What is the voting power of each individual voting member?

b. Would Beloit gain power if its delegation would agree to vote as a bloc while others did not?

c. Would Janesville gain power if its delegation would agree to vote as a bloc while others did not?

d. How would the voting powers change if both Beloit and Janesville agreed to vote as separate blocs while others did not?

e. If we think of the cities of Janesville and Beloit as individual players, the cities can choose to allow board members to act as individuals or require them to act as a bloc. Further, if their cardinal utilities are proportional to their relative share of power as calculated in the previous exercises, then the scenario can be thought of as a strategic game. Analyze this game.

*8.35 Prove Theorems 8.4.2 and 8.4.3.

*8.36 Describe a real-world scenario in which three or more agents make decisions via a voting type process. Model the scenario as a voting game. Analyze the voting game by finding the Shapley-Shubik and/or Banzhaf power indices. Interpret your results.

CHAPTER 9

Resource Allocation

The central mathematical problem in this chapter is that of dividing assets fairly among a group of agents. We suggest several reasonable allocation methods, describe fairness properties in order to characterize some of the allocation methods, and build bargaining, coalition, strategic, and incomplete information game models. We finish with a comparison of the multiple approaches.

In previous chapters we began by defining a game which served as a model for a variety of scenarios. Now that we have the modeling tools associated with a variety of games, we begin with a scenario and investigate how different games can serve as models.

> **Inheritance.** Bob, Carol, and Doug have inherited equal shares in their mother's estate, which consists of a cabin, a car, silverware, and $6,000 in cash. The three siblings are friendly with each other, but it is important that they each believe that the division of property is fair. They have differing views about the worth of each item; however, each thinks that the cabin makes up a majority of the estate's value, and so the person receiving the cabin will need to give money to the other two siblings to even up the final distribution. The siblings must decide how to divide up the estate [41].

Since the siblings are willing to exchange money to help ensure each receives an equal share of their mother's estate, we will assume that each can ascribe a monetary value to each item as shown in Table 9.1.

TABLE 9.1 Inheritance Monetary Valuations

Item/Sibling	Bob	Carol	Doug
Cabin	$66,000	$60,000	$42,000
Car	$10,000	$12,000	$7,000
Silverware	$17,000	$27,000	$11,000
Cash	$6,000	$6,000	$6,000
Estate (sum)	$99,000	$105,000	$66,000

9.1 RESOURCE ALLOCATION PROBLEM

To begin modeling this scenario, we first recognize it as a resource allocation problem.

Definition 9.1.1. A *resource allocation problem* consists of the following:

1. A finite (possibly empty) set L of items.

2. A fixed amount M of money.

3. A finite set N of $n \geq 2$ agents.

4. For each item $i \in L$ and agent $j \in N$, a monetary valuation $v_{ij} \geq 0$.

A resolution to the problem is an *allocation* (x, m) consisting of the following:

1. A matrix $x = [x_{ij}]$ where x_{ij} is the proportion of item $i \in L$ allocated to agent $j \in N$ satisfying

$$x_{ij} \geq 0 \text{ for all items } i \in L \text{ and agents } j \in N, \tag{9.1}$$

$$\sum_{j \in N} x_{ij} = 1 \text{ for all items } i \in L, \text{ and} \tag{9.2}$$

2. A vector m whose entry m_j is the amount of money from M allocated to agent $j \in N$ satisfying

$$\sum_{j \in N} m_j = M. \tag{9.3}$$

Because our agents have assigned monetary valuations for each item, we will assume that they are self-interested, are risk neutral with respect to both items and money, and determine their value of each item independently of the other items. Thus it is reasonable to define the *payoff* to agent $j \in N$ to be

$$u_j(x, m) = \sum_{i \in L} v_{ij} x_{ij} + m_j. \tag{9.4}$$

Agent j's *monetary valuation of the estate* is

$$v_j = \sum_{i \in L} v_{ij} + M. \tag{9.5}$$

Agent j's *share* of the estate is

$$s_j(x, m) = u_j(x, m)/v_j. \tag{9.6}$$

With the addition of the above payoff functions, a resource allocation problem is still not quite a game because the agents have not yet been called upon to act as players and rules of play have not been specified. In this section, we suggest solutions to the resource allocation problem without providing a game model. Sections 9.2 and 9.3 will provide game models in order to suggest solutions.

The Inheritance scenario is a resource allocation problem with $L = \{$Cabin, Car, Silverware$\}$, $M = \$6,000$, $N = \{$Bob, Carol, Doug$\}$, and v defined by Table 9.1, e.g., $v_{\text{Cabin, Doug}} = \$42,000$. The final row in the table is simply the sum of the numbers above, indicating each sibling's monetary valuation for the entire estate.

Implicit in choosing to use Equation 9.4 to model the agents' utilities are the following assumptions. First, agents are self-interested; this excludes the possibility that Carol may enjoy her brothers receiving some of the items. Second, the monetary valuation for an item is considered to be independent of ownership; this excludes the possibility that Bob may value Carol over Doug owning the silverware because Bob thinks Carol will take better care of this item of the family's legacy. Third, payoffs are assumed to be proportional; this excludes the possibility that someone might think that one place setting of silverware is worth less than one-twelfth of twelve place settings of silverware. Fourth, payoffs are assumed to be additive; this excludes the possibility that a person who already has silverware might value receiving a second set more if also receiving the cabin where the silverware could be used. Fifth, each item is considered to be a "good" or worthless; this excludes the possibility that one of the items is a weekly chore to be performed.

It is reasonable to ask why the siblings might have such different valuations of the non-cash items. After all, is not the monetary worth of an item the price agreed to by a willing seller and buyer? Of course, it may be difficult to determine such prices. Even nominally objective real estate appraisers are likely to obtain different appraisals for the worth of the cabin, and *Kelly Blue Book*^{TM}(www.kbb.com) provides three different values (selling to a dealer, buying from another consumer, and trading-in to a dealer) for every car. Carol may attach great sentimental value to the silverware, Bob may place a premium on their antique style and age, and Doug may prefer something more modern and new. In our discussion about preferences in Chapter 2, we made the point that it is important to determine the actual preferences of the players. Here this is reflected by the agents' different monetary valuations.

A possible allocation would be for Bob and Carol to split ownership of the cabin evenly via a timeshare arrangement, Doug to obtain full ownership of the car, Carol to obtain full ownership of the silverware, and Bob and Carol to each give Doug $5,000 in addition to the $6,000 left by their mother. Table 9.2 provides this information in terms of allocation notation.

TABLE 9.2   Timeshare Allocation

x_{ij}	Bob	Carol	Doug
Cabin	0.5	0.5	0
Car	0	0	1
Silverware	0	1	0
m_j	−$5,000	−$5,000	$16,000

Observe that each of the numbers in each item row are non-negative (an agent cannot have negative ownership of an item); each item row sums to 1 (there is exactly one of each item to be allocated); and the money row sums to $M = \$6,000$ (the money is completely allocated). The payoffs to the agents are

$$u_{\text{Bob}}(x,m) = (\$66,000)(0.5) + (\$10,000)(0) + (\$17,000)(0) - \$\ 5,000 = \$28,000$$
$$u_{\text{Carol}}(x,m) = (\$60,000)(0.5) + (\$12,000)(0) + (\$27,000)(1) - \$\ 5,000 = \$52,000$$
$$u_{\text{Doug}}(x,m) = (\$42,000)(0\ \ ) + (\$\ 7,000)(1) + (\$11,000)(0) + \$16,000 = \$23,000.$$

The agents' monetary valuations of the estate are (verifying the last row of Table 9.1)

$$v_{\text{Bob}} = \$66,000 + \$10,000 + \$17,000 + \$6,000 = \$\ 99,000$$
$$v_{\text{Carol}} = \$60,000 + \$12,000 + \$27,000 + \$6,000 = \$105,000$$
$$v_{\text{Doug}} = \$42,000 + \$\ 7,000 + \$11,000 + \$6,000 = \$\ 66,000.$$

Finally, the agents' shares are

$$s_{\text{Bob}}(x,m) = \$28,000/\$\ 99,000 \approx 28.3\%$$
$$s_{\text{Carol}}(x,m) = \$52,000/\$105,000 \approx 49.5\%$$
$$s_{\text{Doug}}(x,m) = \$23,000/\$\ 66,000 \approx 34.8\%.$$

Since Bob shares less than one-third of the estate from his perspective, he could object to this allocation as a violation of his one-third ownership in the estate. This suggests that a fair allocation should provide a share to Bob that is at least one-third, perhaps even as much as Carol's 49.5% share!

We just described an *ad hoc* allocation of items and money. There are many allocation methods for resource allocation problems. We summarize some here and give the resulting allocations for the Inheritance problem in Table 9.3.

TABLE 9.3   Inheritance Problem Allocations

Method/Agent	Bob	Carol	Doug
Ad hoc payoff (share)	$\frac{1}{2}$cabin − $5,000 $28,000 (28.3%)	$\frac{1}{2}$cabin + silver − $5,000 $52,000 (49.5%)	car + $16,000 $23,000 (34.8%)
Equal split payoff (share)	$\frac{1}{3}$estate $33,000 (33.3%)	$\frac{1}{3}$estate $35,000 (33.3%)	$\frac{1}{3}$estate $22,000 (33.3%)
first-price auction payoff (share)	cabin − $29,000 $37,000 (37.4%)	car + silver − $2,000 $37,000 (35.2%)	$37,000 $37,000 (56.1%)
second-price auction payoff (share)	cabin − $29,000 $37,000 (37.4%)	car + silver + $4,000 $43,000 (41.0%)	$31,000 $31,000 (47.0%)
Equal shares payoff (share)	cabin − $25,300 $40,700 (41.1%)	car + silver + $4,167 $43,167 (41.1%)	$27,133 $27,133 (41.1%)
Knaster payoff (share)	cabin − $26,000 $40,000 (40.4%)	car + silver + $3,000 $42,000 (40.0%)	$29,000 $29,000 (43.9%)
Isolated equal shares payoff (share)	cabin − $25,588 $40,412 (40.8%)	car + silver + $4,791 $43,791 (41.7%)	$26,797 $26,797 (40.6%)

The most obvious method is the *equal split method*, which divides ownership of each item and the available money equally to each agent. In the remaining methods, each agent's valuation on an item is taken as a bid for that item and each item is allocated to the highest bidder. In the *first-price auction method*, the highest bidder pays their bid amount to the estate, and the estate in the end divides all the money it holds equally among the agents. The *second-price auction method* requires the highest bidder to pay only an amount equal to the item's second highest valuation. After allocating items to the highest bidders, the *equal shares method* gives money from M or takes money from agents until all agents' shares are equal. *Knaster's method* is a compromise between the equal share and first-price auction outcomes. Each agent is first assured of receiving a $1/n$ share of the estate. Any value remaining is divided equally among the agents. The final method, *isolated equal shares*, treats each item separately, using the equal shares method independently on each item. That is, the item is sold to the highest bidder and proceeds divided so that the amount each agent receives represents an equal share of that item. It's important to note that treating the items independently gives a different allocation than using the equal shares method on the whole estate.

As is often the case, finding reasonable solutions is not too difficult. Selection among the reasonable solutions is more difficult. If the goal is fairness, we can model fairness with formalized properties and endeavor to characterize solutions with these properties.

Examining the *ad hoc* allocation, Bob may object because his share is only 28.3% of the estate even though he is a one-third owner of the estate. This suggests that allocations should be *proportionate*, i.e., each agent's share should be at least $1/n$. Doug's share is more than one-third. Nonetheless, he may object because his valuation of what Carol received $((0.5)(\$42,000) + (1)(\$11,000) - \$5,000 = \$27,000)$ is higher than his valuation of what he received $((1)(\$7,000) + \$16,000 = \$23,000)$. This suggests that allocations should be *envy free*, i.e., each agent should value the items and money received at least as much as that player would value the items and money received by any other agent.

Carol and Doug together could object, because if Carol gives Doug $9,500 and takes the car, both her payoff and Doug's payoff would increase by $2,500. This suggests that allocations should be *efficient*, i.e., no other allocation should increase the payoff for some agents without decreasing the payoff for other agents.

Additionally, we could make the argument that all agents should receive an equal value or equal shares, properties known as *payoff equitable* and *share equitable*.

We noted in the isolated equal shares method that the final payoffs to the agents were different than those given by the equal shares method applied to the whole estate. With some allocation methods, however, the method applied separately to items yields the same payoff as the method applied to the entire estate. We will say that such methods are *additive*.

TABLE 9.4   Resource Allocation Properties and Methods

	Equal Split	First Price Auction	Second Price Auction	Equal Shares	Knaster	Isolated Equal Shares
Proportionate	always	always	always	always	always	always
Envy Free	always	always	always	often	often	often
Efficient	rarely	always	always	always	always	always
Payoff Equitable	rarely	always	rarely	rarely	rarely	rarely
Share Equitable	always	rarely	rarely	always	rarely	rarely
Additive	always	always	always	rarely	always	always

Table 9.4 describes how often each of the resource allocation methods satisfies each of the fairness properties. Clearly, "always" means that the method satisfies the property on every resource allocation problem. By "often," we mean that the method satisfies the property on a positive fraction of the problems, while "rarely" indicates that the method satisfies the property only on very rare special problems.

Ideally, we would want an allocation method to have all of these properties. From the table, we see that none of the methods discussed so far can do this. Is there a different allocation method that would satisfy all six properties? The following theorem characterizes some of the methods based on fairness properties and shows that no method can satisfy all of them.

Theorem 9.1.1 (Allocation Characterization). Consider an allocation of items.

1. The allocation is efficient if and only if the only agents who receive any positive portion of an item are those who value it the most if and only if it maximizes the agent's *payoff sum* $\bar{u}(x,m) = \sum_{j \in N} u_j(x,m)$ over all allocations.

2. The allocation is efficient and payoff equitable if and only if it is a first-price auction allocation.

3. The allocation is efficient and share equitable if and only if it is an equal share allocation.

4. The allocation maximizes the minimum agent payoff if and only if it is a first-price auction allocation.

This characterization theorem can help us decide which allocation method to use based on the fairness properties that are important in a given scenario. There are three general philosophical ideas that often arise in discussions about allocation of resources and more generally about social justice issues. Aristotle's equity principle is that rewards should be proportional to contribution. This corresponds with an allocation being payoff or share equitable since a resource allocation problem assumes equal ownership by the agents. Jeremy Bentham's [10] utilitarianism recommends the outcome that will produce the most social good. Utilitarianism then is formalized by the efficient property. John Rawl's [87] theory of justice recommends that the least well-off member of society should be made as well-off as possible. This theory of justice is partially formalized by the proportionate and envy-free properties and more closely by the last item in the characterization theorem. Further information about allocation methods, properties, and applications can be found in [127] and [126].

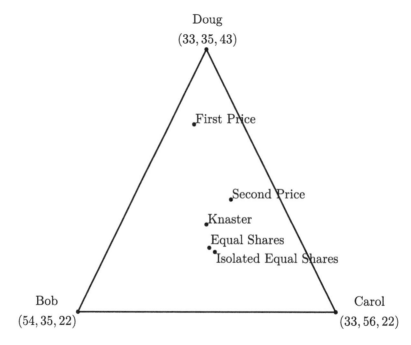

FIGURE 9.1 Inheritance Efficient and Proportionate Payoff Vectors.

We close this section with a return to the Inheritance resource allocation problem. Figure 9.1 provides a geometric view for some possible allocations. The filled-in triangle is the set of efficient and proportionate allocations with the vertices corresponding to the efficient allocations in which one agent receives the entire surplus above the equal split payoff vector $(33, 35, 22)$ where the numbers are in thousands of dollars. The allocations are generally clustered, with first-price most favoring Doug and the isolated equal shares least favoring Doug.

9.2 BARGAINING AND COALITION MODELS

Now that we've investigated important aspects of a resource allocation problem, we can apply game theory models to solve the resource allocation problem. The agents become the players, the payoffs are as defined, and we add rules of play to fit the model we choose. In this section, we will examine how to model a resource allocation problem as a bargaining game and as a coalition game.

Allocations already provide us with outcomes, and the monetary valuations provide us with payoff functions. To fully define a bargaining game, there needs to be a disagreement outcome. In the resource allocation problem, we could choose the equal split allocation, perhaps minus some fees paid to lawyers for adjudicating a final resolution if agreement cannot be reached. Alternatively, the estate could be sold on the open market and the revenue split equally among the players, an option we will call a *monetary split* outcome. This motivates the following definition.

Definition 9.2.1. A *resource allocation bargaining game* consists of a resource allocation problem (L, M, N, v) together with the following:

1. The set of players given by N.

2. Outcomes given by all possible allocations including a disagreement outcome, the partial allocation (x_d, m_d) where

$$x_{ij} \geq 0 \text{ for all items } i \in L \text{ and agents } j \in N,$$

$$\sum_{j \in N} x_{ij} \leq 1 \text{ for all items } i \in L, \text{ and}$$

$$\sum_{j \in N} m_j \leq M.$$

3. The game ends when all players agree upon an allocation, or a pre-specified time limit is reached resulting in the disagreement outcome.

4. Payoff functions $u_j(x, m) = \sum_{i \in L} v_{ij} x_{ij} + m_j$ giving the value of each allocation to each player.

All three of the bargaining game solutions introduced in Chapter 4 (egalitarian, Raiffa, and Nash) are efficient and unbiased. We only need these two properties to find the resource allocation bargaining game solutions using either the equal split or the monetary split disagreement allocation.

Theorem 9.2.1. Suppose equal split is the disagreement allocation. If a resource allocation bargaining game solution method is efficient and unbiased, then it will yield the Knaster allocation.

Proof. For convenience, let $N = \{1, 2, \ldots, n\}$. Given any allocation x of the items satisfying Equations 9.1 and 9.2, the set of possible payoff vectors is

$$\{u(x, m) : \sum_{j \in N} m_j = M\} = \{y \in \mathbb{R}^n : \sum_{j \in N} y_j = \sum_{i \in L} \sum_{j \in N} v_{ij} x_{ij} + M\}$$

since the payoff to a player is equal to their valuation of items i times the proportion x_{ij} of the item they receive and the only constraint on m is Equation 9.3.

The set of all payoff vectors for a fixed allocation x of items is symmetric with respect to any disagreement payoff vector. The set of all possible payoff vectors is the union of these symmetric sets over all possible allocations of the items, hence, it is also symmetric with respect to any disagreement payoff vector. Since the solution method is unbiased, it will yield a payoff vector of the form $d + \lambda(1, \ldots, 1)$ where $d = \frac{1}{n}(v_1, \ldots, v_n)$ is the payoff vector for equal split. Since the solution method is efficient, λ must be chosen to be maximal, which is exactly the formula for Knaster. □

Theorem 9.2.2. Suppose a monetary split is the disagreement allocation. If the resource allocation bargaining game solution method is efficient and unbiased, then it will yield the first-price auction allocation.

> *Proof.* The proof of this theorem follows the proof of the previous theorem except that the disagreement payoff vector is $d = c(1, \ldots, 1)$ for some number c. Thus, the payoff vector chosen must be $(c + \lambda)(1, \ldots, 1)$ where λ is chosen to be maximal. By the Allocation Characterization Theorem, the allocation must be the first-price auction allocation. $\qquad \square$

We can also choose to model the resource allocation problem as a coalition game and apply the solution methods we have for these games, specifically the Shapley and nucleolus allocation methods. To create a resource allocation coalition game, we must first specify the worth of each coalition. The worth of the coalition should represent the proportion of the set N that the coalition S represents. Optimistically, the worth of a coalition should be that proportion of the highest values of the items plus the money. Pessimistically, the worth of a coalition might be that proportion of the equal split allocation.

Definition 9.2.2. An *optimistic resource allocation coalition game* or *pessimistic resource allocation coalition game* consists of a resource allocation problem (L, M, N, v) together with the following:

1. A set of players given by N.

2. For each coalition S of size s, a determined *worth* $w(S) \in \mathbb{R}$. For the optimistic game,

$$w(S) = \frac{s}{n} \left(\sum_{i \in L} \max\{v_{ij} : j \in S\} + M \right).$$

For the pessimistic game, $w(N)$ is the same as for the optimistic game and for $S \neq N$,

$$w(S) = \frac{1}{n} \left(\sum_{i \in L} \sum_{j \in S} v_{ij} \right) + \frac{s}{n} M.$$

3. Any coalition S can come to an agreement to divide $w(S)$ amongst themselves as payoffs. The game ends when there is a coalition structure of agreements, or a pre-specified time limit has passed. In the latter case, each player i that has not entered into a nontrivial coalition receives a payoff of $w(\{i\})$.

4. Utilities u_i given by the payoff received by player i.

Table 9.5 displays the worth of each coalition (in thousands of dollars) for the Inheritance coalition games where players are represented by first initial. Table 9.6 displays the Shapley and nucleolus allocations (in thousands of dollars) for the Inheritance optimistic and pessimistic coalition games. The optimistic coalition game allocations do not correspond to the allocations found using any of the allocation methods of the previous section, but both the Shapley and nucleolus methods applied to the pessimistic coalition game allocations match Knaster allocations for these games. This is generally true.

Theorem 9.2.3. If the pessimistic resource allocation coalition game solution method is efficient, unbiased, and scale invariant, then it will yield the Knaster allocation.

TABLE 9.5   Inheritance Coalition Games

Coalition	Optimistic Worth	Pessimistic Worth
BCD	$66 + 12 + 27 + 6 = 111$	$66 + 12 + 27 + 6 = 111$
BC	$(2/3)(66 + 12 + 27 + 6) = 74$	$33 + 35 = 68$
BD	$(2/3)(66 + 10 + 17 + 6) = 66$	$33 + 22 = 55$
CD	$(2/3)(60 + 12 + 27 + 6) = 70$	$35 + 22 = 57$
B	$(1/3)(66 + 10 + 17 + 6) = 33$	33
C	$(1/3)(60 + 12 + 27 + 6) = 35$	35
D	$(1/3)(42 + 7 + 11 + 6) = 22$	22

TABLE 9.6   Inheritance Coalition Game Allocations

Coalition Game	Method	Bob	Carol	Doug
Optimistic	Shapley	38.5	41.5	31
Optimistic	Nucleolus	37	41	33
Pessimistic	Shapley	40	42	29
Pessimistic	Nucleolus	40	42	29

Proof. Define a new game in which $w'(S) = w(S) - \sum_{j \in S} e_j$ where e_j is j's equal split payoff. Because $w(S)$ came from the pessimistic coalition game, $w'(S) = 0$ for all $S \neq N$. Since the allocation method is efficient and unbiased, the allocation for the new game must give each player $(1/n)(w'(N))$. Since the allocation method is scale invariant, the allocation to player j for the original game is their equal split payoff plus $(1/n)(w'(N))$. This is exactly the Knaster payoff.   □

Figure 9.2 provides a geometric view of the allocations (expressed in thousands of dollars) derived from our analysis of the bargaining and coalition resource allocation games.

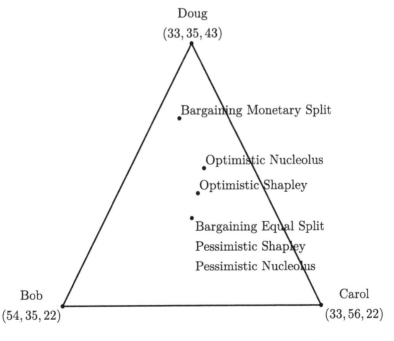

FIGURE 9.2   Inheritance Bargaining and Coalition Game Payoff Vectors.

9.3   COMPLETE AND INCOMPLETE INFORMATION MODELS

In the previous resource allocation game models, we assumed that agent valuations of items were common knowledge. Usually, the only way to know an agent's valuations is by asking; this provides the agents opportunities to be dishonest. We now explore the possibility that agents as players will announce valuations strategically. In our first model, players are still assumed to commonly know everyone's true valuations, but each agent can choose what valuations, which we will call bids, to announce for use in the determination of the allocation.

Definition 9.3.1. A *resource allocation strategic game* consists of a resource allocation problem (L, M, N, v) with integer values of v_{ij} and an allocation method α together with the following:

1. A set of players given by N.

2. Outcomes given by $\alpha(L, M, N, b)$.

3. A strategy for each player $j \in N$ in which they announce their integer bids $b_{ij} \geq 0$ for each item $i \in L$.

4. Each player simultaneously, without knowledge of any other player's choice, chooses a strategy, and the resulting strategy profile determines the outcome $o \in O$.

5. Utility functions equal to the payoff functions

$$u_j(b) = \sum_{i \in L} v_{ij} x_{ij} + m_j$$

   where $(x, m) = \alpha(L, M, N, b)$.

We illustrate with the Inheritance resource allocation problem and a first-price auction method that splits an item evenly among the players that most value the item. If Bob, Carol, and Doug announce their actual valuations, then their payoffs are as calculated in Section 9.1: ($\$37,000, \$37,000, \$37,000$). If Carol were to dishonestly announce the bid $b_{\text{Car,Carol}} = \$18,000$ instead of her valuation $\$27,000$, then she would pay $\$9,000$ less to the estate for receiving the car and each player would receive $\$3,000$ less from the estate, hence, the payoffs would change to ($\$34,000, \$43,000, \$34,000$). The key observation is that by announcing a bid less than her valuation for an item she values the most, Carol was able to increase her payoff. This shows that honest announcements do not form a Nash equilibrium for the Inheritance strategic game.

Consider that instead each player announces the valuations in Table 9.7. Then the first-price auction allocation gives Bob the cabin and he pays $\$64,000$, Carol the car for $\$10,002$, and the silverware for $\$26,000$. In addition, each player gets ($\$64,000 + \$10,002 + \$26,000 + \$6,000)/3 = \$35,334$ in cash for the payoff vector ($\$37,334, \$38,332, \$35,334$).

TABLE 9.7   Inheritance Monetary Valuations

Item/Sibling	Bob	Carol	Doug
Cabin	$64,000	$63,999	$42,000
Car	$10,001	$10,002	$7,000
Silverware	$17,000	$26,000	$25,999

While these payoffs are not as high as in the first-price auction for Doug, no player regrets their announcement given the other players' announcements, as we will show presently.

Recall that player j places a monetary value of v_{ij} on item i and bids b_{ij} on item i. If player j wins item i, then the contribution to their payoff is

$$v_{ij} - b_{ij} + \frac{1}{3}b_{ij} = v_{ij} - \frac{2}{3}b_{ij}.$$

Since this is decreasing in b_{ij}, the player winning an item must bid 1 unit higher than the second highest bid. If player j does not win item i, then the contribution to their payoff is no less than

$$\frac{1}{3}b_{ij}.$$

For player i's best response to involve winning item i, then it must be the case that $v_{ij} - (2/3)b_{ij} \geq (1/3)b_{ij}$, which is equivalent to $v_{ij} \geq b_{ij}$. Therefore any Nash equilibrium will have winning bids no higher than the winner's valuation. Finally, if the highest bid by other players other than i is $b^* \leq v_{ij} - 1$, then player i will obtain their largest payoff by choosing b_{ij} to be $b^* + 1$.

In our example, no one who lost an item would do better by bidding high enough to win the item because to do so would require bids higher than their valuations. The winners of items bid \$1 higher than the second-highest bid; bidding any lower would result in a loss of the difference between their valuation and current bid. Thus, the bids given in Table 9.7 are a Nash equilibrium. Any bids of this type are a Nash equilibrium and *vice versa*.

Theorem 9.3.1. Consider a resource allocation strategic game (L, M, N, v, α) with α the first-price auction allocation and each item i having a unique player $1(i)$ who most values it. We denote by $v_{i2(i)}$ the second highest value for item i ($2(i)$ can refer to any of the players with this second highest value), and assume that $v_{i1(i)} \geq v_{i2(i)} + 2$. The Nash equilibria of the game are announcements b satisfying either

1. $v_{i2(i)} \leq b_{i1(i)} < v_{i1(i)}$ and $b_{ij} < b_{i1(i)}$ for all players $j \neq 1(i)$ with $b_{ik} = b_{i1(i)} - 1$ for at least one player k, or

2. $b_{i1(i)} = v_{i1(i)}$ and $b_{ij} = b_{i1(i)} - 1$ for all players $j \neq 1(i)$.

> *Proof.* We will first determine best responses, next show that announcements satisfying conditions 1 and 2 of the theorem are Nash equilibria, and finally that the only Nash equilibria are those described by conditions 1 and 2.
>
> Our first step is to determine best responses to a given pure-strategy profile b. Because the first-price auction method is additive, we can focus on a single item i separately from the other items. Let c_{ij} be the highest bid on item i by the players other than player j, and let m_{ij} be the number of players other than j who are bidding c_{ij}. Given a pure-strategy profile b, player j's partial payoff due to item i after changing their bid from b_{ij} to a is
>
> $$u_{ij}(a) = \begin{cases} v_{ij} - a + \frac{1}{n}a, & \text{if } a > c_{ij} \\ \frac{1}{m_{ij}+1}(v_{ij} - c_{ij}) + \frac{1}{n}c_{ij}, & \text{if } a = c_{ij} \\ \frac{1}{n}c_{ij}, & \text{if } a < c_{ij} \end{cases}$$
>
> where the first term in the $a = c_{ij}$ case comes from each of the tied winning bidders sharing the item and its cost equally. The expression for $u_{ij}(a)$ is constant for $a < c_{ij}$ and is decreasing with respect to a for $a > c_{ij}$, hence, the best response is $b_{ij} = c_{ij} + 1$,

$b_{ij} = c_{ij}$, or $b_{ij} = c_{ij} - 1$ depending on which of the quantities $u_{ij}(c_{ij} + 1)$, $u_{ij}(c_{ij})$, or $u_{ij}(c_{ij} - 1)$ is largest. With some algebra, we can establish that

$$u_{ij}(c_{ij} + 1) \geq u_{ij}(c_{ij}) \Leftrightarrow v_{ij} \geq c_{ij} + \frac{n-1}{n} \frac{m_{ij} + 1}{m_{ij}}$$

$$u_{ij}(c_{ij} + 1) \geq u_{ij}(c_{ij} - 1) \Leftrightarrow v_{ij} \geq c_{ij} + \frac{n-1}{n}$$

$$u_{ij}(c_{ij}) \geq u_{ij}(c_{ij} - 1) \Leftrightarrow v_{ij} \geq c_{ij}.$$

Therefore,

$$c_{ij} + 1 \text{ is a best response if } v_{ij} \geq c_{ij} + \frac{n-1}{n} \frac{m_{ij} + 1}{m_{ij}}$$

$$c_{ij} \text{ is a best response if } c_{ij} + \frac{n-1}{n} \frac{m_{ij} + 1}{m_{ij}} \geq v_{ij} \geq c_{ij}$$

$$0, 1, \ldots, c_{ij} - 1 \text{ are best responses if } c_{ij} \geq v_{ij}.$$

Since $1 \leq m_{ij} \leq n - 1$, it follows that $n/(n-1) \leq (m_{ij} + 1)/m_{ij} \leq 2$, which implies

$$1 \leq \frac{n-1}{n} \frac{m_{ij} + 1}{m_{ij}} < 2. \tag{9.7}$$

Suppose the announcements b satisfy condition 1:

$$v_{i2(i)} \leq b_{i1(i)} < v_{i1(i)}$$

and

$$b_{ij} < b_{i1(i)}$$

for all players $j \neq 1(i)$ with $b_{ik} = b_{i1(i)} - 1$ for at least one player k. Then $c_{i1(i)} = b_{i1(i)} - 1$ which implies

$$b_{i1(i)} = c_{i1(i)} + 1$$

and

$$v_{i1(i)} \geq b_{i1(i)} + 1 = c_{i1(i)} + 2 > c_{i1(i)} + ((n-1)/n)((m_{ij} + 1)/(m_{ij}))$$

(by Equation 9.7), showing that player $1(i)$ has chosen a best response. For any player $j \neq 1(i)$, it also follows that $c_{ij} = b_{i1(i)} > b_{ij}$, which implies $b_{ij} < c_{ij}$ and $c_{ij} = b_{i1(i)} \geq v_{i2(i)} \geq v_{ij}$, showing that player j has chosen a best response. Thus, b is a Nash equilibrium.

Suppose the announcements b satisfy condition 2:

$$b_{i1(i)} = v_{i1(i)}$$

and

$$b_{ij} = b_{i1(i)} - 1$$

for all players $j \neq 1(i)$. Then $c_{i1(i)} = b_{i1(i)} - 1$ which implies

$$b_{i1(i)} = c_{i1(i)} + 1$$

and

$$v_{i1(i)} = b_{i1(i)} = c_{i1(i)} + 1 = c_{ij} + ((n-1)/n)((m_{ij} + 1)/(m_{ij})),$$

showing that player $1(i)$ has chosen a best response. For any player $j \neq 1(i)$, it also follows that $c_{ij} = b_{i1(i)} > b_{ij}$ which implies $b_{ij} < c_{ij}$ and $c_{ij} = b_{i1(i)} = v_{i1(i)} \geq v_{ij}$, showing that player j has chosen a best response. Thus, b is a Nash equilibrium.

Conversely, suppose the announcements b are a Nash equilibrium. Suppose there is a tie among the two or more players in J for the highest bid b_i^* on item i. Then $c_{ij} = b_i^*$ for all players $j \in N$. For each player $j \in J$, since $b_{ij} = b_i^* = c_{ij}$ is a best response,

$$b_i^* + ((n-1)/n)((m_{ij}+1)/(m_{ij})) \geq v_{ij} \geq b_i^*.$$

For each player $k \notin J$, since $b_{ik} < b_i^* = c_{ik}$ is a best response, $b_i^* = c_{ik} \geq v_{ik}$. The last two sentences imply that $v_{ij} \geq v_{ik}$ for $j \in J$ and $k \notin J$, which implies that the unique player who most values item i must satisfy $1(i) \in J$. Now using Inequalities 9.7 on a previous inequality, we obtain $b_i^* + 2 > v_{ij} \geq b_i^*$ for all $j \in J$. This contradicts the assumption $v_{i1(i)} \geq v_{i2(i)} + 2$.

Thus, there is a unique highest bidder l, that is, $b_{il} > b_{ij}$ for all players $j \neq l$, and so $c_{il} < b_{il}$. Since $b_{il} > c_{il}$ is a best response, $b_{il} = c_{il} + 1$ and

$$v_{il} \geq c_{il} + ((n-1)/n)((m_{il}+1)/(m_{il})) \geq c_{il} + 1$$

(by Inequalities 9.7). For each player $j \neq l$, the definition of l implies that $c_{ij} = b_{il}$, and since $b_{ij} < b_{il} = c_{ij}$ is a best response, $c_{ij} \geq v_{ij}$. The previous two sentences together imply

$$v_{il} \geq c_{il} + 1 = (b_{il} - 1) + 1 = b_{il} = c_{ij} \geq v_{ij}.$$

Thus, $l = 1(i)$.

Substituting the last sentence back into the next to last sentence of the previous paragraph, we have $b_{i1(i)} \geq v_{i2(i)}$. Substituting the last sentence back into the first sentence of the previous paragraph, we have $b_{ij} < b_{i1(i)}$ for all players $j \neq 1(i)$. Since by substituting the last sentence back into the second sentence of the previous paragraph $b_{i1(i)} = c_{i1(i)} + 1$, there is at least one player $k \neq 1(i)$ satisfying $b_{ik} = c_{i1(i)} = b_{i1(i)} - 1$. Substituting this last result back into the second sentence of the previous paragraph, we have

$$v_{i1(i)} \geq b_{i1(i)} - 1 + ((n-1)/n)((m_{il}+1)/(m_{il})).$$

By Inequalities 9.7 and the fact that values and bids are integers, we have either (a) $v_{i1(i)} > b_{i1(i)}$, or (b) $v_{i1(i)} = b_{i1(i)}$ and $m = n - 1$. Results (a) and (b) yield the Nash equilibria described in conditions 1 and 2, respectively, of the theorem statement. □

Figure 9.3 shows a gray parallelogram for the set of payoff vectors (expressed in thousands of dollars) corresponding to Nash equilibria in the resource allocation strategic game for the Inheritance resource allocation problem with the first-price auction method. One vertex corresponds to all players paying for their items with the highest valuations (first-price) and another corresponds to all players paying for their items with the second highest valuations (second-price). The left-most vertex corresponds to Bob paying for the cabin with the second highest valuation and Carol paying for the car and silverware with the highest valuation. The right-most vertex corresponds to Bob paying for the cabin with the highest valuation and Carol paying for the car and silverware with the second highest valuation. This provides a picture of the possible variation in payoffs when players can announce bids different from their actual valuations. Observe that players never announce bids higher than the actual valuations.

Of course, it is highly unrealistic to assume that players' monetary valuations for the items is common knowledge. A more realistic model would have each player only have probabilistic knowledge about other players' monetary valuations.

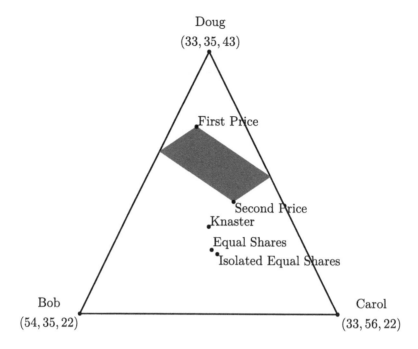

FIGURE 9.3 Inheritance Strategic Game Payoff Vectors.

Definition 9.3.2. A *resource allocation Bayesian game* consists of (L, M, N) from a resource allocation problem and an allocation method α together with the following:

1. A set of players given by N.

2. A set T_j of types for each player $j \in N$ describing their possible valuations $v_{ij} \geq 0$ for each item $i \in L$.

3. A set of actions A_j for each player $j \in N$ in which they announce their real number valuations $b_{ij} \geq 0$ for each item $i \in L$.

4. Outcomes given by $\alpha(L, M, N, b)$ that occur when typed players choose their actions according to the type profiles.

5. Belief functions ϕ_j, where $\phi_j(t_{-j}|t_j)$ is the probability density that the other player types come from the type profile t_{-j} given that player j has type t_j. The function ϕ_j is a probability distribution on the truncated type space $T_{-j} = T_1 \times T_2 \times \cdot T_{j-1} \times T_{j+1} \times \cdots T_n$.

6. Players simultaneously choose actions after privately learning their own types.

7. Utility functions equal to the payoff functions

$$u_j(b) = \sum_{i \in L} v_{ij} x_{ij} + m_j \text{ where } (x, m) = \alpha(L, M, N, b).$$

A strategy for player j is a function from their possible types T_j to their possible actions A_j, that is, knowing what each item is worth personally (i.e., v_{ij} for each item i), player j must indicate what bids to announce (i.e., b_{ij} for each item i). It seems reasonable that in equilibrium, players will announce bids that are less than their valuations and their bid for an item will be an increasing function of their valuation. The following theorem shows that our intuition is correct, at least when the players are interchangeable.

Theorem 9.3.2. Suppose each player's valuation for an item is chosen randomly and independently and the probability density is the same for each player. Then there is a unique symmetric Nash equilibrium in which a player j having a valuation of v_{ij} on item i announces a bid for item i that is the mean of the maximum valuation among all player valuations given that these valuations are no greater than v_{ij}.

Proof. We prove that the strategy profile described in the theorem is a Nash equilibrium and it is the only symmetric Nash equilibrium using a strictly increasing and differentiable bidding function of valuation for each item. The full proof that it is the only symmetric Nash equilibrium is beyond the scope of this book.

We can think of player j's valuation for item i as a random variable V_{ij} having a probability density f_i on the nonnegative real numbers, and the joint density is simply the product of these marginal densities. Since the first-price auction is additive, we can focus on a single item i separately from the other items, will do so in what follows, and will drop the i subscript to unclutter our notation. Hence, player j's valuation for the item is a random variable V_j having a probability density f on the nonnegative real numbers,

$$F(v) = \int_0^v f(t)\, dt$$

is the probability that player j's valuation for the item is no greater than v, $F^n(v)$ is the probability that all players' valuations for the item are no greater than v, and $F^{n-1}(v)$ is the probability that the valuations by players $2, 3, \ldots, n$ for the item are no greater than v. Also observe that since the random variable distributions are described by densities, $\Pr(V_j < v) = \Pr(V_j \le v)$ for all players j.

Suppose $(\beta, \beta, \ldots, \beta)$ is a symmetric Nash equilibrium in which β is a strictly increasing and differentiable function. In order to determine best responses for player 1, suppose players $2, 3, \ldots, n$ use strategy β, player 1 values the item at v, and player 1 announces a bid of b for the item. If $b > \beta(V_2)$, $b > \beta(V_3)$, $\ldots$, and $b > \beta(V_n)$, then player 1 will win the item and obtain a payoff of $v - b + (1/n)(b) = v - ((n-1)/n)(b)$. Since β was assumed to be strictly increasing, a strictly increasing inverse function β^{-1} exists, and the event that player 1 wins can also be described by $V_2 < \beta^{-1}(b)$, $V_3 < \beta^{-1}(b)$, $\ldots$, and $V_n < \beta^{-1}(b)$, hence, the probability that player 1 will win the item is $F^{n-1}(\beta^{-1}(b))$. If $B = \max\{\beta(V_2), \beta(V_3), \ldots, \beta(V_n)\} > b$, then player 1 does not win the item and will obtain a payoff of $(1/n)(B)$. Since β is strictly increasing, the event that player 1 does not win the item can also be described by $T = \max\{V_2, V_3, \ldots, V_n\} > \beta^{-1}(b)$, the resulting payoff is $(1/n)(\beta(T))$, and the probability density differential for $T = t$ is $dF^{n-1}(t) = (n-1)F^{n-2}(t)f(t)\, dt$. Therefore, player 1's expected payoff is

$$u(b) = \left(v - \frac{n-1}{n}b\right)F^{n-1}(\beta^{-1}(b)) + \int_{\beta^{-1}(b)}^{\infty} \frac{1}{n}\beta(t)\, dF^{n-1}(t).$$

The strategy β will be a best response for player 1 if $b = \beta(v)$ maximizes $u(b)$, and a necessary condition for this is that $u'(\beta(v)) = 0$. Using the Fundamental Theorem of Calculus and the product and implicit function derivative rules, we obtain

$$u'(b) = -\frac{n-1}{n}F^{n-1}(\beta^{-1}(b)) + \left(v - \frac{n-1}{n}b\right)\frac{(n-1)F^{n-2}(\beta^{-1}(b))f(\beta^{-1}(b))}{\beta'(\beta^{-1}(b))}$$

$$-\frac{1}{n}\frac{b(n-1)F^{n-2}(\beta^{-1}(b))f(\beta^{-1}(b))}{\beta'(\beta^{-1}(b))}$$

$$=-\frac{n-1}{n}F^{n-1}(\beta^{-1}(b))+(v-b)\frac{(n-1)F^{n-2}(\beta^{-1}(b))f(\beta^{-1}(b))}{\beta'(\beta^{-1}(b))}.$$

The necessary condition $u'(\beta(v)) = 0$ becomes

$$-\frac{n-1}{n}F^{n-1}(v)+(v-\beta(v))\frac{(n-1)F^{n-2}(v)f(v)}{\beta'(v)}=0.$$

After some algebra and the replacement of v with t, we obtain

$$F^n(t)\beta'(t)+nF^{n-1}(t)f(t)\beta(t)=tnF^{n-1}(t)f(t).$$

Recognizing that the left-hand side is the derivative of $F^n(t)\beta(t)$ with respect to t, taking the definite integral of both sides of this equation with respect to t from 0 to v, and recognizing that $F(0) = 0$, we obtain

$$F^n(v)\beta(v)=\int_0^v t\,dF^n(t).$$

This implies that

$$\beta(v)=\int_0^v t\,\frac{dF^n(t)}{F^n(v)} \tag{9.8}$$

which is the mean of the maximum valuation among all player valuations given that these valuations are no greater than v.

We observe from Equation 9.8 that β is strictly increasing (wherever f is positive) and differentiable, as assumed. We have proved that any symmetric Nash equilibrium with increasing and differentiable strategies must have strategies defined by Equation 9.8. To complete the proof that we have found a Nash equilibrium, verify that $b = \beta(v)$ is a maximum (rather than a minimum or inflection) for $u(b)$ by applying the first derivative test. □

This result for a Bayesian game model of the resource allocation problem is not directly comparable to the results from the other game models because the belief functions are not part of nor directly derivable from the resource allocation problem. We explore this further by imagining that each player's valuation for an item is uniformly distributed on an interval. Specifically, for some parameters $\bar{v} \geq \delta > 0$, let each player's valuation for an item be a random variable with probability density $f(v) = 1/2\delta$ if $\bar{v} - \delta \leq v \leq \bar{v} + \delta$. Using Equation 9.8, we can obtain the symmetric Nash equilibrium strategy for bids on this item to be

$$\beta(v)=v-\frac{1}{n+1}(v-\bar{v}+\delta),$$

which clearly shows the general result that players should announce bids that are less than their actual valuations. Since δ is a measure of player uncertainty about the valuations of others, this form of β also shows that the amount by which a player should lower their bid is greater the larger the uncertainty.

One way to compare our Bayesian game model results with previous game model results is to assume that a player j that values item i at v_{ij} believes that the valuations for the other players are drawn independently from a random variable with probability density $f_{ij}(v) = 1/(2\delta_{ij})$ if $v_{ij} - \delta ij \leq v \leq v_{ij} + \delta ij$. Then player j should announce a valuation

$b_{ij} = v - \delta/(n + 2)$. If the actual valuations correspond to those in Table 9.1 for the Inheritance problem, the resulting payoff vector could land within the gray parallelogram depicted in Figure 9.3 or below because for a large enough value for δ_{ij} the winner j of item i may have announced a bid that is smaller than the second highest actual valuation.

Since the bids increase as functions of the valuations in the same manner for each player, each item is given to a player who most values the item. This was also true in the strategic game, and so the allocations obtained from the Nash equilibria of either model yield efficient allocations. This is actually somewhat of a surprising result. In most strategic and Bayesian games, the equilibria are not efficient; Prisoners Dilemma is a canonical example.

Although efficiency is maintained in both the complete and incomplete information models, other fairness properties associated with the first-price auction method may not be maintained. This is because players do not announce their valuations in equilibrium, and so the actual payoff distributions are typically different from what was desired in accordance with the first-price auction method. As is often the case, it may be impossible to create an allocation mechanism that will both have desirable fairness properties and give players incentives to reveal their preferences honestly.

9.4 CONCLUDING REMARKS

This chapter highlights how the assumptions we make about a real-world problem influence the type of game theory model we choose to use and the techniques and tools available to us in our analysis. While we have independently discussed different models in the previous chapters, this chapter demonstrates a pivotal step in the modeling process—selecting the appropriate model for the situation.

We analyzed the common scenario of needing to fairly divide assets among multiple parties. We took this real-life scenario, quantified variables by giving each player's valuation of each asset, turning it into a formal resource allocation problem, and discussed fairness properties we wanted to respect when distributing the assets among the players.

Treating the possible allocations as outcomes, we first modeled this problem as a bargaining game and a coalition game in which players came to a binding agreement on the division of assets. When we introduced the idea that players might be strategic in announcing their valuations of items in order to influence the allocation, we were able to model the resource allocation problem as a strategic game with complete information and a Bayesian game of incomplete information. In all cases we were able to use the techniques and theory of these game theoretic models to analyze the resource allocation problem and see the effects of the type of model we chose on our analysis and conclusions.

Even though we based them on the Inheritance Scenario, all four models were fairly general. This suggests that they are reasonably flexible and might be applied to any resource allocation problem. Because the computations involved in each of the allocation methods are relatively simple, the cost is limited to determining the valuations. These costs might involve acquiring data sets, interviewing agents, or conducting market analysis.

On the other hand, the fidelity of the models is limited somewhat by the ability of the agents to determine their valuations and by their willingness to reveal those valuations. The Bayesian game model is the most sophisticated, however, it requires a costly determination of agent beliefs. It is unlikely that agents would explicitly, or willingly, do what the model suggests. Nonetheless, the results of the model suggest no matter how the agents actually behave, there are reasonable constraints on the final allocations, and the crucial issue is the determination of the two highest valuations for each item.

EXERCISES

9.1 Recall that in a resource allocation problem, having many smaller items makes the allocation process easier. Sometimes we can divide large items into small items before allocating, but we must take care to not violate our assumptions about agent preferences. Each part below describes objects that are part of a resource allocation problem. Discuss, in terms of our assumptions about agent preferences, whether the objects in a particular part should be considered a single item or should be separated into more than one item in a specific manner.

    a. A silverware set containing eight knives, eight forks, and eight spoons.

    b. Checking, savings, and money market accounts.

    c. Five identical standard decks of playing cards.

    d. A completely furnished house.

    e. A dog and two cats that have been living in a house together.

    f. In a forty acre plot of land, the rights to the water, oil, and natural gas that lie underneath.

9.2 Describe scenarios other than an inheritance that could be described as a resource allocation problem.

9.3 Verify the allocations, payoffs, and shares in Table 9.3 for all methods other than Ad hoc.

9.4 As part of a divorce settlement, Tom and Mary must agree upon how to allocate custody of their son and a jointly owned property, business, and $50,000 debt.

    a. Explain why it makes sense to treat $N = \{1, 2\}$.

    b. Determine the set L of items and the amount M. Explain why in this scenario it might make sense to treat the custody of their son or the property as each more than one item. Can any of the items be allocated partially to each person?

    c. Create a table similar to Table 9.1 that gives the valuation of each of the items in L to each of the agents in N. Again, clearly state any assumptions you are making as you determine these valuations.

    d. Which allocation method do you think would be the most appropriate to use? Why?

    e. Which type of game do you think is most appropriate to model the resource allocation problem you have created? What assumptions about the scenario influenced your decision?

9.5 Suppose four siblings have inherited a house, jewelry and $10,000. Model this scenario as a resource allocation problem by determining the following.

    a. Determine the set N of agents, and explain any assumptions you make about whether or not to treat the siblings as individual agents.

    b. Determine the set L of items, and explain any assumptions you made when deciding to either subdivide items or leave them intact.

c. Create a table similar to Table 9.1 that gives the valuation of each of the items in L to each of the agents in N. Again, clearly state any assumptions you are making as you determine these valuations.

d. Which allocation method do you think would be the most appropriate to use? Why?

e. Which type of game do you think is most appropriate to model the resource allocation problem you have created? What assumptions about the scenario influenced your decision?

9.6 Consider a resource allocation problem with no money and a single item. The single item is valued most by only one agent. Without loss of generality, let $L = \{1\}$, $M = 0$, $N = \{1, 2, \ldots, n\}$, and v satisfy $v_1 > v_2 \geq \cdots \geq v_n$ where v_j is player j's valuation of the item.

a. Find formulas for the equal split allocation, payoffs, and shares.

b. Find formulas for the first-price auction allocation, payoffs, and shares.

c. Find formulas for the second-price auction allocation, payoffs, and shares.

d. Find formulas for the equal shares allocation, payoffs, and shares.

e. Find formulas for the Knaster allocation, payoffs, and shares.

9.7 Consider a resource allocation problem in which there is money, but no items. Without loss of generality, let $L = \{\}$, $M > 0$, $N = \{1, 2, \ldots, n\}$, and v has no values. Verify that the equal split, first-price auction, second-price auction, equal shares, Knaster, and isolated equal shares allocate $m_j = \frac{1}{n}M$ for each agent j.

*9.8 Write a computer program that, given a resource allocation problem, calculates the allocation, payoffs, and shares for one or more of the methods described in this chapter.

9.9 The three siblings valuations for each other's Knaster allocations are given in Table 9.8. Looking at the Bob column, we see that Bob thinks what he received is worth

TABLE 9.8   Knaster Allocations

Valuation	Bob	Carol	Doug
Bob's cabin − $26,000	$40,000	$34,000	$16,000
Carol's car + silver + $3,000	$30,000	$42,000	$21,000
Doug's $29,000	$29,000	$29,000	$29,000

($40,000) more than what Carol or Doug received ($30,000 and $29,000). Looking at the Carol column, we see that Carol thinks what she received is worth ($42,000) more than what Bob or Doug received ($34,000 and $29,000). Looking at the Doug column, we see that Doug thinks what he received is worth ($29,000) more than what Bob or Carol received ($16,000 and $21,000). Thus, the Knaster allocation is envy free for the Inheritance problem.

a. Compute each agent's valuation for each agent's *ad hoc* allocation. For which pairs of agents j and k does agent j envy agent k?

    b. Compute each agent's valuation for each agent's equal split allocation. Verify that this allocation is envy free.

    c. Compute each agent's valuation for each agent's first-price auction allocation. Verify that this allocation is envy free.

    d. Compute each agent's valuation for each agent's second-price auction allocation. Verify that this allocation is envy free.

    e. Compute each agent's valuation for each agent's equal shares allocation. Verify that this allocation is envy free.

    f. Compute each agent's valuation for each agent's isolated equal shares allocation. Verify that this allocation is envy free.

*9.10 Explore resource allocation methods and properties: In terms of a general resource allocation problem (L, M, N, v), obtain formal definitions of one or more allocation methods described in the text or of your own design. Obtain formal definitions of one or more of the fairness properties described in the text or of your own design. Then determine how often each method satisfies each property. Finally, prove the Allocation Characterization Theorem or similar conjectures of your own design.

9.11 Choose the allocation method from those described in this chapter that you think is the fairest and explain why. If you think there is a different method, not described in the text, that is fairer, describe the method and explain why it is fairer.

9.12 Consider a resource allocation Bayesian game of incomplete information as described in Theorem 9.3.2. Suppose each player's valuation for an item is a random variable with probability density $f(v) = 1/2\delta$ if $\bar{v} - \delta \leq v \leq \bar{v} + \delta$. Use Equation 9.8 to show that the symmetric Nash equilibrium strategy for valuation announcements on this item is $\beta(v) = v - (1/(n+2))(v - \bar{v} + \delta)$.

*9.13 The text found the Nash equilibria for a general class of resource allocation strategic games using the first-price auction allocation method. Choose a different allocation method and find the Nash equilibria for a general class of resource allocation strategic games using that allocation method. Apply your results to the Inheritance problem. What conclusions do you draw?

*9.14 The text found the symmetric equilibrium for a general class of resource allocation Bayesian games of incomplete information using the first-price auction allocation method. Choose a different allocation method and find the symmetric equilibrium for a general class of resource allocation Bayesian games of incomplete information using that allocation method. Apply your results to the Inheritance problem. What conclusions do you draw?

*9.15 Describe a real-world scenario involving the allocation of resources. Model the scenario as one or more resource allocation games. Analyze the games by using appropriate solution concepts. Interpret your results.

Selected Answers

We urge the reader to make a good faith attempt to solve exercises before looking at the answers found here. Answers provided here are not necessarily full solutions.

1.1a. Which costs less, purchasing gas at a station on our path, or at a station which is out of our way?

1.1b. Gas is cheaper at the further station; the car's fuel economy is constant; we purchase the same amount of gas at either station.

1.1c. The prices of gas at each of the stations, the fuel economy of the car, and the extra miles driven.

1.1d. Calculate S, the savings in gasoline purchased, and T, the cost of driving to the further station. If $S > T$ then we drive to the further station.

1.5 Two of many possible problem statements: What is the best deployment for each of the admirals? Which admiral will win the overall engagement? The players are the two admirals. Outcomes could be which fleets are destroyed. A simple set of rules could be that each admiral simultaneously sends each of their fleets to one of the two ocean areas, where upon opposing fleets pair up and destroy each other (e.g., if Admiral R sends two fleets and Admiral C sends one fleet to the north, then one of Admiral R's fleets and Admiral C's fleet destroy each other, leaving one of Admiral R's fleets unscathed). Presumably, each Admiral would prefer having more fleets left unscathed. Certainly we could envision more complicated rules (e.g., whether a fleet takes on a defensive or offensive role, fleets might choose to surrender rather than be destroyed) and preferences (e.g., the political perception of different outcomes in the two countries could be taken into account).

1.7 No, the bottom two could have been reversed. Or the top two could have been tied and Sophie randomized between the two alternatives. Ray Gardener (www.members.shaw.ca/ray_gardener/essays/posts_cap.htm) suggests a similar choice that we could imagine: the need to rescue two children from different parts of a burning building with the high probability that only one could be saved. He also suggests that since Sophie did not have any coins available for flipping, "we suspect that there must have been some little something that made her choose one child over the other, and she herself suspects that, and the guilt catches up to her."

2.1 See Table A.1.

2.4a. There are at most $m - 1$ subsets since a subset with one element does not require a choice.

TABLE A.1   Subsets of Ice Cream
Parlor Outcomes

Subset	Scarlett's Choice
$\{V, C, S, N\}$	V
$\{V, C, S\}$	V
$\{V, C, N\}$	V
$\{V, S, N\}$	V
$\{C, S, N\}$	C or S
$\{V, C\}$	V
$\{V, S\}$	V
$\{V, N\}$	V
$\{C, S\}$	C or S
$\{C, N\}$	C
$\{S, N\}$	S

2.4b. Since there are 2^m subsets, m singleton subsets, and 1 empty set, there are $2^m - m - 1$ subsets with two or more outcomes.

2.4c. See Table A.2.

TABLE A.2   Construct vs.
Verify Ordinal Preferences

m	$m - 1$	$2^m - m - 1$
3	2	4
4	3	11
5	4	26
6	5	57

2.4d. The formula $2^m - m - 1$ grows much faster than $m - 1$.

2.9a. Suppose the player chooses either outcome if presented with $\{A, B\}$, and the player chooses either outcome if presented with $\{B, C\}$. (a1) Then the player weakly chooses A over B, and the player weakly chooses B over C. By transitivity, the player weakly chooses A over C. (a2) Also by the supposition, the player weakly chooses C over B, and the player weakly chooses B over A. By transitivity, the player weakly chooses C over A. Combining the results (a1) and (a2), we find that the player chooses either outcome if presented with $\{A, C\}$.

2.13a. D, C, B, A.

2.13b. D and C are much more strongly preferred than A and B.

2.13c. The player is indifferent between the outcome C and the lottery $0.1A + 0.9D$, and is indifferent between the outcome B and the lottery $0.8A + 0.2D$.

2.13d.

$$u(L) = u\left(\frac{2}{6}A + \frac{3}{6}B + \frac{1}{6}C\right)$$
$$= \frac{2}{6}u(A) + \frac{3}{6}u(B) + \frac{1}{6}u(C)$$

$$= \frac{2}{6}(0) + \frac{3}{6}(20) + \frac{1}{6}(90)$$
$$= 25$$

2.13e. One approach is to use the outcome-lottery indifferences described in part (b) to obtain the equations

$$v(B) = 0.8v(A) + 0.2v(D)$$
$$v(C) = 0.1v(A) + 0.9v(D).$$

Using the given utilities, we obtain the equations

$$0 = 0.8v(A) + 80$$
$$v(C) = 0.1v(A) + 360$$

whose solution is $v(A) = -100$ and $v(C) = 350$.

2.13f. No. By the Expected Utility Hypothesis, the utility of the given lottery should be $0.5u(A) + 0.5u(C) = 0.5(0) + 0.5(90) = 45 \neq 20$.

2.13g. Possibly. The evidence is consistent with u being a vNM utility function, but we cannot be sure without checking an infinite number of possibilities.

2.20
$$\pi_1(Q_1, Q_2, Q_3) = \left(\frac{a}{b} - \frac{Q_1 + Q_2 + Q_3}{b} \right) Q_1 - cQ_1.$$

2.23 If course grades were an ordinal scale, then it would be meaningful to say that one student performed better than another student, but it would be meaningless to compute average grades. If course grades were an interval scale, then it would be meaningful to compute average grades, but it must be the case that the difference between an A and a B is the same as the difference between a D and an F. If course grades were a ratio scale, then it would be meaningful to say that a student receiving an A performed twice as well as a student receiving a C and infinitely better than a student receiving an F. The primary controversy should be whether grades can be considered an interval scale or merely an ordinal scale.

2.28a. Evan's utility for the lottery is 10. The expected value of the lottery is $1,000. Evan's utility for the expected value of the lottery is approximately 31.6. Since Evan's utility for $1,000 is larger than his utility for a lottery whose expected value is $1,000, Evan is risk adverse.

2.28b. The graph is in Figure A.1. Since the utility curve is concave down, this indicates that Evan is risk adverse.

2.28c. Because $u(x) \geq 10$ implies $x \geq 100$, Evan would choose an offer of $100 or more.

2.28d. Because the utility of this lottery is $\sqrt{4}/2 + \sqrt{4^2}/2^2 + \sqrt{4^3}/2^3 + \cdots + \sqrt{4^k}/2^k + \cdots = 1 + 1 + 1 + \cdots + 1 + \cdots = \infty$, there is no amount of money that can be offered to Evan to dissuade him from choosing the lottery. It seems unreasonable that someone could offer such a lottery to Evan since it is possible that Evan would be given more money than exists on the Earth.

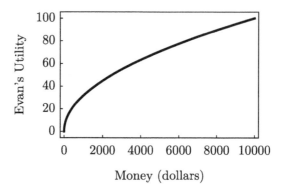

FIGURE A.1 Evan's Utility.

3.1a. Preference order is sharing a stag, catching a hare (whether or not the other hunter catches something), and catching nothing. A hunter catching a hare does not care whether the other hunter has or has not caught a hare, so these outcomes can be tied.

3.1b. In addition to the ordinal utility description, the hunter is indifferent between catching a hare and a 50% chance of catching a stag.

3.1c. No strategies are dominated.

3.1d. Catching a hare is prudential for each hunter with a security level of 1.

3.1e. The Nash equilibria are (Stag, Stag) and (Hare, Hare).

3.1f. The unique efficient strategy profile is (Stag, Stag).

3.1g. Answers will vary. With prior communication, it seems reasonable that the two hunters will choose to hunt for a stag together. If there is no prior communication and at least one hunter is strategically risk adverse, then they may each hunt individually for a hare.

3.5a. For Stag Hunt, let A be Stag and B be Hare. Then

$$R_{12} < R_{21} = R_{22} < R_{11}$$

and

$$C_{21} < C_{12} = C_{22} < C_{11}.$$

It is not absolutely essential that the equalities hold (perhaps a hunter obtaining a hare places a small positive or negative value on the other hunter obtaining a hare). It is important to note that, because linear transformations of vNM utilities or monotonic transformations of ordinal utilities are equivalent, there need not be any relationship between Rose's payoffs and Colin's payoffs.

3.6 If there were only 30 seconds left, Hopkins would hold the ball to the last 10 seconds and then attempt to score without giving Shakopee time to score in return. Shakopee would not attack the ball during that first 20 seconds. That is, both teams would simply wait 20 seconds and turn the game into a 10-second game. Because in the scenario described there is no shot clock, we can extend this solution to the three-minute game: Hopkins should hold the ball until the last 10 seconds, and Shakopee should not attack the ball until that same point in time.

In overtime, if Hopkins wins the tip, then both teams play the proposed strategies for the end of regulation time. If Shakopee wins the tip, it should try to score as quickly as possible to secure the advantage. If they fail to score, then the two teams revert to the proposed strategies for the end of regulation.

3.8 The posited change in the scenario changes the payoff pair for the (Move, Retreat) strategy profile to being very favorable for the Germans and bad for the Americans. The best response payoffs are shown in Table A.3. For the Americans, Hold is now strongly dominant and prudential, and Move is dominated by each of the other two strategies. For the Germans, neither strategy is dominated, but Retreat is prudential. There is a unique Nash equilibrium: (Hold, Retreat). Since this is a constant sum strategic game, all strategy profiles are efficient. Hitler's choice to attack seems even more unreasonable given these game theoretic solutions.

TABLE A.3   Normandy Breakout Payoff Matrix

(American, German)	Attack	Retreat
Reinforce	(3, 4)	(4, 3)
Hold	([6], 1)	([5], [2])
Move	(1, [6])	(2, 5)

3.14 For the strategic game in Table 3.22, Bottom is dominated by both Top and Middle, and Right is dominated by Middle; hence, Bottom and Right can be eliminated in either order. Now Top is dominated by Middle and can be eliminated. Finally, Left is dominated by Middle and can be eliminated. The result is (Middle, Middle).

For the strategic game in Table 3.23, Top is dominated by Middle and Right is dominated by Left; hence, Top and Right can be eliminated in either order. Now Left is dominated by Middle and can be eliminated. Finally, Bottom is dominated by Middle and can be eliminated. The result is (Middle, Middle).

3.22a. See the left half of Figure A.2.

3.22b. See the right half of Figure A.2.

3.22c. See Figure A.3.

3.22d. The best response functions intersect at $(0,0)$, $(1,1)$, and $(5/8, 3/8)$, which describe the three Nash equilibria (Bach, Bach), (Stravinsky, Stravinsky), and $((3/8)\text{Bach} + (5/8)\text{Stravinsky}, (5/8)\text{Bach} + (3/8)\text{Stravinsky})$ with the expected payoff pairs $(10, 8)$, $(8, 10)$, and $(5, 5)$.

3.22e. First, for the expected payoffs to be meaningful, the payoffs in the table must be vNM utilities. It would be easy to implement the mixed strategy: roll an 8-sided die and choose the concert to attend based on the value of the roll. Nevertheless, this seems like a silly equilibrium to achieve since it is not efficient while the pure-strategy Nash equilibria are efficient and better for both players than the completely mixed one.

3.29 Generic Final Jeopardy is symmetric if and only if $p_1 = p_2$.

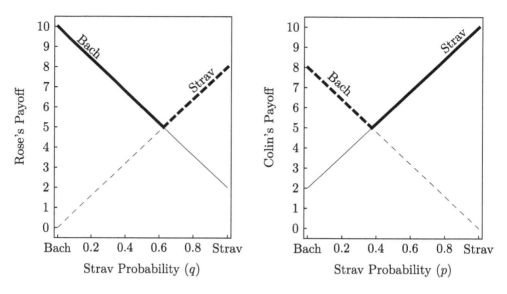

FIGURE A.2 Battle of the Sexes Player Payoff vs. Opponent's Mixed Strategy.

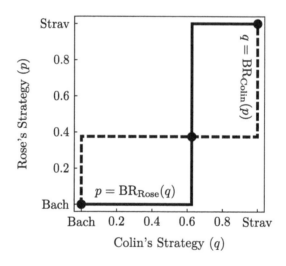

FIGURE A.3 Battle of the Sexes Best Response Functions.

3.36 The Nash equilibria are (Hawk, Dove), (Dove, Hawk), and (s^*, s^*) where $s^* = (14/29)(\text{Hawk}) + (15/29)(\text{Dove})$. The only evolutionarily stable strategy is s^*. Indeed, (s^*, s^*) is the only Nash equilibrium using the same strategy in the strategy profile. Also for any strategy $s = (1 - \lambda)\text{Hawk} + \lambda\text{Dove}$,

$$u_1(s^*, s) - u_1(s, s) = (\frac{14}{29} - (1 - \lambda))(1 - \lambda)(-30)$$
$$+ (\frac{14}{29} - (1 - \lambda))\lambda(40)$$
$$+ (\frac{15}{29} - \lambda)(1 - \lambda)(0)$$

$$+ (\frac{15}{29} - \lambda)\lambda(12)$$
$$= \frac{2}{29}(29\lambda - 15)^2 > 0$$

if $\lambda \neq 15/29$, which shows that s^* is an evolutionarily stable strategy.

3.40 In the Nash equilibrium, $Q_1 = Q_2 = (1/4)a/(1 + bc)$.

4.2 Egalitarian and Raiffa are equal if each player's aspirational payoff is the same amount above the disagreement payoff. All three are equal when the feasible payoff pair region F is symmetric about the line $d + t(1, 1)$.

4.4a. By using these points as payoffs, we are assuming self-interest, additivity, and proportionality. The use of point weights suggest an underlying ratio scale. Answers will vary depending upon one's opinion about how closely these assumptions are met. Since the workers assigned 40 points to having a large wage increase and assigned 20 points to paying a smaller share of the insurance premium, their payoff is $40 + 20 = 60$. Since the owner assigned 50 points to the ability to let workers go, their payoff is 50. In Table A.4, an outcome is represented as (the issues won by workers, the issues won by owners) where each issue is abbreviated with its first letter.

TABLE A.4 Outcomes and Payoffs to Workers and Owners

Outcome	Workers Payoff	Owners Payoff
(WIS,)	100	0
(WI,S)	60	50
(WS,I)	80	40
(IS,W)	60	10
(W,IS)	40	90
(I,WS)	20	60
(S,WI)	40	50
(,WIS)	0	100

4.4b. See Figure A.4. If the two sides maintain the terms of their expiring contract, then the workers will win on the security issue while the owners will win on the wage and insurance issues, leading to the $(40, 50)$ payoff pair.

4.4(c.)i The payoff pair $(40, 50)$ is obtained with (S, WI). Thus, the workers maintain job security, while the owners give small raises and increase the worker share of the insurance premium.

4.4(c.)ii The payoff pair $(20, 95) = (1/2)(0, 100) + (1/2)(40, 90)$ is obtained with $1/2(,\text{WIS})$ $+(1/2)(\text{W,IS})$. Thus, the workers receive a $(1/2)(2\%) + (1/2)(5\%) = 3.5\%$ raise, while the owners win on both insurance and job security issues.

The payoff pair $(84, 32) = (4/5)(80, 40) + (1/5)(100, 0)$ is obtained with $4/5(\text{WS,I})$ $+(1/5)(\text{WIS},)$. Thus, the workers win on the wage and job security issues and their share of the insurance premium increases to $(4/5)(30\%) + (1/5)(20\%) = 28\%$.

4.4(c.)iii The payoff pair $(84, 32) = (4/5)(80, 40) + (1/5)(100, 0)$ is obtained with $4/5(\text{WS,I})$ $+(1/5)(\text{WIS},)$. Thus, the workers win on the wage and job security issues and their share of the insurance premium increases to $(4/5)(30\%) + (1/5)(20\%) = 28\%$.

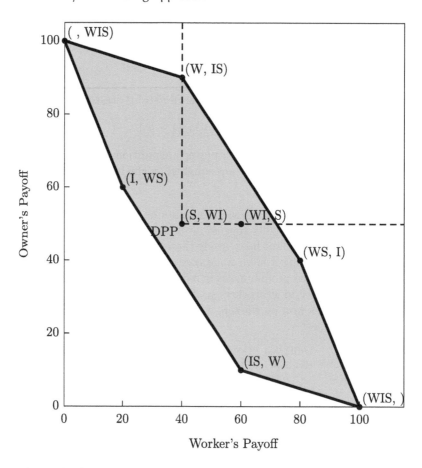

FIGURE A.4   Acme Industrial Feasible Payoff Pairs.

4.4d. The egalitarian payoff pair is the solution of

$$(x, y) = (40, 50) + t(1, 1) = (1 - s)(40, 90) + s(80, 40)$$

yielding

$$(x, y) = (520/9, 610/9) \approx (57.78, 67.78)$$
$$s = 4/9 \approx 0.44$$
$$t = 160/9 \approx 17.78.$$

The corresponding outcome has the workers obtaining the high wage increase, but paying the larger share of the insurance premium. There is a compromise on the security issue significantly favoring the owners.

4.4e. The aspiration payoff for the workers is the x solution to

$$(x, y) = (40, 50) + t(1, 0) = (1 - s)(40, 90) + s(80, 40),$$

and the aspiration payoff for the owners is the y solution to

$$(x, y) = (40, 50) + t(0, 1) = (1 - s)(40, 90) + s(80, 40).$$

Hence, the aspiration payoff pair is

$$a = (72, 90).$$

The Raiffa payoff pair is the solution of

$$(x, y) = (1 - t)(40, 50) + t(72, 90) = (1 - s)(40, 90) + s(80, 40)$$

yielding

$$(x, y) = (56, 70)$$
$$s = 2/5$$
$$t = 1/2.$$

The corresponding outcome has the workers obtaining a 5% raise, paying the increased premium, and obtaining a slightly better compromise on job security than with the egalitarian solution.

4.4f. The Nash payoff pair is the solution of

$$max(x - 40)(y - 50) s.t. (x, y) = (1 - s)(40, 90) + s(80, 40)$$

yielding

$$(x, y) = (56, 70)$$
$$s = 2/5.$$

The corresponding outcome is the same as for the Raiffa solution.

4.9a. The right-most boundary of the feasible payoff pairs is of the form $(1-s)(5, 7)+s(10, 0)$ for $0 \le s \le 1$, and the egalitarian payoff pair is of the form $(\frac{25}{7} + t, 5 + t)$ for some $t \ge 0$. This yields the equations

$$5(1 - s) + 10s = \frac{25}{7} + t$$
$$7(1 - s) + 0s = 5 + t,$$

which have the solution $s = 1/21 \approx 0.048$ and $t = 5/3 \approx 1.67$. Hence, the payoff to Rose is $25/7 + 5/3 = 110/21 \approx 5.24$, the payoff to Colin is $5 + 5/3 = 20/3 \approx 6.67$, and an implementing correlated strategy is $(20/21)((\text{Tails,Tails})) + (1/21)((\text{Heads,Heads}))$.

4.9b. The top-most boundary of the feasible region lies on the line $y = 10 - \frac{3}{5}x$, and the intersection of this line with Rose's disagreement payoff $x = 25/7$ yields Colin's aspiration payoff $10 - (3/5)(25/7) = 55/7 \approx 7.86$. The right-most boundary of the feasible region lies on the line $x = 10 - (5/7)(y)$, and the intersection of this line with Colin's disagreement payoff $y = 5$ yields Rose's aspiration payoff $10 - (5/7)(5) = 45/7 \approx 6.43$. The right-most boundary of the feasible payoff pairs is of the form $(1 - s)(5, 7) + s(10, 0)$ for $0 \le s \le 1$, and the Raiffa payoff pair is of the form $(1 - t)(25/7, 5) + t(45/7, 55/7)$ for some $0 \le t \le 1$. This yields the equations

$$5(1 - s) + 10s = \frac{25}{7}(1 - t) + \frac{45}{7}t$$
$$7(1 - s) + 0s = 5(1 - t) + \frac{55}{7}t,$$

which have the solution $s = 1/21 \approx 0.048$ and $t = 5/3 \approx 1.67$. Hence, the payoff to Rose is $25/7 + 5/3 = 110/21 \approx 5.24$, the payoff to Colin is $5 + 5/3 = 20/3 \approx 6.67$, and an implementing correlated strategy is $(20/21)(Tails, Tails) + (1/21)(Heads, Heads)$.

4.14 This payoff pair is not rational because Beth receives less than she could obtain without making an agreement with Ann. Hence, Beth would not agree to this.

5.2a. See Table A.5.

TABLE A.5 Two-Two-Two Outcomes
and Payoffs

Terminal history h	$u_1(h)$	$u_2(h)$
D	1	0
C, F	3	1
C, E, H	0	0
C, E, G	1	2

5.2b. See Table A.6.

TABLE A.6 Two-Two-Two Player
Function

Subhistory h	Acting Player $P(h)$
$\emptyset$	P1
C	P2
C, E	P1

5.2c. $((D, G), E)$ resulting in the payoffs $(2, 0)$.

5.2d. See Table A.7. Note that we list both (D, H) and (D, G) as strategies because the

TABLE A.7 Two-Two-Two
Strategic Game

(P1, P2)	F	E
(D, H)	(2, $\boxed{0}$)	($\boxed{2}$, $\boxed{0}$)
(D, G)	(2, $\boxed{0}$)	($\boxed{2}$, $\boxed{0}$)
(C, H)	($\boxed{3}$, $\boxed{1}$)	(0, 0)
(C, G)	($\boxed{3}$, 1)	(1, $\boxed{2}$)

definition of strategy requires us to identify an action at each decision point regardless of whether that decision point can be reached.

5.2e. $((D, H), E)$, $((D, G), E)$, and $((C, H), F)$.

5.2f. In $((D, H), E)$ and $((C, H), F)$, Player 1 is willing to choose H instead of G resulting in a lower payoff if Player 1 is able to make the choice. In $((D, H), E)$, Player 2 is willing to choose E instead of F resulting in a lower payoff if Player 2 actually is able to make the choice. Choices such as these are often called "incredible threats," and canonical game theory literature has further information.

5.5 Both might have preference ordering C,A,B: Don't publish so you don't have to write an opinion, but if you write, do so with the least agitation.

5.9a. R can now take at most two steps at once. S can make up to three steps at once.

5.11a. Figure A.5 shows that the backward induction solution is (In, Withdraw).

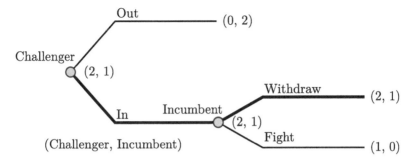

FIGURE A.5 Exercise 5.11(a) Game Tree.

5.12 Table A.8 shows one of the three strategic games with pure-strategy best response payoffs boxed.

TABLE A.8 Primary Strategic Game Alternate (a)

(Challenger, Incumbent)	Withdraw	Fight
In	([2], [1])	([1], 0)
Out	(0, [2])	(0, [2])

5.13 Briefly, at each of a player's action nodes, there is a subgame. The BI solution provided the best response at that node, given the other players' subsequent actions in the subgame. Therefore the player's proposed action is regret-free.

5.20a. Several solutions are possible. One solution is the trees $T_1^{(0)} = \{s1, s3, 15, 23, 45, 5t\}$ and $T_2^{(0)} = \{s2, 13, 14, 24, 35, 4t\}$ as shown in Figure A.6.

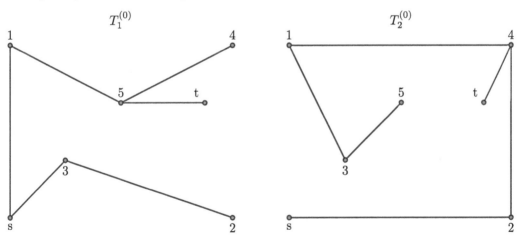

FIGURE A.6 Two Disjoint Trees.

5.20b. Assuming the trees from part (a), edge $s3$ is in tree T_1, and if added to T_2 completes the cycle $s3, s2, 24, 14, 13$. Thus we replace $s3$ in T_1 with $s2$ and mark it claimed in both trees as shown in Figure A.7.

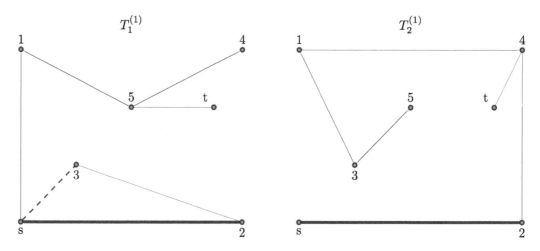

FIGURE A.7 The Two Disjoint Trees after the First Moves.

5.20c. Assuming the trees from part (a), edge $t4$ is in the updated tree T_2, and if added to T_1 completes the cycle $t4, 45, 5t$. Thus we replace $t4$ in T_2 with $5t$ and mark it claimed in both trees as shown in Figure A.8.

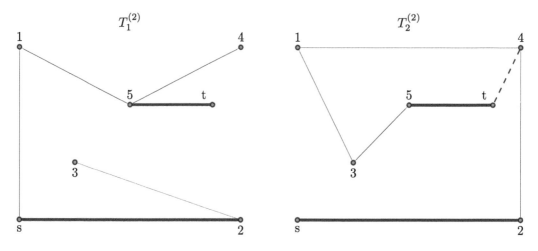

FIGURE A.8 The Two Disjoint Trees after the Second Moves.

5.20d. There are many ways to finish this exercise, even with the given initial trees.

5.26 The managers may decide to take higher-risk decisions with the intent of increasing their bonus, knowing that they have no personal cost if the decision turns out badly for the company.

6.2 For the FoodPro game, Mark has the three pure strategies

$$s_{\text{Mark}}^1(\text{Mark1}) = \text{No},$$
$$s_{\text{Mark}}^2(\text{Mark1}) = \text{Lo, and}$$
$$s_{\text{Mark}}^3(\text{Mark1}) = \text{Hi},$$

and Ben has the four pure strategies

$$s_{\text{Ben}}^1(\text{Ben1}) = \text{No and } s_{\text{Ben}}^1(\text{Ben2}) = \text{No,}$$

$$s_{\text{Ben}}^2(\text{Ben1}) = \text{No and } s_{\text{Ben}}^2(\text{Ben2}) = \text{Hi,}$$

$$s_{\text{Ben}}^3(\text{Ben1}) = \text{No and } s_{\text{Ben}}^3(\text{Ben2}) = \text{No, and}$$

$$s_{\text{Ben}}^4(\text{Ben1}) = \text{Hi and } s_{\text{Ben}}^4(\text{Ben2}) = \text{Hi.}$$

6.5 The assessment (s'', β'') is sequentially rational because (1) given Ben's strategy $s''(\text{Ben1}) = \text{Hi}$ and $s''(\text{Ben2}) = \text{No}$, Mark's best response at Mark1 is Hi since it yields a payoff of 6 versus payoffs of 2 or 4 obtained by choosing No or Lo, (2) Ben's best response at Ben1 is Hi since it yields a payoff of 4 versus a payoff of 2 by choosing No, and (3) given the belief system $\beta''(\text{Ben2}) = (\text{Hi})$, Ben's best response at Ben2 is No since it yields a payoff of $(0)(1) + (1)(6) = 6$ versus a payoff of $(0)(3) + (1)(5) = 5$ by choosing Hi. It also achieves consistency of beliefs because

$$\frac{\Pr((\text{Hi})|G, s'')}{\Pr((\text{Hi})|G, s'') + \Pr((\text{Lo})|G, s'')} = \frac{1}{1+0} = 1 = \beta''((\text{Hi})|\text{Ben2}).$$

The assessment (s''', β''') is sequentially rational because (1) given Ben's strategy $s'''(\text{Ben1}) = \text{Hi}$ and $s'''(\text{Ben2}) = (1/2)(\text{No}) + (1/2)(\text{Hi})$, Mark's best response at Mark1 is any probabilistic mixture of Lo and Hi since they each yield a payoff of $(1/2)(4)+(1/2)(5) = (1/2)(6) + (1/2)(3) = 4.5$ versus a payoff of 2 obtained by choosing No, (2) Ben's best response at Ben1 is Hi since it yields a payoff of 4 versus a payoff of 2 by choosing No, and (3) given the belief system $\beta'''(\text{Ben2}) = (1/3)((\text{Lo})) + (2/3)((\text{Hi}))$, Ben's best response at Ben2 is any probabilistic mixture of No and Hi since they each yield a payoff of $(1/3)(1) + (2/3)(6) = (1/3)(3) + (2/3)(5) = 13/3$. It also achieves consistency of beliefs because

$$\frac{\Pr((\text{Lo})|G, s''')}{\Pr((\text{Lo})|G, s''') + \Pr((\text{Hi})|G, s''')} = \frac{1/3}{1/3 + 2/3} = \frac{1}{3} = \beta'''((\text{Lo})|\text{Ben2}).$$

6.11a. A possible Colin's payoff as a function of the number of tokens won by Colin graph is given in Figure A.9, where the given information is indicated by points. The graph is clearly concave down, which means Colin is risk-adverse with respect to winnings.

6.15a. In the described assessment, the payoffs for the challenger and incumbent are 2 and 4, respectively. If the challenger unilaterally changes to Unready or Ready, her payoff changes to 0 or 1, respectively, both decreases. If the incumbent unilaterally changes to Withdraw, his payoff does not change. Finally, the beliefs are consistent with the strategy since neither node is reached.

6.15c. Consider the arbitrary assessment (s, β) provided by Figure A.10, that is, the challenger's strategy is $s(\text{Challenger}) = (1 - a - b)\text{Out} + a\text{Unready} + b\text{Ready}$, the incumbent's strategy is $s(\text{Incumbent}) = (1 - c)\text{Withdraw} + c\text{Fight}$, and the belief system is $\beta(\text{Incumbent}) = (1 - p)(\text{Unready}) + p(\text{Ready})$.

At his information set, the incumbent's payoffs for adopting the pure strategies Withdraw and Fight are $(1 - p)(2) + p(2) = 2$ and $(1 - p)(3) + p(1) = 3 - 2p$, respectively. Since these payoffs are equal if and only if $p = 1/2$, we consider the three cases $p > 1/2$, $p < 1/2$, and $p = 1/2$.

Case 1. Suppose $p > 1/2$. Then $2 > 3 - 2p$, which implies the incumbent's unique best

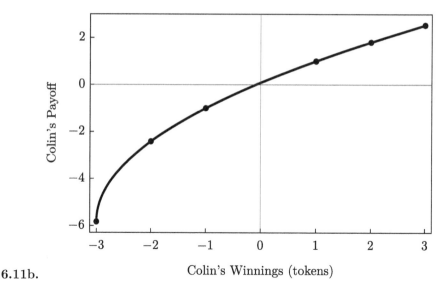

6.11b.

FIGURE A.9 Colin's Utility Function for Modified One-Card Poker.

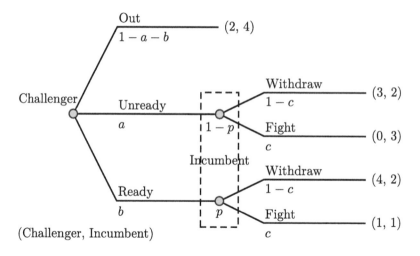

FIGURE A.10 Primary Game Tree General Assessment.

response is Withdraw ($c = 0$), and to this the challenger's unique best response is Ready ($a = 0$ and $b = 1$), and to achieve consistency of beliefs, $p = 1$. This yields a weak sequential equilibrium in which the challenger chooses Ready and the incumbent chooses Withdraw. Since all probabilities are 0 or 1, payoffs need only be interpreted as ordinal utilities.

Case 2. Suppose $p < 1/2$. Then $2 < 3 - 2p$, which implies the incumbent's best response is Fight ($c = 1$), and to this the challenger's unique best response is Out ($a = b = 0$), and achieving consistency of beliefs places no further constraint on p. This yields the weak sequential equilibria in which the challenger chooses Out and the incumbent chooses Fight under the belief that if he needs to choose an action, the challenger has chosen Unready with probability more than $1/2$.

Case 3. Suppose $p = 1/2$. Then $2 = 3 - 2p$, which implies all incumbent strategies are best responses. Now no matter what strategy the incumbent adopts, the challenger's payoff for Ready, $(1 - c)(4) + c(1) = 4 - 3c$, is greater than the challenger's payoff for Unready, $(1 - c)(3) + c(0) = 3 - 3c$, which implies $a = 0$, and now achieving consistency of beliefs demands $b = a = 0$. For Out to be a best response for the challenger, $2 \geq 4 - 3c$, which is equivalent to $c \geq 2/3$. This yields the weak sequential equilibria in which the challenger chooses Out and the incumbent chooses Fight with probability at least $2/3$ under the belief that if he needs to choose an action, the challenger has chosen Unready with probability $1/2$.

6.18 The strategies (D, c, L) and (C, c, R) are both Nash equilibria. The former is not a weak sequential equilibria since player two's action is not optimal given what the other two are doing. The latter strategy is a sequential rationale for players one and two, and because this strategy profile does not reach player three's information set, he is free to assign any belief that he would like. If his belief that player one will select D is no more than $1/3$, then R in preferred over L.

6.22a. At RE2 and RE3, the Romans believe they are at the lower node and so Day (with a payoff of 2) is a best response in comparison with Night (with a payoff of 1). At GT2 and GT3, the Germans find Ambush (with a payoff of 1) is a best response in comparison with Attack (with a payoff of -2). At GT1, the payoffs to Lake and Forest are $(1/2)(1) + (1/2)(2) = 1.5$ and $(1/2)(2) + (1/2)(1) = 1.5$, and so any mixture is a best response. At RE1, the payoffs to Don't, Forest, and Lake are -1, $(1/2)(0) + (1/2)(-2) = -1$, and $(1/2)(-2) + (1/2)(0) = -1$, and so any mixture is a best response.

6.22b. Only the RE1 information set is reached with positive probability, and so the condition is vacuously true. This condition differs from consistency of beliefs because it does not consider every possible subgame.

6.22c. The consistency condition does not hold on the subgames rooted at GT2 and GT3.

6.22d. It is an unreasonable solution because it posits a set of beliefs that are not consistent with best response behavior if GT2 or GT3 were reached with positive probability.

6.26 Figure A.11 is the game tree for the modified scenario. The only change is removal of

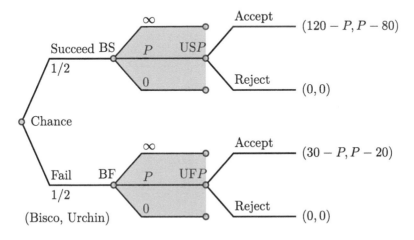

FIGURE A.11 Modified Takeover Game Tree.

the single information set that previously included the nodes labeled BS and BF that are now in separate information sets. Urchin should Accept if Succeed and the offer is at least 80 and if Fail and the offer is at least 20. Bisco should offer 80.

6.31 Suppose Switzerland uses Coordinate on Message. We will show that Coordinate on Message is a best response at each information set for Botswana. Given Switzerland's strategy, the only way for Botswana to change the outcome is to change its first message and what action it chooses. If the first-round messages match, Switzerland will choose H_S, and so Botswana will obtain a positive payoff by choosing H_B and a 0 payoff by choosing L_B; hence, Botswana's best response to matching messages is to choose H_B. Similarly, Botswana's best response to non-matching messages is to choose L_B. Now since Switzerland is sending each message with probability $1/2$, no matter what message Botswana sends, the probability that the messages match is $1/2$, and so Botswana does not change its expected payoff by changing its message; hence, sending the message 1_{B1} with probability $1/2$ is a best response, making Coordinate on Message a best response to Coordinate on Message. The expected payoff to each country when both use Coordinate on Message is $(1/2)(a+1)$, and so the payoff sum is $a+1$, which is the maximum sum for pairs of unaware strategies (as shown in the proof to the previous theorem).

6.34 In order to choose their bids, each bidder uses a deterministic, nonnegative, increasing, and differentiable strategy β that maps possible values of the resource to bids.

7.2 See Table A.9.

TABLE A.9   Exercise 7.2 Actions and Payoffs

Stage	1	2	3	4	5	6	7	Mean
RandomCooperate(0.4)'s Action	D	C	D	D	D	C	C	
OneTitForTwoConsecutiveTats' Action	C	C	C	C	D	D	C	
RandomCooperate(0.4)'s Payoff	5	3	5	5	1	0	3	$22/7 = 3.14$
OneTitForTwoConsecutiveTats' Payoff	0	3	0	0	1	5	3	$12/7 = 1.71$

7.6 When each player uses AllCooperate, the action pair chosen in each stage is CC, hence, the expected payoff per stage to each player is R. If Rose changes her strategy to AllDefect, then the action pair chosen in each stage is DC, hence, the expected payoff per stage to Rose is T. Since $T > R$, AllCooperate is not a best response to AllCooperate.

7.9a. See Figure A.12.

7.9b.

$$
\mathbf{A} = \begin{array}{c} \phantom{\mathbf{A}=} \begin{matrix} CC & CD & DC & DD \end{matrix} \\ \begin{pmatrix} \rho & 0 & 0 & \rho \\ 0 & \rho & \rho & 0 \\ 1-\rho & 0 & 0 & 1-\rho \\ 0 & 1-\rho & 1-\rho & 0 \end{pmatrix} \end{array} \begin{matrix} \text{From/To} \\ CC \\ CD \\ DC \\ DD \end{matrix} \;,\quad
\mathbf{x} = \begin{array}{c} \begin{matrix} \emptyset \end{matrix} \\ \begin{pmatrix} \rho \\ 0 \\ 1-\rho \\ 0 \end{pmatrix} \end{array} \begin{matrix} \text{From/To} \\ CC \\ CD \\ DC \\ DD \end{matrix}
$$

7.9c.

$$
\mathbf{x}^k = \begin{pmatrix} \rho^2 \\ \rho(1-\rho) \\ (1-\rho)\rho \\ (1-\rho)^2 \end{pmatrix}, \text{ for } k = 2, 3, \ldots
$$

7.9d. Rose's expected payoff per stage is

$$
\frac{\delta\rho(1-\rho)S + \delta(1-\rho)^2 P + \rho(1-\delta\rho)R + (1-\rho)(1-\delta\rho)T}{1+\delta(1-2\rho)}.
$$

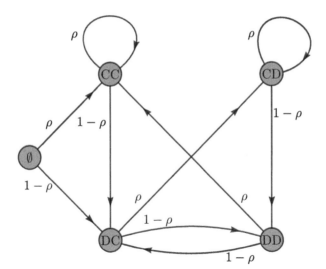

FIGURE A.12   RandomCooperate(ρ) versus Pavlov State Transition Diagram.

Colin's expected payoff per stage is

$$\frac{(1-\rho)(1-\delta\rho)S + \delta(1-\rho)^2 P + \rho(1-\delta\rho)R + \delta\rho(1-\rho)T}{1 + \delta(1-2\rho)}.$$

7.9e. Each player's payoff is 2.44.

7.18 The state transition matrix and initial state probability distribution vector are

$$\mathbf{A} = \begin{matrix} CC & CD & DC & DD & \text{From/To} \\ \begin{pmatrix} 0 & 0 & 0 & 0 \\ 0 & 0 & 0 & 0 \\ 1-\xi & 1-\xi & \xi & \xi \\ \xi & \xi & 1-\xi & 1-\xi \end{pmatrix} & & & & \begin{matrix} CC \\ CD \\ DC \\ DD \end{matrix} \end{matrix} \quad \text{and} \quad \mathbf{x} = \begin{matrix} \emptyset & \text{From/To} \\ \begin{pmatrix} 0 \\ 0 \\ 1 \\ 0 \end{pmatrix} & \begin{matrix} CC \\ CD \\ DC \\ DD \end{matrix} \end{matrix}.$$

The expected payoffs per stage are

$$\mathbf{v} = (1-\delta)\mathbf{\Gamma}(\mathbf{I} - \delta\mathbf{A})^{-1}\mathbf{x} = \begin{matrix} & \text{payoff} & \text{Player} \\ & \begin{pmatrix} T - (T-P)\delta(1-\xi) \\ S + (P-S)\delta(1-\xi) \end{pmatrix} & \begin{matrix} \text{Rose/AllDefect} \\ \text{Colin/TitForTat} \end{matrix} \end{matrix}.$$

7.19 The state transition matrix and initial state probability distribution vector are

$$\mathbf{A} = \begin{matrix} CC & CD & DC & DD & \text{From/To} \\ \begin{pmatrix} \eta(1-\eta) & \eta(1-\eta) & \eta^2 & \eta^2 \\ \eta^2 & \eta^2 & \eta(1-\eta) & \eta(1-\eta) \\ (1-\eta)^2 & (1-\eta)^2 & (1-\eta)\eta & (1-\eta)\eta \\ (1-\eta)\eta & (1-\eta)\eta & (1-\eta)^2 & (1-\eta)^2 \end{pmatrix} & & & & \begin{matrix} CC \\ CD \\ DC \\ DD \end{matrix} \end{matrix} \quad \text{and} \quad \mathbf{x} = \begin{matrix} \emptyset & \text{From/To} \\ \begin{pmatrix} \eta(1-\eta) \\ \eta^2 \\ (1-\eta)^2 \\ (1-\eta)\eta \end{pmatrix} & \begin{matrix} CC \\ CD \\ DC \\ DD \end{matrix} \end{matrix}.$$

The expected payoffs per stage as δ approaches 1 and as η approaches 0 are

$$\mathbf{v} = \lim_{\delta \to 1} \lim_{\eta \to 0} (1 - \delta)\mathbf{\Gamma}(\mathbf{I} - \delta\mathbf{A})^{-1}\mathbf{x} = \begin{pmatrix} P \\ P \end{pmatrix} \begin{matrix} \text{payoff} \\ \text{Rose/AllDefect} \\ \text{Colin/TitForTat} \end{matrix} .$$

7.20 The state transition matrix and initial state probability distribution vector are

$$\mathbf{A} = \begin{pmatrix} 1 - \xi & 1 - \xi & \xi & \xi \\ \xi & \xi & 1 - \xi & 1 - \xi \\ 0 & 0 & 0 & 0 \\ 0 & 0 & 0 & 0 \end{pmatrix} \begin{matrix} CC & CD & DC & DD & \text{From/To} \\ CC \\ CD \\ DC \\ DD \end{matrix} \quad \text{and} \quad \mathbf{x} = \begin{pmatrix} 1 \\ 0 \\ 0 \\ 0 \end{pmatrix} \begin{matrix} \emptyset & \text{From/To} \\ CC \\ CD \\ DC \\ DD \end{matrix} .$$

The expected payoffs per stage are

$$\mathbf{v} = (1 - \delta)\mathbf{\Gamma}(\mathbf{I} - \delta\mathbf{A})^{-1}\mathbf{x} = \begin{pmatrix} R - (R - S)\delta\xi \\ R + (T - R)\delta\xi \end{pmatrix} \begin{matrix} \text{payoff} \\ \text{Rose/AllCooperate} \\ \text{Colin/TitForTat} \end{matrix} .$$

7.24 The replicator evolution mechanism is defined by $\mathbf{F}_i(\mathbf{p}) = p_i v_i/\bar{v} = p_i(\mathbf{Vp})_i/\mathbf{p}^T\mathbf{Vp}$. If $p_i = 0$, then $\mathbf{F}_i(\mathbf{p}) = 0v_i/\bar{v} = 0$. Thus, the replicator evolution mechanism is mutation free. If $p_i > 0$, $p_j > 0$, and $v_i \geq v_j$, then $\mathbf{F}_i(\mathbf{p})/p_i = v_i/\bar{v} \geq v_j/\bar{v} = \mathbf{F}_j(\mathbf{p})/p_j$. Thus, the replicator evolution mechanism is payoff monotone.

7.25 Mathematica commands to find the fixed strategies:

```
V = {{1,1.436,3.10627},{0.891,2.91315,2.85319},{0.473433,2.86506,2.95948}};
Solve[{p (V.p - p.V.p) == 0, Total[p] == 1, p >= {0, 0, 0}}, p]
```

To check whether a strategy distribution $\mathbf{p}$ is Nash, compute $\mathbf{Vp}$ and check whether all entries are no larger than for those for which $p_i > 0$. This was false in only one case: $\mathbf{V}(0,0,1) = (3.11, 2.85, 2.89)$ where we see that $3.11 > 2.89$.

Mathematica commands to determine dynamic stability at $(1,0,0)$:

```
q = {q1, q2, q3};
J = D[(V.q)q/(q.V.q), {q}] /. {q1->1, q2->0, q3->0};
Map[Abs, Eigenvalues[J]]
```

For this case, all three eigenvalues are less than 1, and so $(1,0,0)$ is dynamically stable. The same commands with appropriate number changes in the second command can be used to determine the dynamic stability at the other five fixed strategy distributions. Except for $(0,1,0)$, each has at least one absolute value of an eigenvalue greater than 1.

Since $\mathbf{V}(1,0,0) = (1, 0.891, 0.473)$, we can see that $\mathbf{q}\mathbf{V}(1,0,0) < (1,0,0)\mathbf{V}(1,0,0)$ for all strategy distributions $\mathbf{q} \neq (1,0,0)$, hence, $(1,0,0)$ is evolutionarily stable. Since $\mathbf{V}(0,1,0) = (1.436, 2.913, 2.866)$, we can see that $\mathbf{q}\mathbf{V}(0,1,0) < (0,1,0)\mathbf{V}(1,0,0)$ for all strategy distributions $\mathbf{q} \neq (0,1,0)$, hence, $(0,1,0)$ is evolutionarily stable. For $\mathbf{p} = (0, 0.689, 0.311)$ and $\mathbf{q} = (0,1,0)$, we obtain $\mathbf{q}^T\mathbf{Vp} = \mathbf{p}^T\mathbf{Vp}$ but $\mathbf{p}^T\mathbf{Vq} < \mathbf{q}^T\mathbf{Vq}$, and so $(0, 0.689, 0.311)$ is not evolutionarily stable. Identical conclusions can be drawn for $\mathbf{p} = (0.931, 0.069, 0)$ and for $\mathbf{p} = (0.164, 0.133, 0.703)$, using $\mathbf{q} = (0,1,0)$ again in both cases.

7.26a. Hint: think of a tie as midway between losing and winning.

7.26b. The players are the two people, who we will call Rose and Colin. The strategies are Rock, Paper, and Scissors. The payoffs are given in the symmetric game payoff matrix

$$\mathbf{V} = \begin{array}{c} \\ \text{Rock} \\ \text{Paper} \\ \text{Scissors} \end{array} \begin{array}{ccc} \text{Rock} & \text{Paper} & \text{Scissors} \\ \left(\begin{array}{ccc} a & 0 & 2 \\ 2 & a & 0 \\ 0 & 2 & a \end{array} \right). \end{array}$$

Identify the best responses to each pure strategy to show that there is no pure-strategy Nash equilibrium.

7.26h. First, there are replicator evolution systems for which there is no evolutionarily stable strategy distribution. Second, there are replicator evolution systems for which there is no dynamically stable strategy distribution. Third, an evolutionarily stable strategy distribution need not be dynamically stable. Fourth, some evolutions can be periodic.

7.29d. First, there are games for which there is no evolutionarily stable strategy distribution. Second, a dynamically stable strategy distribution need not be evolutionarily stable. Third, the dynamic stability of a strategy distribution depends upon the payoffs in ways that are not scale invariant.

8.1a. Let E, F, and R be abbreviations for the electricity, flood control, and recreation facilities purposes. We think of these purposes as the players. The worth for each coalition is given in the table.

Coalition	EFR	EF	ER	FR	E	F	R
Worth	120	100	80	60	0	0	0

8.1b. The computations are shown in the following table.

Order	Marginal Contribution		
	E	F	R
EFR	0	100	20
ERF	0	40	80
FER	100	0	20
FRE	60	0	60
REF	80	40	0
RFE	60	60	0
φ_i	50	40	30

8.1c. The computations are shown in the following table.

Coalition	Worths								
EFR	60	+	80	+	100	−	120	=	120
EF	0	+	0	+	100	−	0	=	100
ER	0	+	80	+	0	−	0	=	80
FR	60	+	0	+	0	−	0	=	60
other	0	+	0	+	0	−	0	=	0
Player	Allocations								
E	0	+	40	+	50	−	40	=	50
F	30	+	0	+	50	−	40	=	40
R	30	+	40	+	0	−	40	=	30

8.4a. Starting from the Shapley allocation, we shifted 10 from R to E. The calculations are shown in the following table.

Coalition	Worth	Excess With	
		$(50, 40, 30)$	$(60, 40, 20)$
EF	100	-10	0
ER	80	0	0
FR	60	10	0
E	0	50	60
F	0	40	40
R	0	30	20

If (e, f, r) had a minimum excess greater than 0, then

$$e + f - 100 > 0$$
$$e + r - 80 > 0$$
$$f + r - 60 > 0.$$

Adding the three inequalities together and using the efficient property yields

$$2(e + f + r) - 240 \quad > \quad 0$$
$$0 \quad > \quad 0.$$

This contradiction verifies that the minimum excess cannot be increased. For coalitions EF, ER, and FR to have a 0 excess implies

$$e + f - 100 = 0$$
$$e + r - 80 = 0$$
$$f + r - 60 = 0$$

which implies that $(e, f, r) = (60, 40, 20)$ showing that no more improvements can be made.

8.4b. The reduced game on one of the pairs of players and the corresponding nucleoli are shown in the table.

S	$w^{EF}(S)$
EF	$60 + 40 = 100$
E	$\max\{0, 80 - 20\} = 60$
F	$\max\{0, 60 - 20\} = 40$
i	ν_i
E	60
F	40

8.10 Suppose x is an allocation that is not player rational in the coalition game (N, w). Then $x_i < w(\{i\})$ for some player i. Notice that for each coalition S not containing i, the difference $e(x, S) - e(x, S \cup \{i\}) = w(S \cup \{i\}) - w(S) - x_i \geq w(\{i\}) - x_i$ (by superadditivity) > 0, and so $e(x, S) > e(x, S \cup \{i\})$. Therefore, changing the allocation by taking away small amounts from other players to give to player i will increase the smallest excess. Thus, an allocation that is not player rational cannot be the prenucleolus.

For the game $w(123) = 5$, $w(12) = w(13) = 6$, $w(1) = 3$, and $w(23) = w(2) = w(3) = 1$, the prenucleolus is $x = (5, 0, 0)$ because $e(x, 12) = e(x, 13) = e(x, 23) = e(x, 2) = e(x, 3) = -1 < 2 = e(x, 1)$ and it is impossible to increase the excess of any the 12, 13, 23, 2, or 3 coalitions without decreasing another one of them. The nucleolus is $(3, 1, 1)$ because it is the only efficient and player-rational allocation.

8.20a. If x is an allocation in which $x_j < x_k$ for some $j, k \in \{2, 3, 4, 5\}$, then (Y, y) where $Y = \{1, j\}$, $y_1 = 1 - x_k + (x_k - x_j)/2$, and $y_j = x_j + (x_k - x_j)/2$ is an objection by j against k at x to which there is no counter-objection, because the strongest counter-objection (Z, z) would have $Z = \{1, k\}$, $z_1 + z_k = 1$, $z_1 \geq y_1$, and $z_k \geq x_k$, which implies $1 = z_1 + z_k \geq 1 + (x_k - x_j)/2 > 1$, a contradiction. Thus, if x is a bargaining allocation, then $x = (a, b, b, b, b)$ where $a + 4b = 1$. Since the coalition structure is $\{1, 2345\}$, it follows that the unique bargaining allocation is $x = (0, 1/4, 1/4, 1/4, 1/4)$.

8.25a. A correct answer is any weighted voting system $[q; v_1, v_2, v_3, v_4, v_5]$ that satisfies the following inequalities.

$$
\begin{aligned}
v_1 + v_2 & & & & & \geq q \\
v_1 & + v_3 & & & & \geq q \\
v_1 & & + v_4 & & & \geq q \\
v_1 & & & + v_5 & & \geq q \\
& v_2 + v_3 & + v_4 & + v_5 & & \geq q \\
v_1 & & & & & < q \\
& v_2 + v_3 & + v_4 & & & < q \\
& v_2 + v_3 & & + v_5 & & < q \\
& v_2 & + v_4 & + v_5 & & < q \\
& v_3 & + v_4 & + v_5 & & < q
\end{aligned}
$$

One especially simple answer is $[4; 3, 1, 1, 1, 1]$.

8.27b. $\varphi_1 = (n - r + 1)/n$ and $\varphi_i = (r - 1)/(n(n - 1))$ for all players $i \neq 1$.

8.27c. $\eta_1 = \sum_{k=r-1}^{n-1} \binom{n-1}{k}$ and $\eta_i = \binom{n-2}{r-2}$ for all players $i \neq 1$.

8.27d. $\nu = (1, 0, \dots, 0)$ (why?).

9.1a. Answers may vary, but it would be reasonable to keep the set as a single item rather than 24 items (the individual utensils), 8 items (the sets of a single knife, fork, and spoon), or 3 items (8 knives, 8 forks, 8 spoons). The set is likely to be worth more to a person than the sum of the worths of subsets of the utensils.

9.1b. Answers may vary, but it would be reasonable to combine all three into the money held by the estate. This is the special "item" that is considered finely divisible.

9.3 There are six methods that need to be considered. We consider two here.

1. First-price auction allocates the items to those who value them the most: the cabin to Bob, the car to Carol, and the silverware to Carol. The agents winning the items pay the corresponding valuations to the estate: Bob pays $66,000 and Carol pays $12,000+ $27,000 = \$39,000$. The estate now has $6,000+$66,000+$39,000 = \$111,000$, which is then divided equally among the agents, which results in the following net changes in money: $-\$66,000 + \$111,000/3 = -\$29,000$ for Bob, $-\$39,000 + \$111,000/3 = -\$2,000$ for Carol, and $\$111,000/3 = \$37,000$ for Doug. Agent payoffs are calculated from Equation 9.4:

$$
\begin{aligned}
u_{\text{Bob}}(x, m) &= \$66,000 & - \$29,000 &= \$37,000 \\
u_{\text{Carol}}(x, m) &= \$12,000 + \$27,000 - & \$2,000 &= \$37,000 \\
u_{\text{Doug}}(x, m) &= \$0 & + \$37,000 &= \$37,000.
\end{aligned}
$$

Agent shares are calculated from Equation 9.6:

$$
\begin{aligned}
s_{\text{Bob}}(x, m) &= \$37,000/\ \$99,000 \approx 37.4\% \\
s_{\text{Carol}}(x, m) &= \$37,000/\$105,000 \approx 35.2\% \\
s_{\text{Doug}}(x, m) &= \$37,000/\ \$66,000 \approx 56.1\%.
\end{aligned}
$$

2. Knaster allocates the items to those who value them the most: the cabin to Bob, the car to Carol, and the silverware to Carol. Money is then distributed so that each player receives the same value (λ) above a one-third share of the estate, that is,

$$\begin{aligned}
\$66,000 \quad + \quad m_{\text{Bob}} \quad &= \quad \$99,000/3 \quad + \quad \lambda \\
\$39,000 \quad + \quad m_{\text{Carol}} \quad &= \quad \$105,000/3 \quad + \quad \lambda \\
m_{\text{Doug}} \quad &= \quad \$66,000/3 \quad + \quad \lambda.
\end{aligned}$$

Summing the above equations and using

$$m_{\text{Bob}} + m_{\text{Carol}} + m_{\text{Doug}} = \$6,000,$$

we obtain

$$\$105,000 + \$6,000 = \$90,000 + 3\lambda$$

which implies

$$\lambda = \$7,000$$
$$m_{\text{Bob}} = -\$33,000 + \$7,000 = -\$26,000$$
$$m_{\text{Carol}} = -\$4,000 + \$7,000 = \$3,000$$
$$m_{\text{Doug}} = \$22,000 + \$7,000 = \$29,000.$$

Agent payoffs and shares are calculated from Equations 9.4 and 9.6.

9.4b. A possible solution is $L = \{$primary custody, holiday visitation, house, barn, corn field, central offices, franchises$\}$ and $M = D = -\$50,000$. This shows how there may be separate definable parts to child custody, a piece of property, or a business. Custody of a child can be split by the amount of time spent with each parent or the amount of financial responsibility each parent bears. If it is sufficiently large, property can be meaningfully subdivided into separate lots. If a business is incorporated, ownership becomes finely divisible.

9.6c. The allocation (x, m) satisfies

$$x = (1, 0, \ldots, 0)$$

and

$$m = \frac{1}{n}(-(n-1), 1, \ldots, 1)v_2.$$

Agent j's payoff is

$$u_j(x, m) = \begin{cases} v_1 - v_2 + \frac{1}{n}v_2, & \text{if } j = 1 \\ \frac{1}{n}v_2, & \text{if } j > 1 \end{cases}$$

and share is

$$s_j(x, m) = \begin{cases} 1 - (1 - \frac{1}{n})\frac{v_2}{v_1}, & \text{if } j = 1 \\ \frac{1}{n}\frac{v_2}{v_j}, & \text{if } j > 1. \end{cases}$$

9.9a. Bob envies Carol, and Doug envies Carol.

Bibliography

[1] A. Allais. Le comportement de l'homme rationnel devant le risque: Critique des postulats et axiomes de l'ecole americaine. *Econometrica*, 21:503–546, 1953.

[2] L. Allen. Games bargaining: A proposed application of the theory of games to collective bargaining. *Yale Law Journal*, 65:660–693, 1956.

[3] J. Andreoni and J. Miller. Giving according to GARP: An experimental study of rationality and altruism. *Econometrica*, 70:737–753, 2002.

[4] R. J. Aumann and L. S. Shapley. Long-term competition: A game-theoretic analysis. In N. Megiddo, editor, *Essays in Game Theory in Honor of Michael Maschler*, pages 1–15. Springer, 1994.

[5] R. Axelrod. *The Evolution of Cooperation*. Basic Books, 1985.

[6] J. Banks. *Signaling Games in Political Science*. Harwood Academic Publishers, 1991.

[7] J. F. III Banzhaf. Weighted voting doesn't work: A mathematical analysis. *Rutgers Law Review*, 19:317–343, 1965.

[8] E. Barron. *Game Theory an Introduction*. John Wiley and Sons, Hoboken NJ, 2nd edition, 2013.

[9] M.R. Baye, D. Kovenock, and C.G. de Vries. The all-pay auction with complete information. *Economic Theory*, 8:291–305, 1996.

[10] H. Bentham and J. Bowring (ed.). *The Works of Jeremy Bentham*. Nabu Press, 1838-1843; Reprinted 2012.

[11] E. Berlekamp, J. Conway, and R. Guy. *Winning Ways for Your Mathematical Plays*. A.K. Peters, 2nd edition, 2001.

[12] H. Bierman and L. Fernandez. *Game Theory with Economic Applications, 2nd ed.* Addison Wesley, 1998.

[13] D.T. Bishop, C. Cannings, and M. Smith. The war of attrition with random rewards. *Journal of Theoretical Biology*, 74:377–389, 1978.

[14] M. Block (Host). Aerial Skiing. All Things Considered [Radio Broadcast]. National Public Radio. 2014.

[15] S. Brams and A. Taylor. *Fair Division: From Cake-Cutting to Dispute Resolution*. Cambridge University Press, Cambridge, UK, 1996.

[16] D. Brooks. The Cuomo College fiasco. *New York Times*, April 14, 2017.

[17] M. Broom and A. Rychtar. *Game-Theoretical Models in Biology*. CRC Press, Boca Raton, FL, 2013.

[18] C. Browne. *Connection Games: Variations on a Theme*. A. K. Peters, 2005.

[19] R. Brualdi. Networks and the Shannon switching game. *Delta (Waukesha)*, 4:1–23, 1974.

[20] P. Bump. The great Georgia-Tennessee border war of 2103 is upon us. *The Atlantic*, 2013.

[21] C. Camerer. *Behavioral Game Theory*. Princeton University Press, 2003.

[22] D. Cleveland. Overturning the last stone: The final step in returning precedential status to all opinions. *J. App. Prac. & Process*, 61:?, 2009.

[23] D. Cobb. Florida ruling could have implications in Georgia-Tennessee border dispute. *Times Free Press*, 2017.

[24] G. Codding. *The International Telecommunications Union: An Experiment in International Cooperation*. E.J. Brill, 1952.

[25] B. Cohen. The basketball game that became a stalemate. *The Wall Street Journal*, 2018.

[26] A. Cournot. *Recherches sur les principes mathematiques de la theorie des richeses [translation: Research into the mathematical principles of the theory of wealth, (Bacon, N)]*. McMillan, 1897.

[27] M. Davis and M. Maschler. The kernel of cooperative games. *Naval Research Logistics Quarterly*, 12:223–59, 1965.

[28] R. Dawkins. *The Selfish Gene*. Oxford University Press, Oxford, UK, 1976.

[29] P. Dutta. *Strategy and Games*. The MIT Press, 1999.

[30] Economist Editorial Board. A river runs through it. *The Economist*, 2018.

[31] Economist Editorial Board. France, where the grossly unprepared try for maths degrees. *The Economist*, May 5, 2018.

[32] P. Edelman and J. Chen. The most dangerous justice: The supreme court at the bar of mathematics. *Southern California Law Review*, 70:63:63–111, 1996.

[33] D. Ellsberg. Risk, ambiguity, and the Savage axioms. *Quarterly Journal of Economics*, 75:643–669, 1961.

[34] European Union. Official website of the European Union. `https://europa.eu/european-union/index_en`. [Online; accessed October 2018].

[35] Jeopardy! Fans. J!-archive. `http://j-archive.com`.

[36] D. Felsenthal and M. Machover. *The Measurement of Voting Power: Theory and Practice, Problems and Paradoxes*. Edward Elgar, Cheltenham, 1998.

[37] D. Felsenthal and M. Machover. Voting power measurement: A story of misreinvention. *Soc Choice Welfare*, 25:485–506, 2005.

[38] P. Fishburn. Untitled. Presentation at the Operations Research Society of America meeting, 1987.

[39] D. Friedman and S. Sunder. *Experimental Methods.* Cambridge University Press, 1994.

[40] S. Garfunkel and M. Montgomery, editors. *GAIMME: Guidelines for Assessment and Instruction in Mathematical Modeling Education.* Consortium for Mathematics and Its Applications and Society for Industrial and Applied Mathematics, Philadelphia, 2016.

[41] R. Gillman and D. Housman. *Models of Conflict and Cooperation.* American Mathematical Society, Providence, RI, 2009.

[42] H. Gintis. *Game Theory Evolving: A Problem-Centered Introduction to Modeling Strategic Interaction.* Princeton University Press, Princeton, NJ, 2000.

[43] U. Gneezy and R. Smorodinsky. All-pay auctions—an experimental study. *Journal of Economic Behavior & Organization*, 64:255–275, 2006.

[44] T. Groseclose and J. Milyo. Sincere versus sophosticated voting in Congress: Theory and evidence. *The Journal of Politics*, 72:60–73, 2010.

[45] T. Groseclose and J. Milyo. Sincere versus sophosticated voting when legislators vote sequentially. *Soc. Choice Welf.*, 40:745–751, 2013.

[46] H. Grotte and P. Brooks. Measuring naval presence using blotto games. *International Journal of Game Theory*, 12:225–236, 1983.

[47] J. Hagel and A. Roth. *The Handbook of Experimental Economics.* Princeton University Press, 1995.

[48] O. Haywood. Military decision and game theory. *Journal of Operations Research Society of America*, 2:365–385, 1954.

[49] J. Hofbauer and K. Sigmund. *The Theory of Evolution and Dynamical Systems.* Cambridge University Press, 1988.

[50] E. Hoffman, K. McCabe, K. Shachat, and V.L. Smith. Preferences, property rights and anonymity in bargaining games. *Games and Economic Behavior*, 7:346–80, 1994.

[51] D. Housman. Five player voting games. `http://www2.goshen.edu/~dhousman/gtbook/FivePlayerVotingGames.xls`. Online; accessed December 2018.

[52] D. Housman and L. Clark. Core and monotonic allocation methods. *Internat. J. Game Theory*, 27:611–616, 1998.

[53] A. Hudson, editor. *Catalysing Ocean Finance, Volume II.* United Nations Development Programme, 2012.

[54] A. Johnston. Free public college: Utopian no more. *The Chronicle of Higher Education*, September 25, 2016.

[55] P. Jones. Jeopardy! Wagering in pre-final tie scenarios. Unpublished manuscript. 2016.

[56] E. Kalai and M. Smorodinsky. Other solutions to Nash's bargaining problem. *Econometrica*, 43:513–518, 1975.

[57] S.N. Katz and M.L. Aronson. *Negotiating to Settlement in Divorce.* Prentice Hall Law & Business, 1987.

[58] B. Keijzer, T. Klos, and Y. Zhang. Finding optimal solutions for voting game design problems. *Journal of Artificial Intelligence Research*, 50:105–140, 2014.

[59] S. Kimbrough. *Agents, Games, and Evolution: Strategies at Work and Play.* Cambridge University Press, 2012.

[60] A. Kreighbaum. Democratic platform spurs excitement of advocates of free community college. *Insider Higher Education*, July 28, 2016.

[61] L. Langley. Romance of the seas: Strange mating habits of the seahorse. *National Geographic*, June 25, 2016.

[62] D. Leech and R. Leech. Computer algorithms for voting power analysis. `http://homepages.warwick.ac.uk/~ecaae/`. Accessed: 2018-06-06.

[63] A. Lehman. A solution of the Shannon switching game. *J. Soc. Indust. App. Math.*, 12:687–725, 1964.

[64] W. Lucas. Measuring power in weighted voting systems. In Steven J. Brams, William F. Lucas, and Philip D. Straffin, Jr., editors, *Political and Related Models*, pages 183–238. Springer-Verlag, New York, 1983.

[65] W. F. Lucas, J. C. Maceli, M. Hilliard, and D. Housman. Reapportionment by weighted voting. *Cornell Operations Research and Industrial Engineering Technical Report 533*, 1982.

[66] R. Mansfield. Strategies for the Shannon switching game. *American Mathematical Monthly*, 103:250–252, 1996.

[67] R. Marks. Repeated games and finite automata. In J. Creedy, J. Eichberger, and J. Borland, editors, *Recent Developments in Game Theory*, pages 43–64. Edward Elgar, 1992.

[68] M. Maschler, B. Peleg, and L. Shapley. Geometric properties of the kernel, nucleolus and related solution concepts. *Mathematics of Operations Research*, 4:303–38, 1979.

[69] M. Maschler, E. Solan, and S. Shmuel Zamir. *Game Theory.* Cambridge University Press, Cambridge, UK, 2013.

[70] J. Maynard Smith and G. Price. The logic of animal conflict. *Nature*, 246:15–18, 1973.

[71] R. McKelvey. Gambit: Software tools for game theory, version 15.1.0. `http://www.gambit-project.org`.

[72] J. Morrow. *Game Theory for Political Scientists.* Princeton University Press, 1994.

[73] J. Morrow. Modeling the forms of international cooperation: Distribution vs information. *International Organization*, 48:387–423, 1994.

[74] R. Nagel. Unravelling in guessing games: An experimental study. *American Economic Review*, 1995.

[75] J. Nash. The bargaining problem. *Econometrica*, 18:155–162, 1950.

[76] J. Nash. Non-cooperative games. *Ann. of Math.*, 54:286–295, 1951.

[77] G. Orshan. The prenucleolus and the reduced game property: Equal treatment replaces anonymity. *Internat. J. Game Theory*, 22:241–248, 1993.

[78] M. Osborne. *Introduction to Game Theory*. Oxford University Press, Oxford, 2004.

[79] I. Palacios-Huerta. Professionals play minimax. *Review of Economic Studies*, 70:395–415, 2003.

[80] I. Palacios-Huerta. *Beautiful Game Theory*. Princeton University Press, 2014.

[81] T. Parker. Allocation of the Tennessee Valley Authority projects. *Transactions of the American Society of Civil Engineers*, 108:174–187, 1943.

[82] W. Pech and P. Swicegood. Trust and trustworthiness: A game theory transcontinental experiment. *International Business & Economics Research Journal*, 2013.

[83] B. Peleg and P. Sudholter. *Introduction to the Theory of Cooperative Games*. Springer-Verlag, 2007.

[84] Associated Press. The AP top 25 poll: Poll methodology. `https://collegefootball.ap.org/poll`. [Online; accessed 10-Sep-2018].

[85] H. Raiffa. Arbitration schemes for generalized two-person games. In H. W. Kuhn and A. W. Tucker, editors, *Arbitration Schemes for Generalized Two-Person Games*, number 28 in Ann. Math. Stud., pages 361–387. Princeton University Press, Princeton, NJ, 1953.

[86] J.S. Ransmeier. *The Tennessee Valley Authority: A Case Study in the Economics of Multiple Purpose Stream Planning*. Vanderbilt University Press, 1942.

[87] J. Rawls. *A Theory of Justice*. Harvard University Press, Cambridge, MA, 1971.

[88] P. Rock. Untitled conversation. Virginia Wesleyan University Dept. of Biology.

[89] T. Saaty. *The Analytic Hierarchy Process: Planning, Priority Setting, and Resource Allocation*. McGraw Hill, New York, 1980.

[90] W. Sandholm. *Population Games and Evolutionary Dynamics*. The MIT Press, 2010.

[91] Jeff E. Schaprio. Gillespie, Northam finally agree on something: Secrecy. *Richmond Times-Dispatch*, October 14 2017.

[92] J. Schatz. The Effects of Religious Affiliation on Social Preferences. Unpublished manuscript. 2013.

[93] J. Schiemann. *Does Torture Work?* Oxford University Press, 2016.

[94] D. Schmeidler. The nucleolus of a characteristic function game. *SIAM J. App. Math.*, 17:1163–1170, 1969.

[95] R. Selten. Reexamination of the perfectness concept for equilibrium points in extensive games. *International Journal of Game Theory*, 4:25–55, 1975.

[96] C. Shannon. Game playing machines. *Journal of the Franklin Institute*, 260:447–453, 1955.

[97] L. Shapley. A value for n-person games. In H.W. Kuhn and A.W. Tucker, editors, *Contributions to the Theory of Games*, volume 2 of *Ann. Math. Stud.*, pages 307–317. Princeton University Press, Princeton, NJ, 1953.

[98] L. Shapley and M. Shubik. A method for evaluating the distribution of power in a committee system. *American Political Science Review*, 48:787–792, 1954.

[99] A. N. Siegel. *Combinatorial Game Theory*. American Mathematical Society, 2013.

[100] K. Sigmund. *The Calculus of Selfishness*. Princeton University Press, 2010.

[101] J.M. Smith. *Evolution and the Theory of Games*. Cambridge university press, 1982.

[102] M. Smith. The theory of games and the evolution of animal conflicts. *Journal of Theoretical Biology*, 47:209–21, 1974.

[103] A.I. Sobolev. The characterization of optimality principles in cooperative games by functional equations. In N.N. Vorobev, editor, *Mathematical Methods in Social Sciences*, pages 94–151. Vilnius, 1975.

[104] S. Sorin. Repeated games with complete information. In R. Aumann and S. Hart, editors, *Handbook of Game Theory with Economic Applications, Vol. 1*, pages 71–108. North-Holland, 1992.

[105] M. Spence. Job market signaling. *Quarterly Journal of Economics*, 87(3):355–374, August 1973.

[106] M. Stein. *How the States Got Their Shapes*. HarperCollins, 2008.

[107] P. Straffin. Homogeneity, independence and power indices. *Public Choice*, 30:107–118, 1977.

[108] P. Straffin. The power of voting blocs: An example. *Mathematics Magazine*, 50:22–24, 1977.

[109] P. Straffin. Power indices in politics. In Steven J. Brams, William F. Lucas, and Philip D. Straffin, Jr., editors, *Political and Related Models*, pages 256–321. Springer-Verlag, New York, 1983.

[110] P. Straffin. *Game Theory and Strategy*. Mathematical Association of America, Washington DC, 1993.

[111] D. Tabler. James Camak botches surveying the ga-tn border. twice. `http://www.appalachianhistory.net/2018/07/james-camak-botches-surveying-gatn.html`, 2015.

[112] S. Tadelis. *Game Theory: An Introduction*. Princeton University Press, 2013.

[113] J. Tarentino. School is expensive, is it worth it? *The Wall Street Journal*, April 14, 2018.

[114] P. Taylor and L. Jonker. Evolutionarily stable strategies and game dynamics. *Mathematical Biosciences*, 40:145–156, 1978.

[115] W. Thomson. Cooperative models of bargaining. In R. Aumann and S. Hart, editors, *Handbook of Game Theory with Economic Applications, Vol. 2*, volume 11 of *Handbooks in Economics*, pages 1237–1284. Elsevier, Amsterdam, Netherlands, 1994.

[116] W. Thomson, editor. *Bargaining and the Theory of Cooperative Games: John Nash and Beyond*. Edward Elgar Publishing, 2010.

[117] J.B. Van Huyck, R.C. Battalio, and Cook J. Adaptive behavior and coordination failure. *Journal of Economic Behavior and Organization*, 1997.

[118] W. Vickrey. Counterspeculation, auctions, and competitive sealed tenders. *The Journal of Finance*, 16:8–37, 1961.

[119] J. von Neumann and O. Morgenstern. *Theory of Games and Economic Behavior*. Princeton University Press, 1944.

[120] H. Von Stackelberg. *Martform und Gleichgewicht*. Julius Springer, 1934.

[121] F. Weissing. Evolutionary stability and dynamic stability in a class of evolutionary normal form games. In R. Selten, editor, *Game Equilibrium Models I: Evolution and Game Dynamics*, pages 29–97. Springer, 1991.

[122] Wikipedia. `https://en.wikipedia.org/wiki/Tic-Tac-Toe`. Online; accessed December 2018.

[123] Wikipedia. `https://en.wikipedia.org/wiki/Hex_(board_game)`. Online; accessed December 2018.

[124] Wolfram—Alpha. `http://www.wolframalpha.com/input/?i=eigenvector+of+ %7B%7B1,1%2F3,4%7D,%7B3,1,5%7D,%7B1%2F4,1%2F5,1%7D%7D`. Online; accessed September 2018.

[125] H. Young. Monotonic solutions of cooperative games. *Internat. J. Game Theory*, 14(2):65–72, 1985.

[126] H. P. Young. *Cost Allocation: Methods, Principles, Application*. North-Holland, 1985.

[127] H. P. Young. *Equity in Theory and Practice*. Princeton University Press, 1994.

[128] E. Zermelo. Uber eine anwendung der mengeniehre auf die theorie des schachspiels. In *Proceedings of the Fifth Congress of Mathematicians*, pages 501–504, Cambridge, 1913. Cambridge University Press.

Index

Acme Industrial, 98
Adams-Onis Treaty, 102
additive, 245, 279
adjacent, 126
AHP, 25
all pay auction, 79
AllCooperate, 235
AllDefect, 208
allocation, 239, 276
Allocation Characterization Theorem, 279
analytic hierarchy process, 25
Another Folk Theorem, 218
AP Ranking, 28
arise from a common prior, 164
Ascending Bid Auction, 62
aspiration payoff pair, 86
assessment, 150
Attrition Behavior, 56
aware, 178

babbling, 178
Backward induction, 108
backward induction outcome, 108
backward induction strategy, 108
Banzhaf Characterization Theorem, 266
bargaining allocation, 254
Basketball Show-down, 73
Battle of the Sexes, 71
Bayesian equilibrium, 166
Bayesian game, 164
behavior strategy, 148
belief system, 149
best response, 41, 150
bilateral bargaining game, 81
binding agreement, 81
Biological Auction, 187
Black Sea Cooperation, 6
Bridg-it, 125
Buying a Home, 10

chance, 164
Chicken, 72
child, 108
Chomp, 142

Choose H, 178
Choose L, 178
coalition game, 239
Coalition Game Impossibility Theorem, 251
coalition monotone, 251
coalition structure, 239
coalitions, 239
coextensive, 127
College Choices, 163
College Education, 174
Columbia River Watershed, 97
combinatorial game, 120
Communicate Dishonestly, 184
Communicate Honestly, 184
complete, 14
complete information, 145
compound lottery, 17
consistency of beliefs, 151
consistent, 250
Continental Divide, 69
continuation probability, 212
continuous, 18
convex, 269
convex hull, 90
Cooperate, 43
cooperate, 206
cooperative games, 4
Coordinate on Message, 180
Corporate Home, 252
correlated strategy, 89
corresponding strategic game, 148
cost, 2
counter-objection, 254
County Fair Prize, 82
Cournot Duopoly, 24, 61
cycle, 126

Defect, 43
defect, 206
disagreement outcome, 81
disagreement payoff pair, 81
Divorce Settlement, 99
dominant, 41
dominates, 41, 92

dynamic game, 103

edges, 105
efficient, 42, 92, 241, 245, 279
Egalitarian Characterization Theorem, 93
egalitarian outcome, 84
egalitarian payoff pair, 84
empty history, 103
envy free, 279
EPA, 240
equal shares method, 278
equal split method, 278
Escalation, 63
evolution, 219
evolutionarily stable strategy, 60
evolutionary game theory, 61
excess, 248
Expected Utility Hypothesis, 18
experimental game theory, 71
extensive game, 145

fair, 89
favorable signal, 177
feasible payoff pairs, 81
Fidelity, 2
Final Jeopardy, 48
Finding Nash Equilibria Theorem, 58
Finding Subgame Perfect Equilibria
        Theorem, 115
finite repeated game, 205
first price auction, 79
First Price Sealed Bid Auction, 202
first-price auction method, 278
flexibility, 2
Flow, 208
Folk Theorem, 214
Follow, 185
Follow the Signal, 181
Follow the Signal Mostly, 183
FoodPro, 146
Free College, 133
free disposal, 95

Gambit, 58
game, 4
        Bayesian game, 164
        coalition game, 239
        extensive game, 145
        finite repeated game, 205
        noisy repeated game, 215
        random repeated game, 212

sequential game, 103
        strategic game, 39
        voting game, 258
game theoretic models, 4
game tree, 108
generalizable, 14
graph, 126
GrimTrigger, 208

Hex, 132
Hike, 198
history, 103

Ice Cream Parlor, 13
implementation error, 215
incident, 126
independent of irrelevant payoff pairs, 95
indifferent, 13
individually monotone, 94
Inheritance, 275
International Collaboration, 175
interval scale, 26
isolated equal shares, 278

Job Offers, 25
Joint Police Force, 268

Knaster's method, 278

Lead, 185
least core, 269
Legislative Gridlock, 267
linear program, 136
losing, 258
lottery, 17

marginal contribution, 241
Matching Coins, 88
mathematical modeling, 1
Mating Behavior, 53
Media Streaming, 104
Merit Raise, 37
minimal winning coalitions, 258
mixed prudential strategy, 78
modeling, 1
monetary split, 281
monetary valuation of the estate, 276
monotonic, 18
moral hazard, 133
MostlyCooperative, 235
mutation-free, 227

Nash Characterization Theorem, 95
Nash equilibrium, 41
Nash Equilibrium Existence Theorem, 56
Nash outcome, 86
Nash payoff pair, 86
Nash's Hex Theorem, 132
National Public Radio, 206
Natural Evolution, 218
negative, 131
neutral, 131
Nim, 123
nodes, 104
noisy repeated game, 215
non-cooperative games, 4
non-terminal histories, 103
Normandy Breakout, 44
nucleolus, 249
null players, 272

objection, 254
Office Coffee, 43
Olympic Aerial Skiing, 6
One Card Poker, 153
OneTitForTwoConsecutiveTats, 234
optimistic resource allocation coalition
    game, 282
ordinal preferences, 14
Ordinal Preferences Characterization
    Theorem, 14
ordinal scale, 26
ordinally equivalent, 16
outcome, 4

parent, 108
Patent Race, 111
path, 126
Pavlov, 208
payoff, 276
payoff equitable, 279
payoff matrix, 45
payoff monotone, 227
payoff pair, 41
payoff sum, 279
payoffs, 40
perception error, 215
perfect information, 145
perfect recall, 146
pessimistic resource allocation coalition
    game, 282
pivotal, 259
player function, 103

player rational, 240, 245
Poison, 141
Political Challenger Take 2, 158
Political Challenger, 112
pooling equilibria, 174
positive, 127
Predicting Scaled Averages, 70
preferences
    complete, 14
    continuous, 18
    generalizable, 14
    monotonic, 18
    probabilistic, 18
    substitutable, 18
    transitive, 14
prefers, 13
prenucleolus, 249
Prenucleolus Characterization Theorem,
    251
Prisoner's Dilemma, 40
prisoner's dilemma scenario, 43
probabilistic, 18
proportionate, 279
prudential, 42
Purchasing Gasoline, 10
pure strategy, 148

raffle, 29
Raiffa Characterization Theorem, 95
Raiffa outcome, 86
random repeated game, 212
RandomCooperate(ρ), 208
ranked, 15
ratio equivalent, 26
ratio scale, 26
rational, 84, 92
rational behavior, 7
reduced coalition game, 250
Repeated Social Dilemmas, 206
replicator evolution mechanism, 220
replicator with mutations evolution
    mechanism, 231
representation, 259
resource allocation bargaining game, 281
resource allocation Bayesian game, 288
resource allocation problem, 275
resource allocation strategic game, 284
risk, 30
risk adverse, 30
risk loving, 30
risk neutral, 30

Rock County, 273
Rock Paper Scissors, 237
root, 151
rules, 4

scale invariant, 95, 251
scenario
    Acme Industrial, 98
    Adams-Onis Treaty, 102
    AP Ranking, 28
    Ascending Bid Auction, 62
    Attrition Behavior, 56
    Basketball Show-down, 73
    Battle of the Sexes, 71
    Biological Auction, 187
    Black Sea Cooperation, 6
    Buying a Home, 10
    Chicken, 72
    College Choices, 163
    College Education, 174
    Columbia River Watershed, 97
    Continental Divide, 69
    Corporate Home, 252
    County Fair Prize, 82
    Cournot Duopoly, 24
    Divorce Settlement, 99
    EPA, 240
    Escalation, 63
    Final Jeopardy, 48
    First Price Sealed Bid Auction, 202
    FoodPro, 146
    Free College, 133
    Hike, 198
    Ice Cream Parlor, 13
    Inheritance, 275
    International Collaboration, 175
    Job Offers, 25
    Joint Police Force, 268
    Legislative Gridlock, 267
    Mating Behavior, 53
    Media Streaming, 104
    Merit Raise, 37
    Natural Evolution, 218
    Normandy Breakout, 44
    Office Coffee, 43
    Olympic Aerial Skiing, 6
    One Card Poker, 153
    Patent Race, 111
    Political Challenger Take 2, 158
    Political Challenger., 112
    Predicting Scaled Averages, 70

Prisoner's Dilemma, 40
Purchasing Gasoline, 10
Repeated Social Dilemmas, 206
Rock County, 273
Second Price Sealed Bid Auction, 203
Self-Interest and Other-Interest, 24
Senators Voting, 65
Sequential Duopoly, 118
Stag Hunt, 71
Takeover Bid, 171
Tennessee-Georgia Border, 102
Tonkins County, 257
Tour Bus, 5
Trio, 45
Trivial Bridge, 199
TVA, 267
Twisted Matching Coins, 88
U.S. Appellate Court, 110
U.S. Legislative System, 273
United Nations Security Council, 266
Video Store, 10
Virginia Governor's Race, 73
Voting Order, 116
Warfare, 160
second price auction, 79
Second Price Sealed Bid Auction, 203
second-price auction method, 278
security level, 42
Self-Interest and Other-Interest, 24
Selten's Horse, 198
Senators Voting, 65
send uninformative messages, 178
separating equilibria, 174
Sequential Duopoly, 118
sequential game, 103
sequentially rational, 150
Shannon switching game, 126
Shapley allocation, 241
Shapley Characterization Theorem, 246
Shapley-Shubik Characterization Theorem,
    265
Shapley-Shubik power index, 259
share, 276
share equitable, 279
simple lottery, 17
Simplified Attrition Behavior, 78
sincere strategy, 65
social dilemma, 206
solution, 92
solution method, 92
spanning tree, 127

Stag Hunt, 71
state transition matrix, 210
strategic game, 39
strategy, 39, 165
strategy evolution system, 219
strongly dominant, 41
strongly dominates, 41, 92
strongly monotone, 92
subgame, 115, 151, 228
subgame perfect equilibrium, 115
subgraph, 127
subsidy free, 244, 245
substitutable, 18
superadditive, 240
symmetric, 92
symmetric strategic game, 59

Takeover Bid, 171
Tennessee-Georgia Border, 102
terminal history, 103
theorem
    Allocation Characterization, 279
    Another Folk, 218
    Banzhaf Characterization, 266
    Coalition Game Impossibility, 251
    Egalitarian Characterization, 93
    Finding Nash Equilibria, 58
    Finding Subgame Perfect Equilibria,
        115
    Folk, 214
    Nash Characterization, 95
    Nash Equilibrium Existence, 56
    Nash's Hex, 132
    Ordinal Preferences Characterization,
        14
    Prenucleolus Characterization, 251
    Raiffa Characterization, 95
    Shapley Characterization, 246
    Shapley-Shubik Characterization, 265
    vNM Preferences Characterization, 18
    Weak Sequential Equilibrium
        Existence, 159
    Zermelo's, 108
Tic-Tac-Toe, 120
TitForTat, 208

Tonkins County, 257
Tour Bus, 5
transitive, 14
Trio, 45
Trivial Bridge, 199
TVA, 267
Twisted Matching Coins, 88
TwoForOne, 208
type space, 164
types, 164

U.S. Appellate Court, 110
U.S. Legislative System, 273
Ultimatum scenario, 7
unanimity, 244
unaware, 178
unbiased, 92, 243, 245
unfavorable signal, 177
United Nations Security Council, 266

veto power, 271
Vickery auction, 79
Video Store, 10
Virginia Governor's Race, 73
vNM equivalent, 22
vNM preferences, 18
vNM Preferences Characterization
        Theorem, 18
von Neumann-Morgenstern preferences, 18
voting game, 258
Voting Order, 116

Warfare, 160
weak sequential equilibrium, 152
Weak Sequential Equilibrium Existence
        Theorem, 159
weakly chooses, 14
weighted voting system, 259
winning, 258
worth, 239, 282

Y, 143

Zermelo's Theorem, 108
zero sum, 74